Boundary Integral Equations in Elasticity Theory

SOLID MECHANICS AND ITS APPLICATIONS
Volume 99

Series Editor: G.M.L. GLADWELL
Department of Civil Engineering
University of Waterloo
Waterloo, Ontario, Canada N2L 3GI

Aims and Scope of the Series

The fundamental questions arising in mechanics are: *Why?, How?,* and *How much?* The aim of this series is to provide lucid accounts written by authoritative researchers giving vision and insight in answering these questions on the subject of mechanics as it relates to solids.

The scope of the series covers the entire spectrum of solid mechanics. Thus it includes the foundation of mechanics; variational formulations; computational mechanics; statics, kinematics and dynamics of rigid and elastic bodies: vibrations of solids and structures; dynamical systems and chaos; the theories of elasticity, plasticity and viscoelasticity; composite materials; rods, beams, shells and membranes; structural control and stability; soils, rocks and geomechanics; fracture; tribology; experimental mechanics; biomechanics and machine design.

The median level of presentation is the first year graduate student. Some texts are monographs defining the current state of the field; others are accessible to final year undergraduates; but essentially the emphasis is on readability and clarity.

For a list of related mechanics titles, see final pages.

Boundary Integral Equations in Elasticity Theory

by

A.M. LINKOV

Institute for Problems of Mechanical Engineering,
Russian Academy of Sciences,
St. Petersburg, Russia

KLUWER ACADEMIC PUBLISHERS
DORDRECHT / BOSTON / LONDON

A C.I.P. Catalogue record for this book is available from the Library of Congress.

ISBN 1-4020-0574-1

Published by Kluwer Academic Publishers,
P.O. Box 17, 3300 AA Dordrecht, The Netherlands.

Sold and distributed in North, Central and South America
by Kluwer Academic Publishers,
101 Philip Drive, Norwell, MA 02061, U.S.A.

In all other countries, sold and distributed
by Kluwer Academic Publishers,
P.O. Box 322, 3300 AH Dordrecht, The Netherlands.

Printed on acid-free paper

Printed in the Netherlands.

CONTENTS

PREFACE

by the author to the English edition

The book aims to present a powerful new tool of computational mechanics, *complex variable* boundary integral equations (CV-BIE). The book is conceived as a continuation of the classical monograph by N. I. Muskhelishvili into the computer era.

Two years have passed since the Russian edition of the present book. We have seen growing interest in numerical simulation of media with internal structure, and have evidence of the potential of the new methods. The evidence was especially clear in problems relating to multiple grains, blocks, cracks, inclusions and voids. This prompted me, when preparing the English edition, to place more emphasis on such topics.

The other change was inspired by Professor Graham Gladwell. It was he who urged me to abridge the chain of formulae and to increase the number of examples. Now the reader will find more examples showing the potential and advantages of the analysis.

The first chapter of the book contains a simple exposition of the theory of real variable potentials, including the hypersingular potential and the hypersingular equations. This makes up for the absence of such exposition in current textbooks, and reveals important links between the real variable BIE and the complex variable counterparts. The chapter may also help readers who are learning or lecturing on the boundary element method.

Throughout the entire book, two points are persistently emphasized: (i) the importance of the *hypersingular* equations and (ii) the computational advantages of using *complex variables*. We repeat, hypersingular equations are important because they contain the very values which characterize contact interaction: displacement discontinuities and tractions. This makes them a natural tool for handling multiple discontinuities. The main advantage of the CV-BIE is that they allow easy evaluation of singular and hypersingular integrals over arbitrary curvilinear elements. In real variables this operation is incomparably more difficult.

We increase the usefulness of the hypersingular equations and complex variables by combining them in the complex variable hypersingular boundary integral equations (CVH-BIE). The CVH-BIE are given in Chapters 2-9 for piece-wise homogeneous media with cracks, inclusions and voids, for finite and infinite regions, for periodic and doubly periodic problems and for bonded half-planes.

Chapter 10 contains a simple theory of complex variable hypersingular integrals of arbitrary order. The theory justifies various operations with the integrals: coming to the limit, integration by parts, and differentiation under the integral sign. We hope that this will "deliver the reader from hyper-fear of hyper-singularity". In contrast, the next Chapter 11, containing the theory of the CVH-BIE, is presented for completeness; a reader interested in applications may skip it.

I consider the complex variable boundary element method (CV-BEM) to be the best way to solve problems involving discontinuities on multiple surfaces. For this reason, the method is presented in detail in two chapters. One of them (Chapter 12) contains the stages of the CV-BEM, and simple quadrature rules for hypersingular, singular and proper integrals for ordinary, one-sided and two-sided tip elements. The other (Chapter 13) gives recipes for choosing computational parameters such as the size of boundary elements.

For the readers interested in the method of mechanical quadratures, traditionally employed for isolated cracks, this method is presented in Chapter 14.

The book is addressed to a wide range of readers: graduate students, academics, researchers and engineers. It may be of use to those who wants to calculate stresses, strains, stress intensity factors and effective properties of a medium with internal structure when dealing with problems of material science, fracture mechanics, micromechanics, soil and rock mechanics, geomechanics, civil and mechanical engineering. It may also serve as a textbook on the theory of real potentials (Chapter 1), on the CV-BIE (Chapters 2-10) and on the CV-BEM (Chapters 12, 13).

The theoretical results on the relation between real and complex variable BIE, on the equations for periodic and doubly periodic systems, for bonded half-planes, for circular inclusion and on the theory of hypersingular equations are relatively new. They may be of interest to specialists in the theory of elasticity.

The list of references is short: it includes 163 items. I felt unable to present a comprehensive bibliography for two reasons. First, there are too many publications deserving reference; they include, for instance, over 60 papers by N. I. Ioakimidis and P. S. Theocaris, over 30 papers by Y. Z. Chen, many papers by W. T. Ang, J.-T. Chen, D. L. Clements, A. I. Kalandia, V. V. Panasiuk, M. P. Savruk, Y. V. Veriugski and many other authors. Secondly, some papers were unavailable to me, and some I have missed. Consequently, the list of references contains only a limited number of representative publications, many of a review nature. I apologize to those many colleagues whose papers are not cited.

It is my pleasant duty to express gratitude to my friends and colleagues, Anastasia Dobroskok, Ol'ga Grebenschikova, Vadim Koshelev, Alexei Savitski and Victor Zoubkov for their invaluable help in preparing this book. I am very grateful to Professor Zbigniew Olesiak for his Preface to the Russian edition; this Preface is reproduced below. Many thanks to Doctor Sonia Mogilevskaya for collaboration, which resulted in this book. I am also grateful to Doctors Y. Z. Chen and V. F. Koshelev, who advised me on some misprints in the Russian edition: these and other shortcomings have been removed. The support of the Russian Fund of Fundamental Investigations in publishing the book in Russian is highly appreciated.

My special gratitude is to the scientific editor, Professor Graham Gladwell. I am indebted to him for his editorial work, profound comments, persistence and patience in improving the manuscript and my English, and for interacting in a generous and stimulating manner.

PREFACE

by Professor Zbigniew Olesiak

The achievements of Augustin Louis Cauchy, the great French mathematician, are known not only in many branches of pure mathematics, but also in applications of mathematical analysis. Cauchy's problem, Cauchy's integral, Cauchy's criterion of convergence, and Cauchy's theorems play fundamental roles in pure and applied mathematics. He was one of the founders of the mathematical theory of continuous media, and of the theory of elasticity in particular. His interest in both pure and applied mathematics was remarkable, but he was not the only one interested in both. We could make a long list of great mathematicians interested in pure mathematics, and also inspired by problems in theoretical physics, mechanics, biology, economics, finance, computer sciences, etc.

The Cauchy type integral, the notion of the principal value of such an integral, and the equations of the mathematical theory of elasticity in Navier's form, were introduced almost simultaneously at the beginning of the nineteenth century. Also in that century the method of complex variables was applied to problems of mathematical physics reducible to Laplace's or Poisson's partial differential equations. On the one hand, these equations describe important linear problems of heat conduction, hydrostatics, electrostatics, etc.; on the other hand a single harmonic function is sufficient to solve Dirichlet's or Neumann's boundary value problem. The Laplace integral transform and the operational calculus were devised at that time also.

The application of complex variables to G. B. Airy's biharmonic partial differential equation took longer. Although certain formulae were already known in the nineteenth century, (e.g. E. Goursat's complex variable representation of Airy's partial differential equation, and L. N. G. Filon's formula), it was G. V. Kolosov, in St Petersburg who, at the beginning of the twentieth century, derived and applied complex variables to the solutions of two-dimensional problems of the mathematical theory of elasticity. N. I. Muskhelishvili continued his work and developed Kolosov's idea in a number of papers. The first publications in German and French did not help much in the dissemination of the idea - the formulae were rediscovered a number of times, usually in a form far from complete. The first Russian edition of Muskhelishvili's famous monograph appeared in 1933. It had an enormous impact on the development of two-dimensional problems of the mathematical theory of elasticity and related mathematical theory, first in a number of the mathematical centers in the former Soviet Union, and after its translation into English, in the West. Complex variable methods in the theory of elasticity became internationally known with the publication of I. S. Sokolnikoff's: *Mathematical Theory of Elasticity*, (first edition 1946) and the English edition of Muskhelishvili's

Some Basic Problems of the Mathematical Theory of Elasticity and *Singular Integral Equations* (both in 1953).

Muskhelishvili's monographs were excellent in all respects: the elegance of the presentation, precision of the mathematical proofs, and printing practically without misprints.

Academician A. N. Krylov, in the preface to the monograph *Some Basic Problems*, emphasized the virtues of the monograph, and expressed his opinion that in future editions of the monograph, more numerical results should be included and discussed, and their accuracy estimated.

Meanwhile times have changed. Tedious, lengthy calculations have been replaced by computer programming. Powerful personal computers are freely available to mathematicians and engineers, and are in common use even by humanists. In consequence the excellent books such as those by Muskhelishvili, need continuation, extension and revision. What was almost impossible to achieve half a century ago has become a relatively easy matter.

Today there are technological ways to fulfill Academician Krylov's wish: to have a deeper insight into problems, and to obtain the numerical solutions together with knowledge of their accuracy. The technological developments in electronic computers are not sufficient for this purpose. There are many new problems for mathematicians to solve, and new theorems to prove.

The Author of this monograph belongs to a small group of mathematicians who are able to undertake the difficult task of continuing and extending Muskhelishvili's monographs. First of all, he is the author of many original papers relating to boundary integral equations in complex variables, and their applications to systems of curvilinear cracks, non-periodic (1974), periodic and double-periodic (1976), to blocky systems with cracks and inclusions (1983), and bonded half-planes with inclusions and cracks (1998). He introduced and investigated, together with his former student S. G. Mogilevskaya, the complex hypersingular integrals of arbitrary order. The theory of complex variable hypersingular equations, derived by the author, and recently developed for open arcs, has been extended here to include arbitrary systems of open arcs and closed contours.

The monograph consists of an Introduction and four Parts. In Part I the reader will find the discussion of real variable potentials of the theory of elasticity, discussion of the singular solutions and potentials in complex form, and the direct and indirect approaches to solutions. Part II is devoted to the description and discussion of the methods based on the Kolosov-Muskhelishvili theory. The theory of the hypersingular integral equations of a complex variable belongs to the original investigations by the author and Mogilevskaya, and of course was not discussed in Muskhelishvili's books. The theory of hypersingular integrals and integral equations of a complex variable is presented in Part III. In Part IV the reader will find the presentation of numerical methods for complex variable integral equations. It includes the method of boundary elements (BEM), numerical simulation and experiments, and numerical integration.

The book constitutes a significant scientific contribution, as in the nineteenth century, and as for Muskhelishvili's monographs, it is also important for applications of mathematics in the theory of elasticity; besides it was written in St Petersburg! The reader will find not only the presentation of contemporary methods utilizing the power of computer programming, but also new ideas and excellent exposure to contemporary problems, written in a modern style.

INTRODUCTION

Why complex variables? K. F. Gauss called complex variables "a wonderful golden source". In his letter to Bessel he wrote (Gauss [1]): "Analysis ... loses in beauty and value when imaginary quantities are neglected".

Why?

First of all, because of specific properties of analytic functions of a complex variable. These are functions that are represented by convergent polynomial series. If such a function has first derivatives in a two dimensional region, it has derivatives of an arbitrary order in this region. An integral of it does not depend upon the integration path. This gives magnificent results connecting values of a function on the contour of integration with its values away from the contour. Conformal mapping, analytical continuation, meromorphic and quasi-periodic functions, modules, factorization, complex integral transforms, solutions of Riemann's problem, and many other powerful and elegant mathematical tools emerge from the properties of analytic functions.

We will not dwell on these theories because we are interested primarily in applications to computational elasticity. For our theme, two features of complex analytic functions are especially important: 1) convergence of series; and 2) ability to use contour (one-dimensional) integrals to represent values of functions in two-dimensional regions and also their limiting values. The first property allows us to use finite sums to obtain approximations of arbitrary accuracy for a given or unknown function. The second reduces the dimension of a problem: instead of a two-dimensional problem for a plane region, we may solve a one-dimensional equation along its contour. Conformal mapping of curvilinear elements onto a straight element or onto an element along the circumference opens additional computational options. They reduce numerical evaluation of integrals to calculation of a limited number of standard integrals that can often be expressed analytically.

These mathematical properties are of extreme importance in applied problems. The importance is due to the close relationship between complex analytic functions and real *harmonic* functions, that is, functions satisfying Laplace's equation. Indeed, both real and imaginary part of an analytic function satisfy Laplace's equation. This equation appears in many applied problems in electrostatics, thermodynamics, fluid mechanics, and ground water flow; functions of complex variables became a natural way to solve plane harmonic problems (see, e. g., Hromadka and Lai [1], Clements [1], Ang and Park [1], Ang, Clements and Cook [1]). *Bi-harmonic* problems become amenable too, because a bi-harmonic function may be expressed through two harmonic functions. This opens the way to applying complex variables to two-dimensional bi-harmonic problems, in particular to elasticity problems.

A little history. G. V. Kolosov showed how to use complex variables in elasticity theory in 1909 (Kolosov [1]). He represented displacements and stresses through two analytic functions of a complex variable. This led to the solution of elasticity problems through functions of a complex variable, and gave the impetus to the development of the Russian and Georgian schools of elasticity; these schools obtained results far beyond those of their founders (see, e. g., Lekhnitskii [1], Kupradze [1], Kupradze et al [1]).

The outstanding student of Kolosov, N. I. Muskhelishvili, exploited the potential of this approach. His classical monograph (Muskhelishvili [5]) makes use of the analytical advantages of complex variables. Numerous analytical solutions were obtained in this way. Many of them are difficult or impossible to obtain by using only real variables.

Naturally, the analytical advantages of functions of a complex variable influence numerical methods also. Some of those, such as complex series and conformal mapping were presented in Muskhelishvili's monograph [5]. However, being written in the pre-computer era, this monograph focuses on methods that do not require excessive numerical effort. The same is true with regard to early western approaches to elasticity problems by complex variables methods, for instance, to publications by L. A. Westergaard [1] and A. C. Stevenson [1,2]. The first author represented stresses and displacements through one complex function in the particular case when the shear stresses are zero on the real axis. The second author gave an elegant extension of the method to plane problems involving body forces having a potential.

Why boundary integral equations? Muskhelishvili's monograph contains a review and sections on *complex variable boundary integral equations* (CV-BIE) of elasticity theory S. G. Mikhlin [1], N. I. Muskhelishvili [1,2] and D. I. Sherman [1-5] were the first who derived, studied and used such equations. First they used Fredholm's equations, but later on singular equations were used (see, e.g., Sherman [3], Mandjavidze [1,2]). Muskhelishvili and his school developed a comprehensive theory of complex variable singular integral equations (Muskhelishvili [4], Vekua [1,2]).

At that time, these equations served primarily for studying theoretical questions regarding existence and uniqueness of the solution of boundary value problems. Meanwhile, the computational aspects of CV-BIE became very important. The main stimulus for their implementation is the same as that for real boundary integral equations (BIE), namely, the reduction of the geometrical dimension by one. For plane problems, we may solve one-dimensional equations on the contour of a region instead of equations in the whole two-dimensional region. This simplifies input data, reduces computer memory and calculation time. Besides, solutions for infinite regions are found as easily as for finite regions.

Complex variables provide computational advantages over real BIE. The main computational virtue of CV-BIE is that it simplifies the evaluation of singular (and hypersingular) integrals along arbitrary curvilinear elements. It is very difficult to write analogous quadrature formulae in real variables.

Other computational virtues of complex variables are also important. Since modern computers operate with complex arithmetic as easily as with real, we gain when employing one complex function of a complex variable instead of pairs of real functions and variables. The number of integrals in the CV-BIE is less than in real BIE. Another important advantage stems from the analytical nature of some integrals; we obtain simple formulae to check calculations and to control their accuracy. Besides, for the most important types of boundary elements and approximations, *all* the integrals can be evaluated analytically.

Some more history. The computational advantages of the CV-BIE were so obvious that attempts to employ them started immediately after these equations were derived. Muskhelishvili [1,2] not only derived new complex variable equations in 1934 but also explained in 1937 how to solve them numerically (Muskhelishvili [3]). His idea was implemented by A. Ya. Gorgidze and A. K. Rukhadze [1] in 1940. They used all the attributes of a method that presently is known as the boundary element method. Specifically, they divided a contour into elements; approximated functions within the elements, integrated along the elements applying the chosen approximation, and reduced the problem to an algebraic system for the unknown approximation coefficients. After finding the boundary values of functions, they found the values of the functions in the region. We may consider papers by Muskhelishvili [3], Gorgidze and Rukhadze [1] to be pioneering works on the complex variable BEM employed in a pre-computer era.

It was in 1940 that Tc. O. Levina and S. G. Mikhlin [1] used the CV-BIE for numerical calculations, but in a more traditional way. They considered a plane with two holes. This region was mapped onto an annulus, for which Green's function was known. As a result, they obtained a CV-BIE by approximating its kernel by a nearby degenerate kernel. The resulting equation was solved by reduction to an algebraic system. They spent more than six months of intensive work on numerical calculations!

These experiments were mostly of an illustrative nature. They had no systematic development because only enthusiasts could carry them out in the era of mechanical calculators. It was for this reason that early attempts at solving Fredholm's equations separated real and complex parts in the calculations; the potential of complex integration and complex arithmetic was not exploited.

All early works started from equations that contained functions without explicit physical meaning. Only since the seventies has growing interest in numerical applications of integral equations led to new forms of complex variables integral equations containing physically meaningful values as unknowns. Among the first such works are A. M. Linkov [1-6], N. I. Ioakimidis and P. S. Theocaris [1,2,5], P. S. Theocaris and N. I. Ioakimidis [1,2,6], M. P. Savruk [1-4] and G. B. Kovneristov [1]. These works are concerned with contact and mixed boundary value problems, and problems for cracks with arbitrary *curvilinear* contours. There were also many publications concerned with *particular* cases of straight cracks, and cracks along a circumference or some other arcs. They were commonly based on known analytical solutions by Muskhelishvili [5] for straight cracks along a real axis, and for cracks along a circumference (see, e.g., reviews in the books by Parton and Perlin [1], Panasiuk et al. [1], Savruk [4], Savruk et al. [1] and for non-linear problems in Chernykh [1]). For arbitrary contours, the *complex variable boundary element method* was used occasionally for singular equations of elasticity (see Kovneristov [1], Kovneristov et al. [1], Zoubkov [1], Linkov,

Zoubkov and Mogilevskaya [1], Lee [1]). The other method widely used for complex variable BIE, the *complex variable method of mechanical quadrature*, was consistently applied to arbitrary contours by Ioakimidis and Theocaris [1-5], Theocaris and Ioakimidis [1-6], and by Savruk and co-workers. Savruk presented this method in detail in his books (Savruk [4], Savruk, Osiv and Prokopchuk [1]).

Breakthrough from real BIE to complex equations by Muskhelishvili. For a long time there was no link between the real BIE of elasticity theory obtained by traditional methods of potential theory, and complex variable equations obtained from the Kolosov-Muskhelishvili theory. Only for Sherman's equations [1,2], was it noted (see Muskhelishvili [5]) that they present a complex form of the real equations by G. Lauricella [1]. It seemed that real and complex equations were two independent branches of elasticity theory, the first based on singular solutions for point forces, the second on analytical functions of Kolosov-Muskhelishvili.

Is there any connection between them? The answer obtained recently is "YES", there is a close connection. S. G. Mogilevskaya gave this answer for the *direct* approach of the potential theory of elasticity. In essence, she stated that the complex variable singular equations (Linkov [1,2,6]) and the complex variable hypersingular equations (Linkov [7], Linkov and Mogilevskaya [1-3]), initially derived by using Kolosov-Muskhelishvili formulae, could be obtained from the usual real potentials. Specifically, one should start from the Somigliana identities, but use complex forms of fundamental solutions. Soon we introduced appropriate forms of potentials and revealed an analogous connection for the *indirect* approach. For the direct approach we derived Muskhelishvili type integral equations without using his theory. These results are stated in brief in Mogilevskaya and Linkov [1] and presented in detail in Linkov and Mogilevskaya [5].

It may look surprising that the close link between two theoretical branches remained hidden for more than fifty years. In our opinion, the reason lies in the different traditions of different scientific schools.

Indeed, ever since the publication of Kolosov [1], scientists in Russia and the Former Soviet Union had a strong commitment to complex variables in elasticity theory. However, although convenient for two-dimensional problems, this approach cannot serve for three-dimensional problems. Consequently, three-dimensional problems were studied in terms of the classical theory of real potentials (Mikhlin [2], Kupradze [1], Kupradze et al [1], Mikhlin, Morozov and Paukshto [1]). These works are of theoretical rather than computational nature. Certainly, results of the three-dimensional theory easily cover the two-dimensional case. But such extension being quite elementary did not attract attention because, as mentioned, two-dimensional problems had been successfully explored by the method of Kolosov-Muskhelishvili, at that time more efficient and more practical.

This division existed even when progress in computers made BIE an efficient tool for solving practical problems. In the USSR, for instance, CV-BIE were used mainly with the complex method of mechanical quadrature. The latter method derives far more from the theory of real potentials than the method of boundary elements. This also hindered the rapprochement between the two branches of the theory.

Meanwhile, western specialists, not having such a strong commitment to complex variables, started systematic calculations on computers using real variables and classical potentials (see, e.g., Benerjee and Butterfield [1], Brebbia and Walker [1], Cruse and Rizzo [1]). The numerical method of boundary elements, based on the classical real equations emerged swiftly, and was independent of the two-dimensional theory of Kolosov-Muskhelishvili. As a result, it took time to bring together the techniques of the BEM and complex variables.

It is interesting to note that G. B. Kovneristov [1] (see also Kovneristov et al. [1]) came very close to bringing the two together. He was the first to use *complex variable* forms of singular solutions for point forces when deriving Somigliana's identities. He used these identities as they are used in the theory of real potentials. He solved the complex variable equations for contact problems by using the boundary element method. However, his presentation of the second Somigliana identity was not completely in a complex form; it was "semi-complex". The author used *real* components of traction instead of joining them into a *complex* traction-vector. This forced him to separate real and imaginary parts in his equations early on. This also prevented him from disclosing the fundamental connection between the equations of the direct approach, Muskhelishvili's complex variable equations [1,2], and ours (Linkov [1,2,6]).

Scope of the book. The two branches, a) the complex counterparts of real BIE, and b) Kolosov-Muskhelishvili's functions, comprise the first two parts of the book.

Part I reproduces in a complex variable form the well-known real BIE:

Singular solutions $\Rightarrow$ potentials $\Rightarrow$ equations of the indirect and direct approach

Part II follows the path by Kolosov-Muskhelishvili. It starts from their classical formulae and employs the analytic nature of their functions. These two parts complement each other. We discuss the features and capabilities of the methods, and the relations between them. These two parts can be read independently from each other. Our advice is to read both parts of the book; this will give a clear understanding of the problem as a whole.

Part II contains sections concerned with systems of blocks (grains) with cracks, holes and inclusions and with integral representations of Kolosov-Muskhelishvili's functions, periodic and doubly periodic problems and problems for bonded half-planes and for a circular inclusion.

Equations for periodic and doubly periodic problems are derived and studied on the basis of Kolosov-Muskhelishvili theory. Here again we see the advantages of this theory over the method of the first part. In particular, extreme difficulties, arising when building singular solutions for doubly periodic system of point forces, disappear completely (Such solutions are needed in Part I). The equations for doubly periodic problems are very attractive for obtaining dependence between *average* strains and *average* stresses accounting for crack growth and interaction on the boundaries of internal grains, blocks and/or inclusions. In other words, they are convenient for numerical modeling macroscopic constitutive equations by tracing deformations on the structural level.

Finally, in problems for bonded half-planes and for a circular inclusion in a plane, Kolosov-Muskhelishvili's method immediately provides general results. They include as simple particular cases all the previously known BIE for these problems. The equations are simple even in the general case.

We see that the approach of Part I is far from being universal. Sometimes, it is difficult or impossible to use. In addition, Kolosov-Muskhelishvili's theory significantly increases the abilities of the first approach, in particular, when obtaining singular solutions.

In the first two parts, we consistently underline the importance of *complex hypersingular* equations. These equations have been obtained recently (Linkov [1], Linkov and Mogilevskaya [1-3]). Naturally, they were not presented in Muskhelishvili [5]. Consequently, it is worthwhile to explain their important advantages.

Hypersingular equations appeared in response to progress in computational mechanics and applied sciences. The main virtue of hypersingular integrals and hypersingular equations (HSIE) both in two- and three-dimensional problems is that they serve as a convenient way to account for *discontinuities* in physical fields. The discontinuities arise on surfaces of thin regions such as cracks or contacts between grains and inclusions in elasticity problems, on screens in problems of electric plating, on thermal isolating layers in heat transfer problems, on low permeability walls in problems of ground water flow. In elasticity, the HSIE serve to account for displacement discontinuities on surfaces of cracks or geological faults, on boundaries of grains in microstructure, on boundaries of blocks of rock in mines and the Earth's crust, on contacts of elements of constructions. These "multi-contact" problems are attracting ever-growing attention in micromechanics, material science, fracture mechanics, rock and soil mechanics, and geophysics. Hypersingular equations provide a unique tool for tackling these problems. In contrast with singular and Fredholm equations that also may serve for these problems, the hypersingular equations contain the very values which characterize contact interaction, displacement discontinuities and tractions.

The interest in hypersingular equations as a powerful new tool for solving important new problems is reflected in the last two parts of the book.

Part III contains the theory of complex variable integral equations. We focus on the recently developed theory of complex variable hypersingular integrals and integral equations (Linkov and Mogilevskaya [3-5]).

Part IV concentrates on numerical implementations of complex variable integral equations. This last part of the book, in its core, reflects the tremendous changes in our scientific environment, the changes generated by the progress in computers and informatics. We wanted to bring the book closer to challenges of modern computational mechanics and information technologies. An emphasis is placed on the boundary element method as the most popular and flexible in employing the advantages of the BIE. In part, this reflects our personal preference; we developed this method in its complex variable form. Examples illustrate the wide applicability and high efficiency of the complex variable method of boundary elements (CV-BEM).

PART I

METHOD OF POTENTIALS

In this part we follow the usual way of deriving *real* boundary integral equations, by employing the classical theory of potentials including hypersingular potentials. We decided to present the basics of the theory even though it is not the subject of this book. By the questions they have asked, we realize that students and colleagues have difficulty finding a simple modern presentation of the theory of potentials. There is no single paper or book that fills this need. In our opinion, the book by A. G. Ugodchikov and N. M. Khutorianski [1] and the chapter by Hartmann in the book [1] are useful, but they do not employ hypersingular integrals of the Hadamard type, and they consider only closed contours. The excellent books by J. H. Kane [1] and P. K. Benerjee [1] published recently, contain hypersingular potentials but only for the direct approach and in a form insufficient for our goals. On the other hand, there is a swiftly growing flux of papers employing real hypersingular potentials (see, e. g. Ioakimidis [2], Hong and J.-T. Chen [1], Khrishnasamy et al. [1], Linkov [8] and the review by Tanaka et al. [1]) but they do not present the theory as a whole.

For these reasons, we decided to present the theory of real potentials in the first chapter to simplify comprehension of our theme, and to give a complete exposition. The following three chapters, dealing with complex variable equations, become a complex variable restatement of the first chapter.

For convenience we will present theoretical results on real potentials for two dimensions. The results can easily be extended to 3D problems with obvious minor changes such as summing over indices 1, 2, 3 instead of 1, 2 or using a 3D singular solution instead of its 2D counterpart. We employ global Cartesian co-ordinates x_1,

x_2: a point with the co-ordinates x_1, x_2 is denoted by the bold letter **x**. Similarly, a vector with the components u_1, u_2 is denoted by **u**.

Chapter 1

REAL POTENTIALS OF ELASTICITY THEORY

§ 1. PROBLEM FORMULATION

Consider the elastostatics equations for a region D, finite or infinite:

$$\frac{\partial \sigma_{ij}}{\partial x_j} = 0, \qquad i = 1,2,3, \qquad (1.1)$$

where σ_{ij} are components of a stress tensor in the global co-ordinates which are assumed to be Cartesian. Here we use Einstein's summation rule. The components of the stress tensor are connected to the components of the strain tensor by Hooke's law:

$$\sigma_{ij} = c_{ijkl}\varepsilon_{kl}, \qquad i,j = 1,2,3, \qquad (1.2)$$

where $c_{ijkl} = c_{ijlk} = c_{klij}$ are elastic constants; ε_{kl} are the components of the strain tensor:

$$\varepsilon_{kl} = \frac{1}{2}\left(\frac{\partial u_k}{\partial x_l} + \frac{\partial u_l}{\partial x_k}\right), \qquad (1.3)$$

u_k are the components of the displacement vector **u**. Substitution of (1.3) into (1.2) and the result into (1.1) leads to the complete system of elasticity theory in terms of displacements:

$$c_{ijkl}\frac{\partial^2 u_k}{\partial x_j \partial x_l} = 0. \qquad (1.4)$$

In (1.4) we have placed the elasticity moduli outside the differentiation symbol because we assume the medium to be homogeneous. For an isotropic medium we have

$$\sigma_{ij} = \lambda\delta_{ij}\varepsilon_{kk} + 2\mu\varepsilon_{ij}, \tag{1.5}$$

where λ and μ are Lame's coefficients; δ_{ij} is Kroneker's delta: $\delta_{ij} = 1$ if $i = j$, $\delta_{ij} = 0$ if $i \neq j$. The coefficient λ may be expressed in terms of the shear modulus μ and Poisson's ratio ν: $\lambda = 2\mu\nu/(1-2\nu)$. Its substitution into (1.5) gives Hooke's law for an isotropic medium in the form

$$\sigma_{ij} = 2\mu\left(\frac{\nu}{1-2\nu}\delta_{ij}\varepsilon_{kk} + \varepsilon_{ij}\right), \tag{1.6}$$

while the equilibrium equations (1.4) become

$$\frac{\partial^2 u_k}{\partial x_i \partial x_k} + (1-2\nu)\frac{\partial^2 u_i}{\partial x_k^2} = 0. \tag{1.7}$$

In 3D the Latin indices in (1.1)-(1.7) run through the values 1, 2, 3. However, (1.6) and (1.7) also hold for plane strain, when the state of a body does not vary along the x_3 axis. In this case

$$\varepsilon_{33} = 0, \quad \frac{\partial \sigma_{ij}}{\partial x_3} = 0, \quad \frac{\partial \varepsilon_{ij}}{\partial x_3} = 0,$$

and consequently the sums in (1.6) and (1.7) run through the indices 1, 2 only.

Formulae (1.6), (1.7) may also serve for *plane stress* ($\sigma_{31} = \sigma_{32} = \sigma_{33} = 0$), if we change ν to $\nu/(1 - \nu)$. They take a universal form if we use Muskhelishvili's parameter χ:

$$\chi = \begin{cases} 3-4\nu & \text{for plane strain} \\ (3-\nu)/(1+\nu) & \text{for plane stress.} \end{cases} \tag{1.8}$$

In particular, Hooke's law (1.6) takes the same form both for plane strain and plane stress:

$$\begin{aligned} 8\mu\varepsilon_{11} &= (\chi+1)\sigma_{11} - (3-\chi)\sigma_{22}, \\ 8\mu\varepsilon_{22} &= (\chi+1)\sigma_{22} - (3-\chi)\sigma_{11}, \\ 2\mu\varepsilon_{12} &= \sigma_{12}. \end{aligned} \tag{1.9}$$

In the following sections, to obtain results valid both for plane strain and plain stress, we will use only the constants μ and χ.

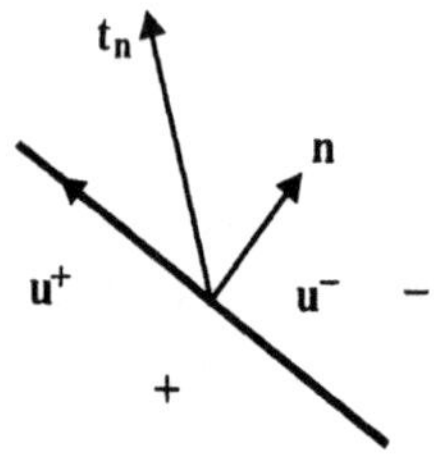

Fig. 1.

The traction vector $\mathbf{t_n}$ on an area with the normal $\mathbf{n}$ (Figure 1) is defined by the stress tensor as

$$t_i = \sigma_{ij} n_j. \tag{1.10}$$

Using (1.2) and (1.3) in (1.10) we obtain

$$\mathbf{t_n} = \mathbf{T_n u}, \tag{1.11}$$

where $\mathbf{T_n}$ is the traction operator; it gives us the traction vector corresponding to a displacement vector $\mathbf{u}$:

$$(\mathbf{T_n u})_i = c_{ijkl} \frac{\partial u_k}{\partial x_l} n_j. \tag{1.12}$$

For an isotropic medium

$$(\mathbf{T_n u})_i = 2\mu \left[\frac{\nu}{1-2\nu} \frac{\partial u_k}{\partial x_k} n_i + \frac{1}{2} \left(\frac{\partial u_i}{\partial x_j} + \frac{\partial u_j}{\partial x_i} \right) n_j \right]. \tag{1.13}$$

For plane problems, we need to change the multiplier $\nu/(1 - 2\nu)$ on the r. h. s. of (1.13) to $\frac{1}{2}(3 - \chi)/(\chi - 1)$, while the indices take only two values: 1 and 2.

In elasticity theory, the traction operator plays the same role that the normal derivative plays in the theory of Laplace's equation. It involves the derivatives of a field $\mathbf{u}$, but this time the derivatives are taken both in the normal and *tangential* directions.

In boundary value problems we seek the solution of the system (1.4) (or its particular form (1.8)) in a region D under given conditions on its boundary. These may be conditions of prescribed displacements

$$\mathbf{u} = \mathbf{f(x)}, \quad \mathbf{x} \in L_D \tag{1.14}$$

or traction

$$\mathbf{t_n} = \mathbf{g}(\mathbf{x}), \quad \mathbf{x} \in L_D, \tag{1.15}$$

or dependence, linear for instance, between traction and displacements

$$\mathbf{Au} + \mathbf{Bt_n} = \mathbf{h}(\mathbf{x}), \quad \mathbf{x} \in L_D, \tag{1.16}$$

where $\mathbf{A}$ and $\mathbf{B}$ are given matrices and $\mathbf{h(x)}$ is a given vector at the points of the boundary.

One may prescribe combinations of these conditions by taking (1.14) at some points of the boundary, (1.15) at other, and (1.16) at the remaining points. For open surfaces (open arcs in plane problems), we may prescribe similar conditions on the jumps in displacements and traction:

$$\Delta\mathbf{u} = \mathbf{f}(\mathbf{x}), \tag{1.17}$$

$$\Delta\mathbf{t_n} = \mathbf{g}(\mathbf{x}), \tag{1.18}$$

$$\mathbf{A}\Delta\mathbf{u} + \mathbf{B}\Delta\mathbf{t_n} = \mathbf{h}(\mathbf{x}), \tag{1.19}$$

where $\Delta\mathbf{u} = \mathbf{u}^+ - \mathbf{u}^-$, $\Delta\mathbf{t} = \mathbf{t_n}^+ - \mathbf{t_n}^-$. For definiteness, we will, as usual, refer the upper index "plus" ("minus") to the side with respect to which the normal is outward (inward). The (1.17)-(1.19) may serve us for closed boundaries as well, if we assume the normal to be outward and apply $\mathbf{u}^- = \mathbf{0}$, $\mathbf{t_n}^- = \mathbf{0}$. Henceforth, we shall use this convention and refer (1.17)-(1.19) both to open and closed boundaries.

The problem of solution (1.4) under boundary conditions (1.17)-(1.19) may be reduced to integral equations on the boundary itself. Singular solutions serve this purpose.

§ 2. INITIAL SINGULAR SOLUTIONS

Suppose that at a point ξ of a region D, finite or infinite, we have a unit point force $\mathbf{F}^1$ acting in the direction of the x_1 axis (Figure 2). The vector displacement field $\mathbf{U}_l^1(\mathbf{x},\xi)$ induced by this force has the components $U_{l1}^1(\mathbf{x},\xi)$, $U_{l2}^1(\mathbf{x},\xi)$ along the axes x_1, x_2. Writing $\mathbf{U}_l^1(\mathbf{x},\xi)$ as a column vector we have

$$\mathbf{U}_l^1(\mathbf{x},\xi) = \begin{pmatrix} \mathrm{U}_{l1}^1(\mathbf{x},\xi) \\ \mathrm{U}_{l2}^1(\mathbf{x},\xi) \end{pmatrix}.$$

In accordance with our theme, we write vectors for the two-dimensional case although all the conclusions of this chapter are valid for three dimensions.

For the traction vector $\mathbf{J}_S^1(\mathbf{x},\xi)$, corresponding to the displacement field $\mathbf{U}_l^1(\mathbf{x},\xi)$, from (1.11), (1.12) we obtain:

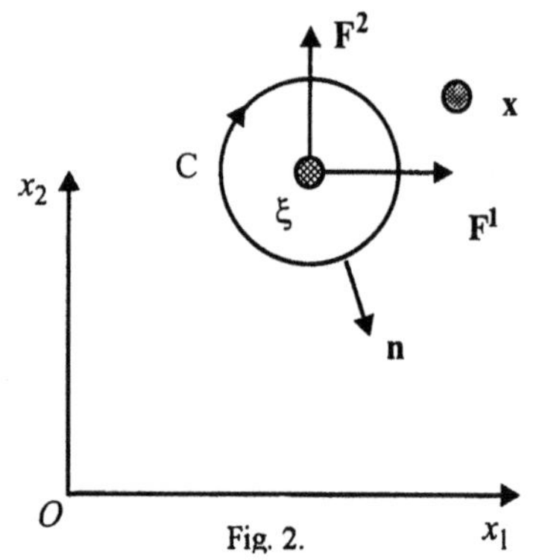

Fig. 2.

$$\mathbf{J}_S^1(\mathbf{x},\xi) = \begin{pmatrix} J_{S1}^1(\mathbf{x},\xi) \\ J_{S2}^1(\mathbf{x},\xi) \end{pmatrix} = \mathbf{T}_{\mathbf{nx}}\mathbf{U}_l^1(\mathbf{x},\xi),$$

where the index $\mathbf{x}$ in the notation of the traction operator presumes differentiation with respect to the co-ordinates of the argument $\mathbf{x}$. It is necessary because the singular solution $\mathbf{U}_l^1(\mathbf{x},\xi)$ depends on the source point ξ as well.

Encircle the point ξ by a small contour C (Figure 2) and take the normal to be outward with respect to the circle containing ξ. Then by the definition of a unit point force in the direction x_1 we have

$$\int_C J_{Si}^1(\mathbf{x},\xi)ds_x = -\delta_{1i}, \; i = 1,2. \tag{2.1}$$

The minus sign on the r. h. s. of (2.1) accounts for the assumed direction of the normal. Hence, the direction of the force acting on the contour C, is opposite to the force $\mathbf{F}^1$: the former equilibrates the latter.

Similarly, for a unit point force $\mathbf{F}^2$ acting in the direction of the x_2 axis we have: the displacement field

$$\mathbf{U}_l^2(\mathbf{x},\xi) = \begin{pmatrix} U_{l1}^2(\mathbf{x},\xi) \\ U_{l2}^2(\mathbf{x},\xi) \end{pmatrix},$$

the traction field

$$\mathbf{J}_S^2(\mathbf{x},\xi) = \begin{pmatrix} J_{S1}^2(\mathbf{x},\xi) \\ J_{S2}^1(\mathbf{x},\xi) \end{pmatrix} = \mathbf{T}_{\mathbf{nx}}\mathbf{U}_l^2(\mathbf{x},\xi)$$

and the equation of equilibrium of a region including the point where the force is applied,

$$\int_C J_{Si}^2(\mathbf{x},\xi)ds_x = -\delta_{2i}, \; i = 1,2.$$

We use the singular solutions $\mathbf{U}_l^1(\mathbf{x},\xi)$, $\mathbf{U}_l^2(\mathbf{x},\xi)$ as columns of the matrix $\mathbf{U}_l(\mathbf{x},\xi)$ of *initial* singular solutions:

$$\mathbf{U}_l(\mathbf{x},\xi) = (\mathbf{U}_l^1(\mathbf{x},\xi), \mathbf{U}_l^2(\mathbf{x},\xi)) = \begin{pmatrix} U_{l1}^1 U_{l1}^2 \\ U_{l2}^1 U_{l2}^2 \end{pmatrix}. \tag{2.2}$$

The simplest initial singular solutions, as usual, are those for an infinite plane. They are termed *fundamental* solutions. For an isotropic plane, the matrix of fundamental solutions corresponds to the well-known Kelvin's solution (see, e. g. Brebbia, Walker [1, § 5.6]). In 2D its components are

$$U_{lj}^k(\mathbf{x},\xi) = \frac{1}{2\pi\mu(\chi+1)}\left(-\chi\delta_{kj}\ln r + \frac{\partial r}{\partial x_k}\frac{\partial r}{\partial x_j}\right), \tag{2.3}$$

where r is the distance between a field point $\mathbf{x}$ and a source point ξ, and χ is Muskhelishvili's parameter (1.8). From (2.3) we see that the initial singular solutions in plane problems have a logarithmic singularity at the point $\mathbf{x}=\xi$. This is marked with the index l in their notation. The corresponding traction, being obtained by differentiation of the displacements given by the columns in (2.2), has a strong singularity. It is marked with the index S. In plane problems it is of the $1/r$ type.

Introduce the matrix of traction with columns $\mathbf{J}_S^1(\mathbf{x},\xi)$, $\mathbf{J}_S^2(\mathbf{x},\xi)$:

$$\mathbf{J}_S(\mathbf{x},\xi) = (\mathbf{J}_S^1(\mathbf{x},\xi), \mathbf{J}_S^2(\mathbf{x},\xi)) = \mathbf{T}_{\mathbf{nx}}\mathbf{U}_l(\mathbf{x},\xi). \tag{2.4}$$

The r. h. s. of (2.4) is understood as the result of applying the traction operator to each column of the matrix of singular solutions $\mathbf{U}_l(\mathbf{x},\xi)$. The emerging components of the traction vectors are sums of derivatives from *various* components of the displacement vector. Being composed of such derivatives, the columns of the traction matrix $\mathbf{J}_S(\mathbf{x},\xi)$ do not satisfy the equations of elasticity theory (1.4).

§ 3. SINGULAR SOLUTIONS OF HIGHER ORDERS

Clearly we may differentiate a whole column of the matrix $\mathbf{U}_l(\mathbf{x},\xi)$ with respect to any co-ordinate of $\mathbf{x}$ or ξ and obtain a new singular solution of equations (1.4) for displacements. But if we differentiate different terms in a column with respect to different co-ordinates, the resulting columns will not satisfy these equations. This holds also for linear combinations of derivatives from different components of the columns. In particular, as mentioned, this holds for the columns of the matrix $\mathbf{J}_S(\mathbf{x},\xi)$.

Meanwhile, there are non-trivial combinations of the derivatives, which give columns satisfying the displacement equations (1.4). Consider the most important of them corresponding to the kernel of the potential of the double layer. Note first, that the matrix of initial singular solutions satisfies the reciprocal equation

$$\mathbf{U}_l(\mathbf{x},\xi) = [\mathbf{U}_l(\xi,\mathbf{x})]^{\mathrm{T}}. \tag{3.1}$$

Here and henceforth the upper symbol T means transposition. Equation (3.1) means that the displacement at the point **x** in the direction of the axis x_i, induced by a unit force applied at the point ξ in the direction of the axis x_j, is equal to the displacement in the direction x_j at the point ξ induced by a unit force at the point **x** in the direction of the axis x_i. This property is clear from (2.3) for the fundamental solution, and follows from the reciprocal theorem in a general case (see § 8).

We conclude from (3.1) that *rows* of the matrix $\mathbf{U}_l(\mathbf{x},\xi)$ taken as functions of ξ satisfy the elasticity equations, because the columns taken as functions of **x** satisfy these equations. Consider a row of $\mathbf{U}_l(\mathbf{x},\xi)$ as a function of ξ. By differentiating it over any coordinate x_k and multiplying by an arbitrary multiplier, we obtain a row that also satisfies the elasticity equations as a function of ξ. By taking another row and making similar operations for any x_k and an arbitrary multiplier, we again obtain a solution of these equations. A linear combination of such solutions is also a solution being taken as a function of ξ.

It is easy to see that the traction operator in (2.4) carries out just these operations on the rows of the matrix $\mathbf{U}_l(\mathbf{x},\xi)$. Hence, the *rows* of the matrix $\mathbf{J}_S(\mathbf{x},\xi)$, defined by (2.4), satisfy elasticity equations taken *as functions of* ξ. As a result, interchanging the symbols **x** and ξ, we obtain the matrix

$$\mathbf{U}_S(\mathbf{x},\xi) = [\mathbf{J}_S(\xi,\mathbf{x})]^{\mathrm{T}}, \tag{3.2}$$

with the *columns* satisfying the elasticity equations (1.4) *as functions of* **x**. This allows us to multiply (3.2) from the right by an arbitrary column-vector that does not depend on **x**. The result again satisfies (1.4) when $\mathbf{x} \neq \xi$. Note that interchanging the symbols **x** and ξ in (2.4), we have

$$\mathbf{U}_S(\mathbf{x},\xi) = [\mathbf{T}_{n\xi}\mathbf{U}_1(\xi,\mathbf{x})]^{\mathrm{T}}. \tag{3.3}$$

The matrix $\mathbf{U}_S(\mathbf{x},\xi)$ is the kernel of the potential analogous to the potential of the double layer in the theory of Laplace's equation. For this reason, we will term it a *kernel of the potential of the double-layer*. Substitution of (3.3) into (3.2) yields

$$\mathbf{U}_S(\mathbf{x},\xi) = [\mathbf{T}_{n\xi}\mathbf{U}_l(\xi,\mathbf{x})]^{\mathrm{T}}. \tag{3.4}$$

The kernel $\mathbf{U}_S(\mathbf{x},\xi)$ has a strong singularity what is marked with the symbol S. In plane problems, this singularity is of the type $1/r$. The columns of $\mathbf{U}_S(\mathbf{x},\xi)$ serve as a field of displacements at points $\mathbf{x}$ other than the singular point ξ.

The traction operator (1.12), applied to the columns of the matrix $\mathbf{U}_S(\mathbf{x},\xi)$, leads to the traction corresponding to these displacements. We obtain the *kernel of the hypersingular potential*

$$\mathbf{J}_H(\mathbf{x},\xi) = \mathbf{T}_{\mathbf{nx}}\mathbf{U}_S(\mathbf{x},\xi). \tag{3.5}$$

At the point $\mathbf{x} = \xi$ it has a hypersingularity, in plane problems of the type $1/r^2$. The index H marks this feature. Insertion (3.4) into (3.5) gives

$$\mathbf{J}_H(\mathbf{x},\xi) = \mathbf{T}_{\mathbf{nx}}\left[\mathbf{T}_{\mathbf{n}\xi}\mathbf{U}_l(\xi,\mathbf{x})\right]^{\mathrm{T}}. \tag{3.6}$$

Equation (3.6) readily yields a reciprocal property, this time for traction:

$$\mathbf{J}_H(\mathbf{x},\xi) = \left[\mathbf{J}_H(\xi,\mathbf{x})\right]^{\mathrm{T}}. \tag{3.7}$$

As agreed, in the kernels (2.2), (2.4), (3.2), (3.5), we consider $\mathbf{x}$ to be a field point while ξ is a source point. These are singular solutions of the elasticity equations: $\mathbf{U}_l(\mathbf{x},\xi)$, $\mathbf{U}_S(\mathbf{x},\xi)$ for displacements; $\mathbf{J}_S(\mathbf{x},\xi)$, $\mathbf{J}_H(\mathbf{x},\xi)$ for traction.

§ 4. LINEAR COMBINATIONS OF SINGULAR SOLUTIONS. POTENTIALS

The system (1.4) is linear. Hence, a linear combination of m singular solutions $\mathbf{F}_1(\mathbf{x},\xi^1),\ldots,\ \mathbf{F}_m(\mathbf{x},\xi^m)$ with a set of singular points $\xi^1,\ldots,\ \xi^m$ is also a solution of (1.4). In particular, if we multiply the initial singular matrix $\mathbf{U}_l(\mathbf{x},\xi)$ from the right by a vector-column $\mathbf{w}$, we obtain a vector which satisfies (1.4). Thus, the vector

$$\mathbf{U}_l(\mathbf{x},\xi^1)\mathbf{w}^1 + \ldots + \mathbf{U}_l(\mathbf{x},\xi^m)\mathbf{w}^m,$$

where $\mathbf{w}^1,\ldots,\ \mathbf{w}^m$ are constant vectors, also satisfies (1.4) as a function of $\mathbf{x}$ when $\mathbf{x} \neq \xi^1,\ldots,\xi^m$.

Analogously, the integral over an arbitrary contour L (Figure 3),

$$\mathbf{u}_l(\mathbf{x}) = \int_L \mathbf{U}_l(\mathbf{x},\xi)\,\mathbf{w}_l(\xi)\,ds_\xi \tag{4.1}$$

satisfies (1.4) at points $\mathbf{x}$ not on L. In other words, (4.1) represents a field of displacements outside L. Here $\mathbf{w}_l$ is a vector-column depending on the point ξ on the contour L. The integral (4.1) is called a *potential of a simple layer*.

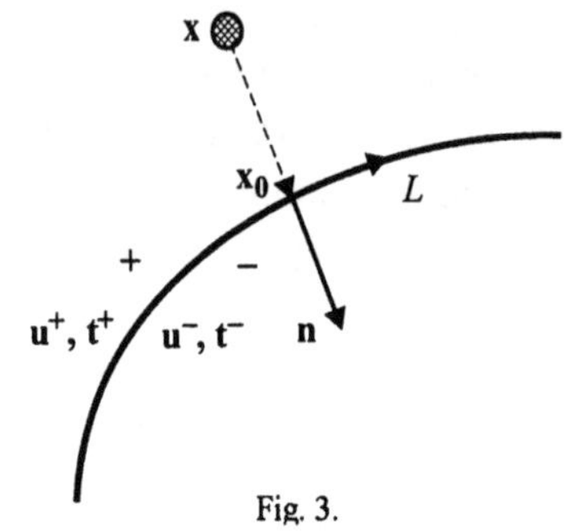

Fig. 3.

Applying the traction operator $\mathbf{T_n}$ to (4.1) we obtain the traction at a field point $\mathbf{x}$ not on L:

$$\mathbf{t}_S(\mathbf{x}) = \mathbf{T}_{\mathbf{nx}}\mathbf{u}_l(\mathbf{x},) = \int\limits_L \mathbf{J}_S(\mathbf{x},\xi)\,\mathbf{w}_l(\xi)ds_\xi. \qquad (4.2)$$

Here we have used the definition of the kernel (2.4). The integral (4.2) may be termed the potential of the traction induced by the potential of a simple layer. It is analogous to the potential of the normal derivative but, in contrast to the latter, it involves differentiation in both normal and tangential directions. This makes integral equations arising from this potential, singular.

In the same way, we may use the kernel (3.4) to obtain a field of displacements at points $\mathbf{x}$ not on L:

$$\mathbf{u}_S(\mathbf{x}) = \int\limits_L \mathbf{U}_S(\mathbf{x},\xi)\,\mathbf{w}_S(\xi)ds_\xi, \qquad (4.3)$$

where the density $\mathbf{w}_S$ is a column vector depending on the point ξ on the contour. The potential (4.3) is termed a *potential of a double layer*. Applying the traction operator $\mathbf{T_n}$ to it, we obtain the corresponding field of traction:

$$\mathbf{t}_H(\mathbf{x}) = \mathbf{T}_{\mathbf{nx}}\mathbf{u}_S(\mathbf{x}) = \int\limits_L \mathbf{J}_H(\mathbf{x},\xi)\,\mathbf{w}_S(\xi)ds_\xi. \qquad (4.4)$$

We have used the definition (3.5) when writing this potential. The potential (4.4) is termed a *hypersingular potential*.

The potentials $\mathbf{u}_l$, $\mathbf{u}_S$, denoted with the symbol $\mathbf{u}$, serve as fields of displacements. The potentials $\mathbf{t}_S$, $\mathbf{t}_H$, denoted with the symbol $\mathbf{t}$ serve as fields of traction. By taking the normal $\mathbf{n_x}$ in the direction of each coordinate axis we obtain all the components of the stress tensor at a field point $\mathbf{x}$. Hence, the potentials (4.2) and (4.4) provide stress fields outside the contour L. The index l, or S, or H marks the type of singularity: l – weak (logarithmic in plane problems), S – strong ($1/r$ in plane problems) and H hypersingular ($1/r^2$ in plane problems). The index l or S in a density marks its origin: $\mathbf{w}_l$ is from the potential of the simple layer, while $\mathbf{w}_S$ is from the potential of the double layer.

§ 5. LIMIT VALUES OF POTENTIALS. PHYSICAL MEANING OF DENSITIES

For sufficiently smooth parts of L and sufficiently smooth densities, the potentials (4.1)-(4.4) have limiting values as $\mathbf{x}$ tends to a point on L. Hence, these potentials may serve to find the fields that satisfy prescribed boundary conditions, for instance, (1.17)-(1.19) on L.

The main terms of the kernels are similar to those in the potentials for Laplace's equation. From this, it follows that the potential of the simple layer is continuous through the contour L:

$$\mathbf{u}_l^+(\mathbf{x}_0) = \mathbf{u}_l^-(\mathbf{x}_0).$$

The same holds for the hypersingular potential if the normal $\mathbf{n}_\mathbf{x}$ is taken in the direction of the normal to L at the point $\mathbf{x}_0$ to which a field point $\mathbf{x}$ tends:

$$\mathbf{t}_H^+(\mathbf{x}_0) = \mathbf{t}_H^-(\mathbf{x}_0).$$

The potentials with the index S, that are $\mathbf{t}_S$ and $\mathbf{u}_S$, in general, have different limits as $\mathbf{x}$ tends to L from different sides. The discontinuities of these potentials are connected with their densities by formulae:

$$\Delta\mathbf{t}(\mathbf{x}_0) = \mathbf{t}_S^+(\mathbf{x}_0) - \mathbf{t}_S^-(\mathbf{x}_0) = \mathbf{w}_l(\mathbf{x}_0), \tag{5.1}$$

$$\Delta\mathbf{u}(\mathbf{x}_0) = \mathbf{u}_S^+(\mathbf{x}_0) - \mathbf{u}_S^-(\mathbf{x}_0) = -\mathbf{w}_S(\mathbf{x}_0). \tag{5.2}$$

From (5.1) we conclude that the density $\mathbf{w}_l$ presents the *traction discontinuity* $\Delta\mathbf{t} = \mathbf{t}_S^+ - \mathbf{t}_S^-$ on the contour L. Hence, the potentials (4.1), (4.2) may serve in problems for open or closed contours with the discontinuities of such type. For instance, they may serve to account for stringers welded to plates. Loads applied to the stringers generate traction discontinuities in a plate, while displacements are continuous.

From (5.2) we see that the density $\mathbf{w}_S$ gives the *displacement discontinuity* $\Delta\mathbf{u} = \mathbf{u}_S^+ - \mathbf{u}_S^-$ on the contour L. Hence, the potentials (4.3), (4.4) may serve in problems for open arcs with displacement discontinuities. These are problems involving cracks and/or blocks (grains) interacting on their boundaries.

We have mentioned that the displacements (4.1) and the traction (4.4) are continuous through L, while (4.2) and (4.3) are discontinuous. Consequently, in a general case, when both traction and displacements are discontinuous, one may use a linear combination of the potentials (4.1) and (4.3) for displacements. Then the corresponding traction is presented by the analogous combination of the potentials

(4.2), (4.4). In particular, this provides equations for cracks when the applied traction is different on opposite sides of a crack.

§ 6.CONNECTION BETWEEN LIMIT AND DIRECT VALUES OF POTENTIALS

The limiting values of the potentials may be expressed by means of the direct values of the integrals. The resulting formulae are:

$$\mathbf{u}_l^{\pm}(\mathbf{x}_0)=\int_L \mathbf{U}_l(\mathbf{x}_0,\xi)\mathbf{w}_l(\xi)ds_\xi, \tag{6.1}$$

$$\mathbf{t}_S^{\pm}(\mathbf{x}_0)=\pm 1/2\,\mathbf{w}_l(\mathbf{x}_0)+\int_L \mathbf{J}_S(\mathbf{x}_0,\xi)\mathbf{w}_l(\xi)ds_\xi, \tag{6.2}$$

$$\mathbf{u}_S^{\pm}(\mathbf{x}_0)=\mp 1/2\,\mathbf{w}_S(\mathbf{x}_0)+\int_L \mathbf{U}_S(\mathbf{x}_0,\xi)\mathbf{w}_S(\xi)ds_\xi, \tag{6.3}$$

$$\mathbf{t}_H^{\pm}(\mathbf{x}_0)=\int_L \mathbf{J}_H(\mathbf{x}_0,\xi)\mathbf{w}_S(\xi)ds_\xi. \tag{6.4}$$

The integral on the r. h. s. of (6.1) is not singular: its kernel has a weak singularity. The integrals on the r. h. s. of (6.2), (6.3) are singular. They are understood as *principal value* (Cauchy) integrals. The integral on the r. h. s. of (6.4) do not exist in a usual sense. It is understood as a *finite-part* (Hadamard) integral. It means that to evaluate it, one should exclude a small ε-vicinity of the point $\mathbf{x}_0$, and delete all the terms in the integral tending to infinity when $\varepsilon\to 0$, i. e. keep only the *finite part* of the integral, and after that come to the limit for $\varepsilon\to 0$. Practical integration based on this definition is quite simple; this becomes especially evident when integration is performed in complex variables (see part III).

These formulae are easy to prove for the simplest initial solution (2.3) by taking L as a straight element. Their extensions to other singular solutions immediately follow since these solutions differ from the fundamental solution (2.3) only by regular terms which do not generate singularities and discontinuities. Extension to arbitrary smooth contours involves only common mathematical techniques. In part III we shall prove formulae of the (6.1) – (6.4) type for the more general case of complex functions.

§ 7. EQUATIONS OF THE INDIRECT APPROACH

The potentials (4.1), (4.3) satisfy elasticity equations for displacements and have limits when a field point $\mathbf{x}$ tends to a point $\mathbf{x}_0$ on the boundary L. The potentials (4.2), (4.4), obtained from (4.1), (4.3) by applying the traction operator, give the corresponding fields of stresses, and also have limits. Hence, we may use these potentials to find the density $\mathbf{w}_l$ or $\mathbf{w}_S$ which makes the boundary conditions of the types (1.17)-(1.19) satisfied. By taking the limit when a field point $\mathbf{x}$ tends to a point $\mathbf{x}_0$ on the contour, and using the boundary conditions, we arrive at boundary integral equations.

The potential of the simple layer (4.1) leads to a Fredholm equation of the first kind. This makes it inconvenient if displacements are prescribed on the boundary. Meanwhile, the corresponding traction, defined by the potential (4.2), does not have this shortcoming. As a result, the potential (4.2) serves for *closed* contours when the condition for traction (1.18) is prescribed on the boundary. By taking the limit from the region D we obtain from (4.2):

$$\lim_{\mathbf{x}\to\mathbf{x}_0^+} \int_L \mathbf{J}_S(\mathbf{x},\xi)\mathbf{w}_l(\xi)ds_\xi = \mathbf{t}_n(\mathbf{x}_0). \tag{7.1}$$

For the given displacements (1.17) on the *closed* boundary, it is reasonable to use the potential of the double layer (4.3). For it, by taking the limit from the region and satisfying the boundary condition we have

$$\lim_{\mathbf{x}\to\mathbf{x}_0^+} \int_L \mathbf{U}_S(\mathbf{x},\xi)\mathbf{w}_S(\xi)ds_\xi = \mathbf{u}(\mathbf{x}_0). \tag{7.2}$$

The boundary integral equations (7.1), (7.2) are written by using the limit *after* integration. The corresponding numerical procedure does not involve evaluation of the principal value integrals.

These equations may also be written in terms of the limit *before* integration. By using (6.2) and (6.3) for the upper signs we obtain

$$\frac{1}{2}\mathbf{w}_l(\mathbf{x}_0) + \int_L \mathbf{J}_S(\mathbf{x}_0,\xi)\mathbf{w}_l(\xi)ds_\xi = \mathbf{t}_n(\mathbf{x}_0), \tag{7.3}$$

$$-\frac{1}{2}\mathbf{w}_S(\mathbf{x}_0) + \int_L \mathbf{U}_S(\mathbf{x}_0,\xi)\mathbf{w}_S(\xi)ds_\xi = \mathbf{u}(\mathbf{x}_0). \tag{7.4}$$

These are *singular* boundary equations: they contain principal value integrals.

For *open* arcs, the potential (4.2) in accordance with the physical meaning of its density, which represents traction discontinuity, may serve to find this discontinuity under given displacements on an arc. Naturally, since the displacements defined by (4.1) are continuous through L, the values of displacements $\mathbf{u}(\mathbf{x}_0)$ should be the same on the both sides of L. Again one may use either the form (7.1) not involving principal value integrals or the form (7.3) involving such integrals.

The potential (4.3) is also of limited use for *open* arcs because, as follows from (5.2), its density contains displacements on both sides of L. Meanwhile, the traction defined by this potential and given by the hypersingular potential (4.4) is very useful when studying discontinuous fields of displacements. In this case, the continuity of the traction through L expressed by (6.4) requires the same prescribed values of the traction on both sides of L. By taking the limit in (4.4) and accounting for (5.2), we arrive at the boundary hypersingular equation

$$\lim_{\mathbf{x}\to\mathbf{x}_0}\int_L \mathbf{J}_H(\mathbf{x},\xi)[-\Delta\mathbf{u}(\xi)]ds_\xi = \mathbf{t}_n(\mathbf{x}_0). \tag{7.5}$$

It is written in the form that corresponds to taking the limit *after* integration. By applying (6.4) we may also write it the form corresponding to coming to the limit *before* integration:

$$\int_L \mathbf{J}_H(\mathbf{x}_0,\xi)[-\Delta\mathbf{u}(\xi)]ds_\xi = \mathbf{t}_n(\mathbf{x}_0). \tag{7.6}$$

Now we consider the field point belonging to the contour from the beginning. This requires a special treatment of the divergent integral on the l. h. s. of (7.6). It is understood as a finite-part (Hadamard) integral. One may use either the form (7.5) not involving the finite-part integrals or the form (7.6) based on such integrals.

The contour in (7.5) and (7.6) may be closed as well. In this case, we actually consider also the region D^- complementing the region D in the whole plane. The density $-\Delta\mathbf{u}$ represents the displacement discontinuity for these two regions under the same traction on their contour.

The contour in (7.5), (7.6) may consist of a finite number of open and closed parts. The hypersingular equations (7.5) or (7.6) serve as a very convenient means for solving problems for cracks and contact interaction: they contain the very values that describe contact interaction, specifically, the traction and the displacement discontinuities.

The equation (7.5) provides a basis of the displacement discontinuity method in its original form suggested by S. L. Crouch (Crouch [1], see also Crouch and Starfield [1]). Applying only *straight elements* to represent a contour and using only *constant density* (displacement discontinuity) on each element, he did not write down the hypersingular equation explicitly. He came to the limit *after* integration over such elements. The direct values of the integrals were not employed. This approach corresponds to zero order approximation both of the contour and the density in (7.5). Presently, we may treat the displacement discontinuity method more broadly, as a general method employing displacement discontinuities. Then (7.5) and (7.6) are hypersingular equations of the displacement discontinuity method for the indirect approach. In § 9, we will see that this method may be extended to the direct approach.

Note two features of the indirect approach. First, as follows from the starting formulae (4.1)-(4.4), the properties of a medium are supposed to be the same on both sides of the integration contour. Secondly, the indirect approach is applicable both for closed and open contours.

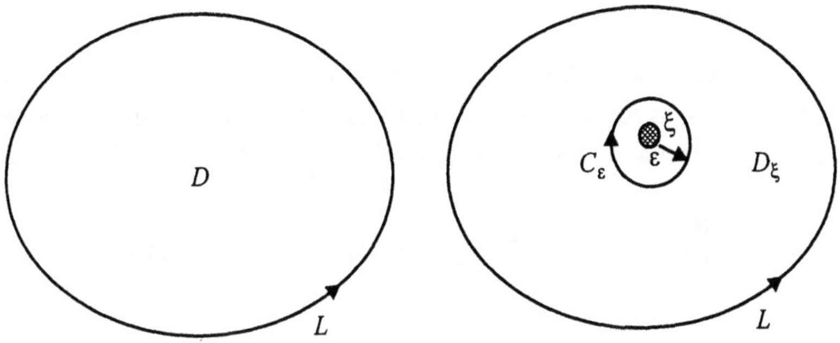

Fig. 4.

§ 8. EQUATIONS OF THE DIRECT APPROACH

The integral equations of the direct approach are based on an integral equation expressing the reciprocal theorem. In elasticity theory this theorem is called Betti's formula. Its proof is simple and can be found in textbooks on elasticity theory (see, e. g. Timoshenko and Goodier [1]). Consider a region D with the contour L. Let $\mathbf{u}(\mathbf{x})$ and $\mathbf{u}^*(\mathbf{x})$ be two sufficiently smooth fields of displacements (Figure 4). Denote the corresponding tractions on L by $\mathbf{t}_n(\mathbf{x})$ and $\mathbf{t}_n^*(\mathbf{x})$. Then Betti's formulae is

$$\int_L \left[\mathbf{t}_n^*(\mathbf{x})\right]^{\mathrm{T}} \mathbf{u}(\mathbf{x})\, ds_x = \int_L \left[\mathbf{u}^*(\mathbf{x})\right]^{\mathrm{T}} \mathbf{t}_n(\mathbf{x})\, ds_x. \tag{8.1}$$

The normal to the boundary is assumed to be outward with respect to the region D.

Note that Betti's formula (8.1) is true only if the fields $\mathbf{u}(\mathbf{x})$ and $\mathbf{u}^*(\mathbf{x})$ do not have singularities within D. Hence, if we take the singular solution $\mathbf{U}_l^{\mathbf{k}}(\mathbf{x},\xi)$ corresponding to the k-th column of the matrix $\mathbf{U}_l(\mathbf{x},\xi)$ as the field $\mathbf{u}^*(\mathbf{x})$ when the source point ξ belongs to D, we need to exclude the point $\mathbf{x} = \xi$ from the considered region (Figure 4). In particular, we may subtract from D a circle with the center at the point ξ and a small radius ε. Denote the remaining region by D_ξ; its contour consists of L and the circumference C_ε of the circle. The normal to C_ε is directed toward the center of the circle, that is, outward from D_ξ, as is supposed in (8.1).

Assume now that $\mathbf{u}(\mathbf{x})$ and $\mathbf{t}_n(\mathbf{x})$ are fields corresponding to the solution of a boundary value problem for the region D, while the fields $\mathbf{u}^*(\mathbf{x})$ and $\mathbf{t}_n^*(\mathbf{x})$ are defined by the k-th column of the matrix $\mathbf{U}_l(\mathbf{x},\xi)$, that is $\mathbf{u}^* = \mathbf{U}_l^{\mathbf{k}}(\mathbf{x},\xi)$, $\mathbf{t}_n^* = \mathbf{J}_S^{\mathbf{k}}(\mathbf{x},\xi)$. Suppose first that the source point ξ belongs to D. Then for the contour $L + C_\varepsilon$ of the region D_ξ Betti's formula (8.1) gives

$$\int_L \left[\mathbf{J}_S^{\mathbf{k}}(\mathbf{x},\xi)\right]^{\mathrm{T}} \mathbf{u}(\mathbf{x})\, ds_x + \int_{C_\varepsilon} \left[\mathbf{J}_S^{\mathbf{k}}(\mathbf{x},\xi)\right]^{\mathrm{T}} \mathbf{u}(\mathbf{x})\, ds_x =$$

$$= \int_L \left[\mathbf{U}_l^{\mathbf{k}}(\mathbf{x},\xi)\right]^{\mathrm{T}} \mathbf{t}_n(\mathbf{x})\, ds_x + \int_{C_\varepsilon} \left[\mathbf{U}_l^{\mathbf{k}}(\mathbf{x},\xi)\right]^{\mathrm{T}} \mathbf{t}_n(\mathbf{x})\, ds_x. \tag{8.2}$$

In the limit when the radius ε tends to zero, the integral from the second term on the l. h. s. of (8.2), as follows from (2.1), tends to $u_k(\xi)$; the integral from the second term on the r. h. s. tends to zero. As a result, for any point ξ *within D* we obtain

$$\int_L \left\{ \left[\mathbf{U}_l^{\mathbf{k}}(\mathbf{x},\xi)\right]^{\mathrm{T}} \mathbf{t}_{\mathbf{n}}(\mathbf{x}) - \left[\mathbf{J}_S^{\mathbf{k}}(\mathbf{x},\xi)\right]^{\mathrm{T}} \mathbf{u}(\mathbf{x}) \right\} ds_x = u_k(\xi). \tag{8.3}$$

Comment. If we take as $\mathbf{u(x)}$ a column j of the matrix $\mathbf{U}_l(\mathbf{x},\xi_1)$ with another source point ξ_1, then we must similarly exclude this point as well. Take the matrix of initial singular solutions $\mathbf{U}_l(\mathbf{x},\xi)$ as Green's matrix satisfying homogeneous boundary conditions on L. As a result, (8.3) takes the form of the reciprocity relation between the components $\mathbf{U}_l(\mathbf{x},\xi)$:

$$U_{lj}^k(\xi,\xi_1) = U_{lk}^j(\xi_1,\xi).$$

By taking the letter $\mathbf{x}$ instead of ξ_1 we may write this equation as

$$\mathbf{U}_l(\mathbf{x},\xi) = \left[\mathbf{U}_l(\xi,\mathbf{x})\right]^{\mathrm{T}}. \tag{8.4}$$

This proves (3.1) for an arbitrary Green's matrix $\mathbf{U}_l(\mathbf{x},\xi)$.

Return to the case when $\mathbf{u(x)}$ is not a singular solution but the solution of a boundary value problem for the region D. Take the source point outside D and L. Then $\mathbf{U}_l(\mathbf{x},\xi)$ does not have singularity within D, and there is no need to cut out a circle with the circumference C_ε. Hence,

$$\int_L \left\{ \left[\mathbf{U}_l^{\mathbf{k}}(\mathbf{x},\xi)\right]^{\mathrm{T}} \mathbf{t}_{\mathbf{n}}(\mathbf{x}) - \left[\mathbf{J}_{\mathrm{S}}^{\mathbf{k}}(\mathbf{x},\xi)\right]^{\mathrm{T}} \mathbf{u}(\mathbf{x}) \right\} ds_x = \begin{cases} u_k(\xi), & \xi \in D, \\ 0, & \xi \notin D+L. \end{cases}$$

Interchange the symbols $\mathbf{x}$ and ξ to have, as usual, the integration variable denoted by ξ:

$$\int_L \left\{ \left[\mathbf{U}_l^{\mathbf{k}}(\xi,\mathbf{x})\right]^{\mathrm{T}} \mathbf{t}_{\mathbf{n}}(\xi) - \left[\mathbf{J}_S^{\mathbf{k}}(\xi,\mathbf{x})\right]^{\mathrm{T}} \mathbf{u}(\xi) \right\} ds_\xi = \begin{cases} u_k(\mathbf{x}), & \mathbf{x} \in D, \\ 0, & \mathbf{x} \notin D+L. \end{cases}$$

These equations are true for any column of the matrix of the initial singular solutions $\mathbf{U}_l(\mathbf{x},\xi)$. Hence, we may use the matrices $\mathbf{U}_l(\mathbf{x},\xi)$ and $\mathbf{J}_S(\mathbf{x},\xi)$ themselves and we arrive at the Somigliana identities for displacements:

$$\int_L \left\{ \left[\mathbf{U}_l(\xi, \mathbf{x}) \right]^T \mathbf{t}_n(\xi) - \left[\mathbf{J}_S(\xi, \mathbf{x}) \right]^T \mathbf{u}(\xi) \right\} ds_\xi = \begin{cases} \mathbf{u}(\mathbf{x}), & \mathbf{x} \in D, \\ 0, & \mathbf{x} \notin D + L. \end{cases} \quad (8.5)$$

At last, taking into account (8.4) and the definition (3.2) of the kernel $\mathbf{U}_S(\mathbf{x},\xi)$, we obtain the final Somigliana identities for *displacements*:

$$\int_L \left[\mathbf{U}_l(\mathbf{x},\xi)\mathbf{t}_n(\xi) - \mathbf{U}_S(\mathbf{x},\xi)\mathbf{u}(\xi) \right] ds_\xi = \begin{cases} \mathbf{u}(\mathbf{x}), & \mathbf{x} \in D, \\ \mathbf{0}, & \mathbf{x} \notin D + L. \end{cases} \quad (8.6)$$

We see that the l. h. s. of (8.6) is a linear combination of the potentials of the simple (4.1) and the double (4.3) layer. Naturally, it gives a field of displacements. Moreover, the densities in these potentials have the prescribed physical meaning: they represent the limiting values of the *actual* displacements and the *actual* traction. Consequently, the combination is such that it gives the *actual* displacements *within* the region D and *zero* displacements *outside* it.

Apply the traction operator to (8.6) and account for the definitions (2.4), (3.5) of the singular and hypersingular traction potentials. Then we arrive at the Somigliana identities for *traction*:

$$\int_L \left[\mathbf{J}_S(\mathbf{x},\xi)\mathbf{t}_n(\xi) - \mathbf{J}_H(\mathbf{x},\xi)\mathbf{u}(\xi) \right] ds_\xi = \begin{cases} \mathbf{t}_n(\mathbf{x}), & \mathbf{x} \in D, \\ \mathbf{0}, & \mathbf{x} \notin D + L. \end{cases} \quad (8.7)$$

We see that the l. h. s. of (8.7) is a linear combination of the potentials of the singular (4.2) and hypersingular (4.4) traction potential. Naturally, it gives a field of traction. Since, as mentioned, the densities of these potentials have the prescribed physical meaning, it appears that the combination is such that it gives the *actual* traction *within* the region D and *zero* traction *outside* it.

Note that the meaning of the densities is in a complete agreement with their physical interpretation discussed in § 5. Indeed, we may consider the second lines in (8.6), (8.7) as zero displacements and traction outside the region D. Hence, the difference of the limiting values from the left and the right of L is simply equal to the limit from the left side. Then from (8.6), accounting for (6.1), (6.3), we see that in accordance with (5.2)

$$\mathbf{u}_S^+ - \mathbf{u}_S^- = \Delta\mathbf{u}\,.$$

Analogously, from (8.7) accounting for (6.2), (6.4), we see that in accordance with (5.1)

$$\mathbf{t}_S^+ - \mathbf{t}_S^- = \Delta\mathbf{t}\,.$$

Take the limits in the first lines of (8.6), (8.7) when the field point tends to the contour from *within* the region D. Then we arrive at the boundary integral equations of the direct approach, corresponding to coming to the limit *after* integration:

$$\lim_{\mathbf{x}\to\mathbf{x}_0^+}\left[\int_L \mathbf{U}_l(\mathbf{x},\xi)\mathbf{t}_n(\xi)\,ds_\xi - \int_L \mathbf{U}_S(\mathbf{x},\xi)\mathbf{u}(\xi)\,ds_\xi\right] = \mathbf{u}(\mathbf{x}_0), \tag{8.8}$$

$$\lim_{\mathbf{x}\to\mathbf{x}_0^+}\left[\int_L \mathbf{J}_S(\mathbf{x},\xi)\mathbf{t}_n(\xi)\,ds_\xi - \int_L \mathbf{J}_H(\mathbf{x},\xi)\mathbf{u}(\xi)\,ds_\xi\right] = \mathbf{t}_n(\mathbf{x}_0). \tag{8.9}$$

Equation (8.8) serves for elasticity problems under prescribed values of displacements on the closed contour of a region. Equation (8.9) is useful under given values of traction. These forms do not contain the direct values of integrals; they use the limits of usual proper integrals.

Just as in the case of the indirect approach, we may apply formulae (6.1)-(6.4) for the limiting values of the potentials. Then the boundary integral equations take the form corresponding to coming to the limit *before* integration:

$$\int_L \mathbf{U}_l(\mathbf{x}_0,\xi)\mathbf{t}_n(\xi)\,ds_\xi - \int_L \mathbf{U}_S(\mathbf{x}_0,\xi)\mathbf{u}(\xi)\,ds_\xi = \frac{1}{2}\mathbf{u}(\mathbf{x}_0), \tag{8.10}$$

$$\int_L \mathbf{J}_S(\mathbf{x}_0,\xi)\mathbf{t}_n(\xi)\,ds_\xi - \int_L \mathbf{J}_H(\mathbf{x}_0,\xi)\mathbf{u}(\xi)\,ds_\xi = \frac{1}{2}\mathbf{t}_n(\mathbf{x}_0). \tag{8.11}$$

Now in (8.10), (8.11), in contrast to (8.8), (8.9), the integrals with the kernels marked by the index S are understood to be principal value (Cauchy) integrals. In (8.11), in contrast to (8.9), the integral with the kernel $\mathbf{J}_H(\mathbf{x},\xi)$ is understood to be a finite-part (Hadamard) integral.

§ 9. EQUATIONS FOR BLOCKY STRUCTURES AND OPEN ARCS

The direct approach is initially applicable only to closed surfaces of homogeneous regions. Meanwhile, it may be extended to open surfaces and, in plane problems, to piece-wise homogeneous regions. For the latter problems, the extended procedure was suggested by Linkov [6].

Consider a system of n interacting elastic parts, which we will term blocks (Figure 5). Mark with the superscript j the values referring to the j–th bock. Write down equations for *displacements* (8.10) and the second lines of (8.6) for the j-th block:

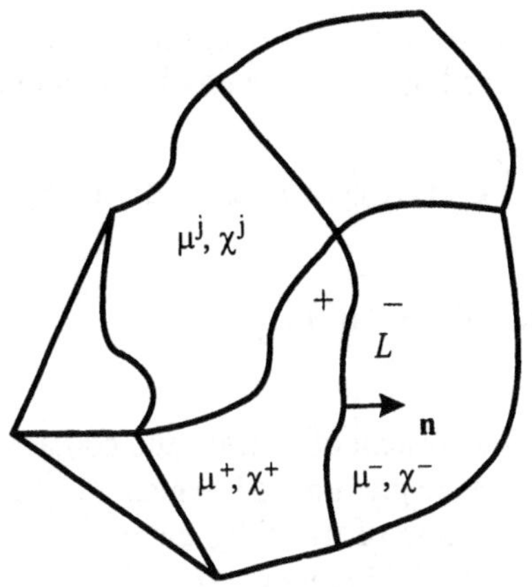

Fig. 5.

$$\int_{L^j} \mathbf{U}_l^{(j)}(\mathbf{x},\xi)\mathbf{t}_n^j(\xi)ds_\xi - \int_{L^j} \mathbf{U}_S^{(j)}(\mathbf{x},\xi)\mathbf{u}^j(\xi)ds_\xi = \begin{cases} \frac{1}{2}\mathbf{u}(\mathbf{x}), & \mathbf{x} \in L^j, \\ \mathbf{0}, & \mathbf{x} \notin D^j + L^j. \end{cases} \tag{9.1}$$

For convenience, we have omitted the index "zero" in the notation of the field point when it belongs to the contour L^j.

Assume that each of the blocks is homogeneous and take Kelvin's fundamental solutions as an initial matrix of the fundamental solutions. For plane problems its components are given by (2.3). From (2.3) it is evident that the matrix $\mathbf{U}_l^{(j)}(\mathbf{x},\xi)$ has the common multiplier $1/(2\mu^j)$. (From dimension considerations, the same is true also in 3D, which allows us to refer the results to 3D). In accordance with the definition (3.4), the matrix $\mathbf{U}_S^{(j)}(\mathbf{x},\xi)$ is obtained from $\mathbf{U}_l^{(j)}(\mathbf{x},\xi)$ by applying the traction operator (1.13), which includes the common multiplier $2\mu^j$. Hence, the matrix $\mathbf{U}_S^{(j)}(\mathbf{x},\xi)$ does not depend on the shear modulus. Suppose that Poisson's ratio is the same for all the blocks. Then the matrices $\mathbf{U}_l^{(j)}(\mathbf{x},\xi)$ differ only by constant multipliers $1/(2\mu^j)$, while the matrices $\mathbf{U}_S^{(j)}(\mathbf{x},\xi)$ are the same for all the blocks. Hence, we may drop the identification symbol j in the product $\mu^j\mathbf{U}_l^{(j)}(\mathbf{x},\xi)$ and in $\mathbf{U}_S^{(j)}(\mathbf{x},\xi)$ writing simply $\mu\mathbf{U}_l(\mathbf{x},\xi)$ and $\mathbf{U}_S(\mathbf{x},\xi)$. Multiplying (9.1) by μ^j we have

$$\int_{L^j} \mu\,\mathbf{U}_l(\mathbf{x},\xi)\mathbf{t}_n^j(\xi)ds_\xi - \int_{L^j} \mu^j\,\mathbf{U}_S(\mathbf{x},\xi)\mathbf{u}^j(\xi)ds_\xi = \begin{cases} \frac{1}{2}\mu\,\mathbf{u}(\mathbf{x}), & \mathbf{x} \in L^j, \\ \mathbf{0}, & \mathbf{x} \notin D^j + L^j. \end{cases}$$

Arbitrarily fix the direction of the normal at each point of the boundaries of blocks in contact. Then at the points of L^j where the normal is inward, we have to take the expressions under the integral with a minus sign. Having this in mind, we sum the equations over all the blocks. Boundaries in contact will be traveled twice with opposite signs of the expressions under the integrals; since the l. h. s. is equal to zero when **x** does not belong to the boundary of a block, we obtain

$$\int_L \mu \mathbf{U}_l(\mathbf{x},\xi)\Delta\mathbf{t}(\xi)ds_\xi - \int_L \mathbf{U}_S(\mathbf{x},\xi)\left[\mu^+\mathbf{u}^+(\xi) - \mu^-\mathbf{u}^-(\xi)\right]ds_\xi =$$
$$= \frac{1}{2}\left[\mu^+\mathbf{u}^+(\mathbf{x}) + \mu^-\mathbf{u}^-(\mathbf{x})\right], \quad \mathbf{x}\in L, \tag{9.2}$$

where L is the total boundary of the system of blocks (the contact between surfaces of adjacent blocks is treated as a single line on which mechanical values may experience discontinuities); the direction of traveling along the contact is arbitrary, the normal is directed to the right of this direction; $\Delta\mathbf{t} = \mathbf{t_n}^+ - \mathbf{t_n}^-$ is the discontinuity of traction on L. As usual, the index "plus" ("minus") refers to a value from the left (right) of the traveling path. On the external boundaries of the blocks we consider $\mathbf{t_n}^- = \mathbf{0}$, $\mathbf{u}^- = \mathbf{0}$, $\mu^- = 0$.

Equation (9.2) is convenient for problems with *traction discontinuities* on the boundaries of parts that may have different shear moduli. In particular, these are problems for plates with stringers welded along contacts. In such problems, one may consider displacements to be continuous along the contacts: $\mathbf{u}^+ = \mathbf{u}^- = \mathbf{u}$. Then (9.2) becomes

$$\int_L \mu \mathbf{U}_l(\mathbf{x},\xi)\Delta\mathbf{t}(\xi)ds_\xi - \int_L \left(\mu^+ - \mu^-\right)\mathbf{U}_S(\mathbf{x},\xi)\mathbf{u}(\xi)ds_\xi = \frac{\mu^+ + \mu^-}{2}\mathbf{u}(\mathbf{x}), \quad \mathbf{x}\in L. \tag{9.3}$$

On the parts of L where the traction is also continuous ($\Delta\mathbf{t} = \mathbf{0}$), while shear moduli are the same ($\mu^+ = \mu^- = \mu$), both integrals on the r. h. s. of (9.3) vanish. Hence, we may exclude such parts from the contour L. This means that the contour L may contain open arcs. In the case when $\mu^+ = \mu^- = \mu$ in the whole plane, (9.3) coincides with equation (6.1) of the indirect approach, where $\mathbf{u}_l = \mathbf{u}$ and, in accordance with (5.1), $\mathbf{w}_l = \Delta\mathbf{t}$.

Apply analogous reasoning to the equations of the direct approach for *traction*. For the j-th block we have from (8.11) and the second line of (8.7):

$$\int_{L^j} \mathbf{J}_S^{(j)}(\mathbf{x},\xi)\mathbf{t}_\mathbf{n}^j(\xi)ds_\xi - \int_{L^j} \mathbf{J}_H^{(j)}(\mathbf{x},\xi)\mathbf{u}^j(\xi)\,ds_\xi = \begin{cases}\mathbf{t_n}(\mathbf{x})/2, & \mathbf{x}\in L^j,\\ \mathbf{0}, & \mathbf{x}\notin D^j + L^j.\end{cases} \tag{9.4}$$

For isotropic blocks, the matrix $\mathbf{J}_S^{(j)}(\mathbf{x},\xi)$ does not depend on the shear modulus μ^j while the matrix $\mathbf{J}_H^{(j)}(\mathbf{x},\xi)$ contains μ^j as a common multiplier. *The peculiarity of plane problems is that the matrix $J_H^{(j)}(x,\xi)$ contains Poisson's ratio* ν *only in a* *common multiplier* $1/(\chi^j+1)$, where χ is Muskhelishvili's parameter (1.8). Hence, $\frac{\chi^j+1}{2\mu^j}\mathbf{J}_H^{(j)}(\mathbf{x},\xi)$ does not depend on both μ and ν.

These properties of $\mathbf{J}_H^{(j)}(\mathbf{x},\xi)$ allow us to write $\frac{\chi^j+1}{2\mu^j}\mathbf{J}_H^{(j)}(\mathbf{x},\xi)$ without the identification symbol j as $\frac{\chi+1}{2\mu}\mathbf{J}_H(\mathbf{x},\xi)$. By multiplying (9.4) by $\frac{\chi^j+1}{2\mu^j}$ and summing (9.4) over all the blocks we obtain

$$\int_L\left[\frac{\chi^++1}{2\mu^+}\mathbf{J}_S^+(\mathbf{x},\xi)\mathbf{t}_\mathbf{n}^+(\xi)-\frac{\chi^-+1}{2\mu^-}\mathbf{J}_S^-(\mathbf{x},\xi)\mathbf{t}_\mathbf{n}^-(\xi)\right]ds_\xi-\int_L\frac{\chi+1}{2\mu}\mathbf{J}_H(\mathbf{x},\xi)\Delta\mathbf{u}(\xi)\,ds_\xi=$$

$$=\frac{1}{2}\left(\frac{\chi^++1}{2\mu^+}\mathbf{t}_\mathbf{n}^+(\mathbf{x})+\frac{\chi^-+1}{2\mu^-}\mathbf{t}_\mathbf{n}^-(\mathbf{x})\right),\quad \mathbf{x}\in L, \tag{9.5}$$

where $\Delta\mathbf{u}=\mathbf{u}^+-\mathbf{u}^-$ is the displacement discontinuity; on the external boundaries of the blocks we assume $\mathbf{t}_\mathbf{n}^-=\mathbf{0}$, $\mathbf{u}^-=\mathbf{0}$, $1/\mu^-=0$.

Equation (9.5) is convenient for problems with *displacement discontinuities* on the contacts of interacting blocks, which may have different shear moduli, and also on the surfaces of cracks. In such problems, one may consider traction to be continuous through the contacts: $\mathbf{t}_\mathbf{n}^+=\mathbf{t}_\mathbf{n}^-=\mathbf{t}_\mathbf{n}$. Then (9.5) becomes

$$\int_L\left[\frac{\chi^++1}{2\mu^+}\mathbf{J}_S^+(\mathbf{x},\xi)-\frac{\chi^-+1}{2\mu^-}\mathbf{J}_S^-(\mathbf{x},\xi)\right]\mathbf{t}_\mathbf{n}(\xi)\,ds_\xi-\int_L\frac{\chi+1}{2\mu}\mathbf{J}_H(\mathbf{x},\xi)\Delta\mathbf{u}(\xi)\,ds_\xi=$$

$$=\frac{1}{2}\left(\frac{\chi^++1}{2\mu^+}+\frac{\chi^-+1}{2\mu^-}\right)\mathbf{t}_\mathbf{n}(\mathbf{x}),\quad \mathbf{x}\in L. \tag{9.6}$$

In a general case, the system of equations (9.2), (9.5) allows us to solve problems for a media with internal structure when discontinuities on the boundaries of structural elements occur both in displacements and in tractions. They may serve also when structural blocks have their own surfaces of discontinuities and/or inclusions (solid lines L_0 in Figure 6). To see this, note that if we have $\Delta\mathbf{u}=\mathbf{0}$, $\Delta\mathbf{t}_n=\mathbf{0}$ and the same elastic properties on both sides of some lines C, shown in Figure 6 with the dashed arcs, then the integrals in the l. h. s. of (9.2), (9.5) are zero over C. Consequently, we may add such curves to the contour L or drop them.

Then, representing a block D^j, containing arcs L_0 as a sum of sub-blocks D^j_1, D^j_2, D^j_3, D^j_4 (Figure 6) and referring the equations (9.2), (9.5) to the new system of

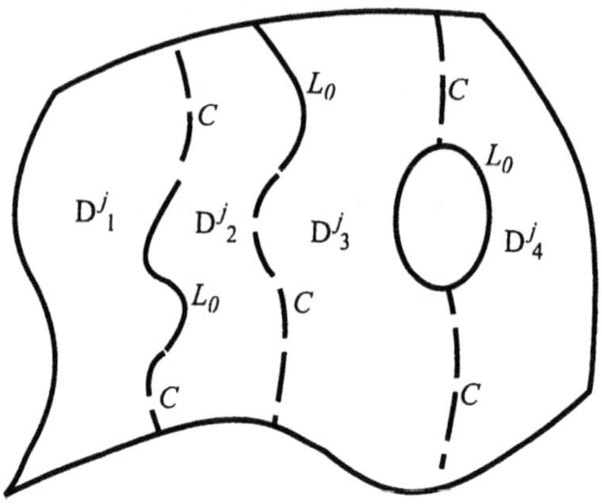

Fig. 6.

blocks, we may keep in the total contour L only those parts of contacts which include L_0. In other words, it is sufficient to include the internal arcs L_0 of discontinuities into the contour L. This means that the contour L may contain open arcs, in particular, internal cracks, and cracks terminating on the boundaries of blocks.

Equations (9.5), (9.6) are attractive for the numerical solution of problems involving internal structure of a medium. They contain the very values, displacement discontinuities and traction, which characterize contact interaction on the surfaces of blocks (grains) and cracks. Involving *displacement discontinuities*, they provide a basis for extension of the displacement discontinuity method by Crouch [1] to the direct approach of the BEM. On the other hand, we will see that for a homogeneous medium they reduce to the equations of the indirect approach.

§ 10. CONNECTION BETWEEN THE INDIRECT AND DIRECT APPROACH

Suppose we have only two "blocks" with the same properties. The first of them represents the given region D, the second is the region D^- complementing D in the whole plane (Figure 7). Then $\mu^+ = \mu^- = \mu$, $\mathbf{J}_S^+(\mathbf{x},\xi) = \mathbf{J}_S^-(\mathbf{x},\xi)$.

Consider two cases: 1) the traction is the same on the common boundary $\mathbf{t}_n^+ = \mathbf{t}_n^- = \mathbf{t}_n$ and 2) the displacements are the same on the common boundary $\mathbf{u}^+ = \mathbf{u}^- = \mathbf{u}$.

For common *traction*, accounting for the equal elastic properties of the blocks, we may write (9.2) and (9.5) as

$$\int_L \mathbf{U}_S(\mathbf{x},\xi)[-\Delta\mathbf{u}(\xi)]ds_\xi = \frac{1}{2}[\mathbf{u}^+(\mathbf{x}) + \mathbf{u}^-(\mathbf{x})], \tag{10.1}$$

$$\int_L \mathbf{J}_H(\mathbf{x},\xi)[-\Delta\mathbf{u}(\xi)]ds_\xi = \mathbf{t}_n(\mathbf{x}), \tag{10.2}$$

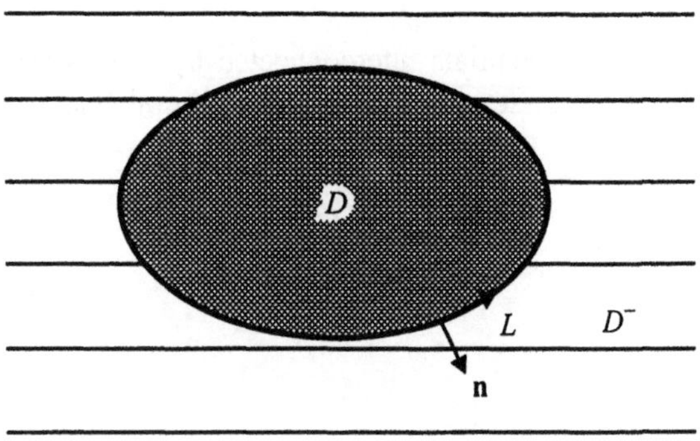

Fig. 7.

where, as previously, $\Delta\mathbf{u} = \mathbf{u}^+ - \mathbf{u}^-$, hence, $\mathbf{u}^- = \mathbf{u}^+ - \Delta\mathbf{u}$. Substitution of the last expression into (10.1) yields

$$\frac{1}{2}\Delta\mathbf{u}(\mathbf{x}) + \int_L \mathbf{U}_S(\mathbf{x},\xi)[-\Delta\mathbf{u}(\xi)]ds_\xi = \mathbf{u}^+(\mathbf{x}). \tag{10.3}$$

Comparing the equation of the direct approach (10.3) with the equation (7.4) of the indirect approach shows that these equations coincide if we take into account the physical meaning of the density $\mathbf{w}_S$: in accordance with (5.2) it represents negative displacement discontinuity. Similarly, the equation of the direct approach (10.2) coincides with the equation (7.6) of the indirect approach.

For common *displacements*, (9.5) becomes

$$\frac{1}{2}\Delta\mathbf{t}_n(\mathbf{x}) + \int_L \mathbf{J}_S(\mathbf{x},\xi)\Delta\mathbf{t}_n(\xi)ds_\xi = \mathbf{t}_n^+(\mathbf{x}), \tag{10.4}$$

where we used $\mathbf{t}_n^- = \mathbf{t}_n^+ - \Delta\mathbf{t}$. Comparing the equation of the direct approach (10.4) with the equation (7.3) of the indirect approach, we see that these equations coincide if we take into account the physical meaning of the density $\mathbf{w}_l$: in accordance with (5.1) it represents the traction discontinuity.

The fact that equations (10.2), (10.4) obtained on the basis of the direct approach coincide with equations (7.6), (7.3) of the indirect approach, was noted in the paper by Pierce, Spottiswoode and Napier [1]. Although the authors discussed Kelvin's solution, their conclusions hold for an arbitrary initial singular solution.

We may exclude from the contour of integration those parts through which both displacements and traction are continuous. Then the stated equivalence of (10.4) with (7.3), (10.3) with (7.4) and (10.2) with (7.6) is extended to open arcs.

We conclude that the equations of the indirect approach obtained in § 7 may be considered as particular cases of the equations of the direct approach. Hence, in the form of equations for blocky systems, the direct method is more general than the indirect approach. It is important that under the mentioned conditions, the direct approach includes cases of blocks with different physical properties. However, the direct approach employs the potentials interconnected by Betti's formula (8.1). For the indirect approach, one may try to employ more general linear combinations of the potentials.

Chapter 2

SINGULAR SOLUTIONS AND POTENTIALS IN COMPLEX FORM

In this Chapter we will follow the procedure described in Chapter 1. In fact, we will simply rewrite the equations in complex variable form. Such a rewriting gives significant benefit for the theoretical analysis and computations. We will need some results on complex variables and functions of complex variables that are given in the first section.

§ 11. PREREQUISITES

Notation. We assume that the reader is familiar with the basics of the theory of the functions of a complex variable (see, e. g. Carrier et al. [1], Churchill [1]). As in the previous chapter, the co-ordinates are taken in a global *right-handed* Cartesian system (Figure 8). However, now we will use the complex variable rather than tensor notation for vectors and multiplication. Thus, as is common in the theory of functions of a complex variable, we will denote the *Cartesian* co-ordinates by x and y, rather than x_1, x_2, which were convenient in Einstein's summation rule; we no longer use this rule. For a source point we will use ξ, η instead of ξ_1, ξ_2.

The complex co-ordinate z of a field point (x, y) is defined as

$$z = x + iy, \tag{11.1}$$

where $i = \sqrt{-1}$ is the imaginary unit. Complex conjugates are marked with an over-bar:

$$\bar{z} = x - iy. \tag{11.2}$$

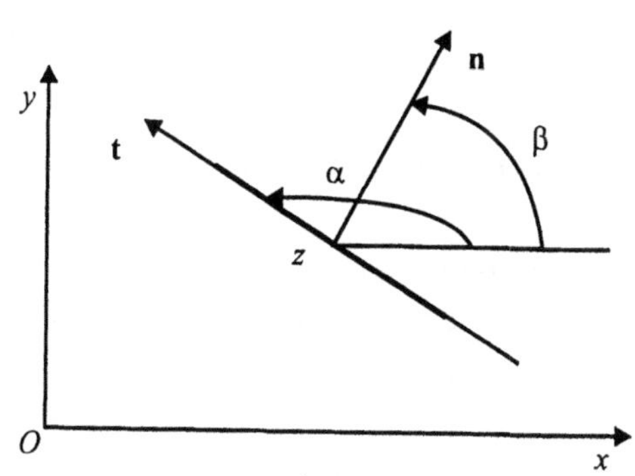

Fig. 8.

From (11.1) and (11.2) it follows that the Cartesian co-ordinates x, y are expressed through the complex variables z and $\bar{z}$ as

$$x = \frac{z+\bar{z}}{2}, \quad y = \frac{z-\bar{z}}{2i}. \tag{11.3}$$

Equations (11.3) imply that a function $g(x, y)$ of two *real* co-ordinates x and y may be considered as a function of two *complex* co-ordinates z and $\bar{z}$: $g_1(z, \bar{z}) = g((z + \bar{z})/2, (z - \bar{z})/(2i))$. For a source point (ξ, η) we have the complex co-ordinate τ and its complex conjugate: $\tau = \xi + i\eta$, $\bar{\tau} = \xi - i\eta$. Hence, $\xi = (\tau + \bar{\tau})/2$, $\eta = (\tau - \bar{\tau})/(2i)$. This implies that any singular solution $P(\mathbf{x}, \xi)$, given as a function of the real co-ordinates of the field point $\mathbf{x}$ and the source point ξ, may be considered as a function of the complex co-ordinates z, $\bar{z}$ of the field point and the complex co-ordinates τ, $\bar{\tau}$ of the source point.

There is a one-to-one relation between a vector $\mathbf{a}$ with real components a_x, a_y, and the complex number a, given by

$$a = a_x + ia_y. \tag{11.4}$$

In the same fashion there is a one-to-one relation between a vector field $\mathbf{a}(z, \bar{z}) = (a_x(x,y), a_y(x,y))^{\mathrm{T}}$ and the complex function

$$a_*(z, \bar{z}) = a_x + ia_y,$$

of the complex co-ordinates z and $\bar{z}$. The star subscript used here and in a few places below denotes that the variables z and $\bar{z}$ are independent. Actually, we need to consider co-ordinates z, $\bar{z}$ to be *independent* only at the very primary step: when transforming real functions of real variables into complex ones. (This will be the case when (i) deriving a complex variable form of the traction operator and (ii) presenting a complex form of Kelvin's solution). After this, the complex variable z completely defines the location of a point (x, y) and consequently its conjugate $\bar{z}$. Now the latter *depends* on z; the function a_* becomes a common composite complex function and may be written as a function $a(z)$ of the one complex variable z:

$$a(z) = a_*(z, \bar{z}(z)) = a_x + ia_y. \tag{11.5}$$

Of course, in general, this function is not holomorphic.

For a displacement field $\mathbf{u}(u_x, u_y)$, by writing it in the form (11.5), we have its usual expression (Muskhelishvili [5])

$$u(z) = u_x + iu_y. \tag{11.6}$$

Almost always we will express a *displacement* field in terms of its components (u_x, u_y) given in the *global* system (x, y). In contrast, the components of a *traction* field will be usually prescribed in a *local* Cartesian system (**n**, **t**) at the field point $z = x + iy$, or at the source point $\tau = \xi + i\eta$.

The local co-ordinates will be always introduced similarly for a field and a source point as follows (Figure 8). For a field point, we fix the *normal* **n** as a unit outward vector normal to an element on which we find the traction. Its direction has the angle β with the x-axis. The tangent vector **t** is directed to the left of **n**. For a source point, we fix the *tangent* **t** in the direction of travel along the element; accordingly, the normal **n** is directed to the right of **t**. In both cases the angle α between the tangent and the x-axis is $\alpha = \pi/2 + \beta$.

The pair (**n**, **t**) comprises a *right-handed* Cartesian system. In this system a vector **b** with co-ordinates b_n, b_t may be written as a complex vector:

$$b = b_n + ib_t. \tag{11.7}$$

To obtain the vector in the global system, we need to multiply (11.7) by $\exp(i\beta)$. As $\beta = \alpha - \pi/2$, we have

$$b_x + ib_y = -i\exp(i\alpha)b. \tag{11.8}$$

From (11.6)-(11.8) it follows that a complex vector of displacements $u(z) = u_x + iu_y$ in the global system is connected to the complex vector of displacements $v(z) = v_n + iv_t$ in the local system by the equation

$$u(z) = -i\exp(i\alpha)v(z). \tag{11.9}$$

The complex vector of the traction σ on an element with the normal **n** and the tangent **t** will be normally given in the local system (**n**, **t**) shown in Figure 8:

$$\sigma(z) = \sigma_{nn} + i\sigma_{nt}. \tag{11.10}$$

In the global system (x, y), we have from (11.8)

$$\sigma_{nx} + i\sigma_{ny} = -i\exp(i\alpha)\sigma(z). \tag{11.11}$$

The complex traction operator. The traction operator T_z defines the connection between a complex field of displacements $u(z)$ and the corresponding complex field of traction $\sigma(z)$. This operator accounts for Hooke's law and the direction of the normal to an element at a point z:

$$\sigma(z) = T_z u(z). \tag{11.12}$$

In accordance with the accepted conventions, we consider that the displacement vector is given in the global, and the traction vector in the local system. The index z in (11.12) shows the variable with respect to which the traction is found. This distinction is necessary because below we will use fields depending on two different complex variables: z of a field point and τ of a source point.

For an isotropic medium, from the definitions (11.10), (11.11), (1.13) one may obtain the following explicit expression for the complex traction operator in plane elasticity (see, e.g. Rabotnov [1]):

$$T_z u(z) = 2\mu\left[\left(\frac{\partial u_*}{\partial z} + \frac{\partial \overline{u_*}}{\partial \overline{z}}\right)\frac{1}{\chi - 1} - \exp(-2i\alpha)\frac{\partial \overline{u_*}}{\partial \overline{z}}\right], \tag{11.13}$$

where χ is Muskhelishvili's parameter, defined by (1.8). Note that here we are writing a real formula in its complex form. Consequently, the r. h. s. of (11.13) presumes that the field u is a function of two *independent* complex variables z and $\overline{z}$; this is marked by the subscript "star"; when applying (11.13), in contrast with further derivations, we assume $\partial \overline{z}/\partial z = \partial z/\partial \overline{z} = 0$.

Equation (11.13) gives all the components of a stress tensor in the global system. Indeed, taking $\alpha = \pi/2$, we have $\sigma_{nn} = \sigma_{xx}$, $\sigma_{nt} = \sigma_{xy}$; taking $\alpha = 0$, we have $\sigma_{nn} = \sigma_{yy}$, $\sigma_{nt} = -\sigma_{xy}$.

Resultant force. In plane elasticity problems, solved in terms of complex variables, we employ an additional value not used in real variables. This is the resultant force applied to an arc. The resultant force is very useful in the theory developed by Muskhelishvili because it allows us to formulate the problems under given boundary traction in a form similar to that for given displacements (see Muskhelishvili [5, § 41, 41a]).

The resultant force is defined as a force applied to the positive side of the arc which connects a fixed point z_0 with the point z (Figure 9). Remember that the positive side is considered to be on the left of the travel direction from z_0 to z. Thus, by definition, the complex resultant force is

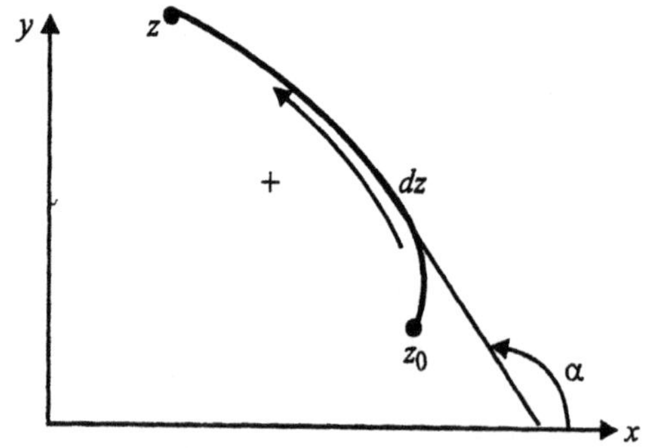

Fig. 9.

$$h(z)=\int_0^s(\sigma_{nx}+i\sigma_{ny})ds, \tag{11.14}$$

where s is the length of the contour of integration from the point z_0; σ_{nx}, σ_{ny} are the components of traction, in this case, in the global system (x, y). The result in (11.14) does not depend on a particular integration path when the contours between z_0 and z belong to the same simply connected region and there are no body forces (Muskhelishvili [5]).

Note that the differential $dz = dx + idy$ of an arc is connected with the increment of its length by equation

$$dz=\exp(i\alpha)ds, \tag{11.15}$$

where α is the angle between the traveling direction and the axis x (Figure 8).

Normally, in plane elasticity we do not use the resultant force itself, but the function $f(z)$ which differs from $h(z)$ by the multiplier i;

$$f(z)=ih(z)=i\int_0^s(\sigma_{nx}+i\sigma_{ny})ds. \tag{11.16}$$

For brevity, when there may be no confusion, we will call this value the resultant force. Using (11.11), (11.15), we can write (11.16) as

$$f(z)=\int_{z_0}^{z}\sigma(z)dz. \tag{11.17}$$

Note that the traction under the integral sign is taken in *local* co-ordinates, connected with the integration point. Integration in (11.17) presumes that in general σ depends on z both explicitly and through the dependence of $\bar{z}$ on z. Vice versa, having the resultant force given in a form of a composite function $f(z) = f_*(z, \bar{z}(z))$, we obtain the traction $\sigma(z)$ by differentiating the traction in accordance with the formula for a composite function:

$$\sigma(z)=\frac{df}{dz}=\frac{\partial f_*}{\partial z}+\frac{\partial f_*}{\partial \bar{z}}\frac{d\bar{z}}{dz}. \tag{11.18}$$

Take into account that along the integration path the arguments z and $\bar{z}$ are connected by the simple dependence following from (11.15):

$$\frac{\partial \bar{z}}{\partial z}=\exp(-2i\alpha). \tag{11.19}$$

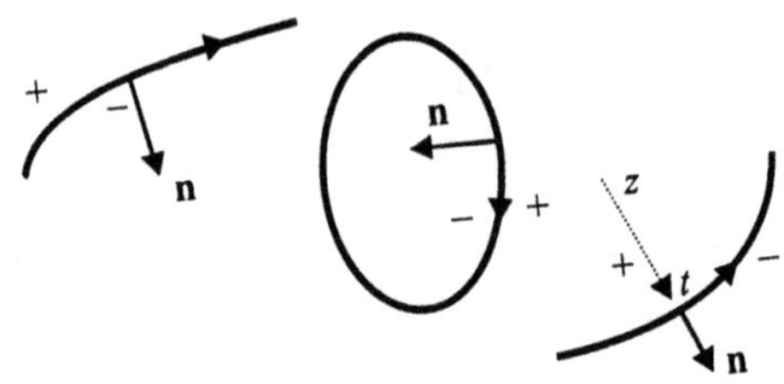

Fig. 10.

Inserting (11.19) into (11.18) we obtain

$$\sigma(z) = \frac{df}{dz} = \frac{\partial f_*}{\partial z} + \frac{\partial f_*}{\partial \bar{z}} \exp(-2i\alpha). \qquad (11.20)$$

Cauchy and Hadamard type integrals. Consider a piece-wise smooth contour L, open, or closed, or comprised of a finite number of open and closed non-intersecting parts (Figure 10). In particular for our applications it is sufficient to assume that these parts have a continuous tangent. Suppose that $g(t)$ is a complex function of the complex co-ordinate t on the contour L; it satisfies Holder's condition (condition H) on each smooth part of L. This means that for two points t_1 and t_2 on such a part, the equality

$$|g(t_2) - g(t_1)| \le A|t_2 - t_1|^{\lambda}, \qquad (11.21)$$

holds, where A and λ are positive constants and $0 < \lambda \le 1$. The constant A is termed Holder's constant; λ is Holder's exponent. If a function $g(t)$ satisfies the condition (11.21) on some part of a contour, we write $g(t) \in H$ on this part. If the function $g(t)$ satisfies (11.21) on the whole contour L excluding a finite number of points, where it may have logarithmic or integrable singularity, it belongs to the class H^* ($g(t) \in H^*$). If it has finite limits at p of such points, it belongs to the class h_p ($g(t) \in h_p$).

In applications of complex variables to plane problems for Laplace's and elasticity equations, the following *Cauchy type integral* is important:

$$G(z) = \frac{1}{2\pi i} \int_L \frac{g(\tau) d\tau}{\tau - z}, \qquad (11.22)$$

where z does not belong to the integration contour L. The function $g(t)$ is called a density, the function $1/(\tau - z)$ is a kernel. It is essential that, for sufficiently smooth L and $g(t)$, the function $G(z)$ is *piece-wise holomorphic*, that is it is holomorphic in any simply connected part of the whole plane, if this part does not include the contour L.[1] This property is the most attractive feature of the Cauchy type integral. From (11.22) we see also that $G(\infty)=0$.

We are also interested in the *Hadamard type integral*:

$$F(z)=\frac{1}{2\pi i}\int_L \frac{g(\tau)d\tau}{(\tau-z)^2}. \tag{11.23}$$

When applying (11.23), we will suppose that the function $g(t)$ has a derivative of the H class on each smooth part of L. The function $F(z)$ is also piece-wise holomorphic in the whole plane and is zero at infinity. It is evident that $F(z) = dG(z)/dz$. For brevity, we will sometimes term the integrals (11.22), (11.23) simply Cauchy and Hadamard integrals.

Limiting values of Cauchy and Hadamard type integrals. In applied problems we need to satisfy prescribed boundary conditions on the boundary of a region. Hence if we want to use Cauchy or/and Hadamard type integrals as a tool for solving boundary value problems, we must have their limiting values when a field point z tends to the contour L from one side (positive or negative). These limits

$$G^{\pm}(t)=\lim_{z\to t^{\pm}}\frac{1}{2\pi i}\int_L \frac{g(\tau)d\tau}{\tau-z}, \quad t\in L, \tag{11.24}$$

$$H^{\pm}(t)=\lim_{z\to t^{\pm}}\frac{1}{2\pi i}\int_L \frac{g(\tau)d\tau}{(\tau-z)^2}, \quad t\in L \tag{11.25}$$

always exist for a point t on a smooth part of L.

To evaluate these limits we may first approximate the density and the contour, evaluate the integrals and *after* this come to the limit. Such a procedure employs only proper integrals. In does not involve the *direct values* of the integrals (11.22), (11.23), that is, the values obtained when we take a point t of the contour as the field point z.

For $z = t$ the integrals (11.22), (11.23) do not exist as proper integrals (unless the density itself is zero at the point t). To use the direct value of integrals we need special definitions explaining how to understand the integration process. In brief, we carry out the following operations: (i) exclude a small ε-vicinity of the point t on L, (ii) evaluate the integral over the remaining part of L, (iii) take the limit as ε tends to zero. Formally, it is sufficient to know that the definitions allow us to differentiate the direct values of the integrals (11.22) and (11.23) under the integration sign and to apply the usual rule of integration by parts.

For applications,. it is important that the definitions provide a means to evaluate the limits (11.24), (11.25). The needed connection between the limiting and the

[1] The function is *holomorphic* in a region if it is *analytic*, that is, it is represented by a convergent power series over z, and is *single valued* in this region.

direct values of the complex integrals by Cauchy and Hadamard is given by the formulae

$$G^{\pm}(t) = \pm\frac{1}{2}g(t) + \frac{1}{2\pi i}\int_L \frac{g(\tau)d\tau}{\tau - t}, \tag{11.26}$$

$$H^{\pm}(t) = \pm\frac{1}{2}g'(t) + \frac{1}{2\pi i}\int_L \frac{g(\tau)d\tau}{(\tau - t)^2}. \tag{11.27}$$

The first pair are the well-known Sokhotski-Plemelj formulae (see, e. g. Muskhelishvili [4,5]). The integral on the r. h. s. of (11.26) is understood as a principal value integral. The second formulae are in fact a corollary of the first (see Linkov, Mogilevskaya [2,3] and Chapter 10). The integral on the r. h. s. of (11.27) is understood as a finite-part integral. Equations (11.26), (11.27) serve as a basis for the alternative numerical procedure of coming to the limit *before* integration.

Holomorphicity theorem. In this part we do not employ Muskhelishvili's theory developed for plane elasticity problems, but we will compare some equations derived below with those initially obtained by employing this theory. To do this, we need the following theorem for piece-wise holomorphic functions.

Theorem. *Functions $\varphi^+(t)$, $\varphi^-(t)$ of H^* class on L represent the limiting values of a piece-wise holomorphic outside L function $\varphi(z)$, equal to zero at infinity, if and only if the equation*

$$\varphi^+ + \varphi^- = \frac{1}{\pi i}\int_L \frac{(\varphi^+ - \varphi^-)d\tau}{\tau - t} \tag{11.28}$$

is fulfilled. The function $\varphi(z)$ is given by the Cauchy type integral:

$$\varphi(z) = \frac{1}{2\pi i}\int_L \frac{(\varphi^+ - \varphi^-)d\tau}{\tau - z}. \tag{11.29}$$

Initially, this theorem was stated for a system of open arcs (Linkov [1]). It is also true in a more general case; its proof is given in § 26.

For a single closed contour we have $\varphi^+(t) = \varphi(t)$, $\varphi^-(t) = 0$ (see, e. g. Muskhelishvili [5, § 73]). Then the condition (11.28) becomes

$$\varphi(t) = \frac{1}{\pi i}\int_L \frac{\varphi(\tau)d\tau}{\tau - t}. \tag{11.30}$$

This case is well known (see Muskhelishvili [5, § 73]).

§ 12. SINGULAR SOLUTIONS IN COMPLEX VARIABLE FORM

We already have initial and higher order singular solutions in *real variable* form. To obtain them in the *complex variable* form, it is sufficient to change real variables into complex ones by using formulae (11.3). The next step consists of composing the resulting values into the complex vectors of displacements (11.6) and traction (11.10). As a result, we arrive at purely complex equations.

Complex initial singular solutions for displacements. Suppose we have the matrix of initial singular solutions in real variables (2.2):

$$\mathbf{U}_l(\mathbf{x},\xi) = \begin{pmatrix} U_{l1}^1 U_{l1}^2 \\ U_{l2}^1 U_{l2}^2 \end{pmatrix}. \tag{12.1}$$

Its columns

$$\mathbf{U}_l^1(\mathbf{x},\xi) = \begin{pmatrix} U_{l1}^1 \\ U_{l2}^1 \end{pmatrix}, \quad \mathbf{U}_l^2(\mathbf{x},\xi) = \begin{pmatrix} U_{l1}^2 \\ U_{l2}^2 \end{pmatrix} \tag{12.2}$$

give the displacement field generated by a unit point force, applied at the point ξ in the direction of the axis $x_1 = x$ (the first column) and $x_2 = y$ (the second column). For definiteness, one may assume that these solutions are given by Kelvin's matrix with the coefficients (2.3).

Use (11.3) to change the real arguments x, y and ξ, η to the complex arguments z, $\bar{z}$ and τ, $\bar{\tau}$. Then, the definition (11.6) of the complex displacement vector in the global system allows us to write the vectors (12.2) as

$$\begin{aligned} U_l^1(z,\tau) &= U_{l1}^1(z,\tau) + iU_{l2}^1(z,\tau), \\ U_l^2(z,\tau) &= U_{l1}^2(z,\tau) + iU_{l2}^2(z,\tau). \end{aligned} \tag{12.3}$$

These are initial singular solutions written in the complex form. They have a logarithmic singularity at $z = \tau$ and serve for the *logarithmic* kernel of the complex potential of the simple layer.

Note that by using the reciprocal relation (3.1) we obtain from (12.3) after some algebra:

$$\begin{aligned} U_l^1(z,\tau) + iU_l^2(z,\tau) &= U_l^1(\tau,z) + iU_l^2(\tau,z), \\ U_l^1(z,\tau) - iU_l^2(z,\tau) &= \overline{U_l^1(\tau,z)} + i\overline{U_l^2(\tau,z)}. \end{aligned} \tag{12.4}$$

Complex singular solutions for traction. We may obtain the complex forms of traction in two ways. The first reproduces for the matrix $\mathbf{J}_S$ the procedure used for the matrix $\mathbf{U}_l$. It employs the traction operator in the real form (1.13). Another approach consists in applying the traction operator in the complex form (11.13) to the initial solutions for displacements written in the complex form (12.3). In this way we have the traction in the *local* rather than in the global system. By employing this procedure, we obtain the tractions $\sigma_S^1(z,\tau)$ and $\sigma_S^2(z,\tau)$:

$$\begin{aligned} \sigma_S^1(z,\tau) &= T_z U_l^1(z,\tau), \\ \sigma_S^2(z,\tau) &= T_z U_l^2(z,\tau). \end{aligned} \tag{12.5}$$

The index S denotes that these tractions have a *strong* (Cauchy type, $1/|z - \tau|$) singularity. To write these complex vectors in the global system, it is sufficient, in accordance with (11.11) to multiply (12.5) by $-i\exp(i\alpha)$. Denoting the complex traction vectors in the *global* system by the symbol j, we have

$$\begin{aligned} j_S^1(z,\tau) &= -i\exp(ia)\sigma_S^1(z,\tau), \\ j_S^2(z,\tau) &= -i\exp(ia)\sigma_S^2(z,\tau). \end{aligned} \tag{12.6}$$

These fields serve for the kernels of the complex traction potential.

Complex singular solutions of higher order for displacements. In real variables, such solutions are defined by (3.2) where $\mathbf{J}_S$ is given by (2.4). We see that to obtain the needed solutions we have to apply the traction operator in the global system to the columns of the matrix (12.1), to interchange the arguments and to transpose the resulting matrix. In complex variables, we need to interchange z and τ in (12.6) and then interchange the imaginary part of the first vector and the real part of the second vector:

$$\begin{aligned} U_S^1(z,\tau) &= \mathrm{Re}[-i\exp(ia)\sigma_S^1(\tau,z)] + i\,\mathrm{Re}[-i\exp(ia)\sigma_S^2(\tau,z)], \\ U_S^2(z,\tau) &= \mathrm{Im}[-i\exp(ia)\sigma_S^1(\tau,z)] + i\,\mathrm{Im}[-i\exp(ia)\sigma_S^2(\tau,z)]. \end{aligned} \tag{12.7}$$

These are complex fields of displacements in the *global* system. They have a *strong* (Cauchy type, $1/|z - \tau|$) singularity and serve for the kernel of the complex singular potential of the double layer.

Complex hypersingular solutions for traction. In accordance with (3.5), the hypersingular solutions for traction are obtained by applying the traction operator to the singular solutions $\mathbf{U}_S^1$ and $\mathbf{U}_S^2$ for displacements. In complex variables we may apply the complex traction operator (11.13) to the complex fields of displacements (12.7). As a result we obtain the complex traction in the *local* system:

$$\begin{aligned} \sigma_H^1(z,\tau) &= T_z U_S^1(z,\tau), \\ \sigma_H^2(z,\tau) &= T_z U_S^2(z,\tau). \end{aligned} \tag{12.8}$$

The index H denotes that these tractions have a hypersingular (Hadamard type, $1/|z - \tau|^2$) singularity at the point $z = \tau$.

We have obtained the kernels of the complex potentials of: 1) simple layer (12.3), 2) singular traction (12.5), 3) double layer (12.7), 4) hypersingular traction (12.8). Two of them, denoted by the symbol U, represent the complex fields of displacements in global co-ordinates, while two others denoted by the symbol σ, represent the fields of traction (in the local co-ordinates). Each of these kernels is nothing more than a common kernel, known in real variables but written in a complex form.

Complex solutions for resultant forces. In addition to these traditional kernels, we will present two more pairs of kernels that are not used in real variables, but play significant roles in the Kolosov-Muskhelishvili theory. In accordance with (11.17), integration of (12.5) over z provides the field of the resultant force with a *logarithmic* singularity:

$$F_l^1(z,\tau) = \int_{z_0}^{z} \sigma_S^1(z,\tau)dz, \quad F_l^2(z,\tau) = \int_{z_0}^{z} \sigma_S^2(z,\tau)dz, \tag{12.9}$$

Analogously, integration of (12.8) provides the field of the resultant force with a *strong* (Cauchy type, $1/(|z - \tau|)$) singularity:

$$F_S^1(z,\tau) = \int_{z_0}^{z} \sigma_H^1(z,\tau)dz, \quad F_S^2(z,\tau) = \int_{z_0}^{z} \sigma_H^2(z,\tau)dz. \tag{12.10}$$

In (12.9) and (12.10), z_0 is an arbitrarily fixed point and the result does not depend on the integration contour, provided that this contour does not encircle the point τ.

Particular case of Kelvin's solution. We have already mentioned that Kelvin's solution (2.3) for point forces at an infinite plane has the simplest matrix of initial singular solutions. It provides illuminating and simple results. We will use it to write down the complex singular solutions explicitly.

Start with equations (2.3) for the components of the matrix $\mathbf{U}_l(\mathbf{x},\xi)$. For its first column $\mathbf{U}_l^1(\mathbf{x},\xi)$, from (2.3) it follows:

$$\mathbf{U}_l^1(\mathbf{x},\xi) = \frac{1}{2\pi\mu(\chi+1)}\begin{pmatrix} -\chi \ln r + (\partial r/\partial x)^2 \\ (\partial r/\partial x)(\partial r/\partial y) \end{pmatrix}.$$

From the definition (12.3) we have:

$$U_l^1(z,\tau) = \frac{1}{2\pi\mu(\chi+1)}\left[-\chi \ln r + \left(\frac{\partial r}{\partial x}\right)^2 + i\left(\frac{\partial r}{\partial x}\right)\left(\frac{\partial r}{\partial y}\right)\right]. \tag{12.11}$$

Apply the complex variables $z = x + iy$ and $\tau = \xi + i\eta$. Then

$$r = \sqrt{(z-\tau)(\bar{z}-\bar{\tau})}, \quad \frac{\partial r}{\partial x} = \frac{\partial r}{\partial z}\frac{\partial z}{\partial x} + \frac{\partial r}{\partial \bar{z}}\frac{\partial \bar{z}}{\partial x} = \frac{1}{2}\left(\sqrt{\frac{\bar{z}-\bar{\tau}}{z-\tau}} + \sqrt{\frac{z-\tau}{\bar{z}-\bar{\tau}}}\right),$$

$$\frac{\partial r}{\partial y} = \frac{\partial r}{\partial z}\frac{\partial z}{\partial y} + \frac{\partial r}{\partial \bar{z}}\frac{\partial \bar{z}}{\partial y} = \frac{1}{2}i\left(\frac{\bar{z}-\bar{\tau}}{r} - \frac{z-\tau}{r}\right).$$

Substitution of these equations into (12.11) gives

$$4\pi\mu(\chi+1)U_i^1(z,\tau) = -\chi\ln[(z-\tau)(\bar{z}-\bar{\tau})] + \frac{z-\tau}{\bar{z}-\bar{\tau}} + 1.$$

The unit on the r. h. s. influences only rigid translation. Omitting it, we may write:

$$4\pi\mu(\chi+1)U_i^1(z,\tau) = -2\chi\ln(z-\tau) + \chi\ln\frac{z-\tau}{\bar{z}-\bar{\tau}} + \frac{z-\tau}{\bar{z}-\bar{\tau}}.$$

Here we have used

$$\ln(\bar{z}-\bar{\tau}) = \ln(z-\tau) - \ln\frac{z-\tau}{\bar{z}-\bar{\tau}}$$

In the same way we obtain the second field, generated by a unit point force at the point τ in the direction of the y axis:

$$4\pi\mu(\chi+1)U_i^2(z,\tau) = i\left[-2\chi\ln(z-\tau) + \chi\ln\frac{z-\tau}{\bar{z}-\bar{\tau}} - \frac{z-\tau}{\bar{z}-\bar{\tau}}\right].$$

Now the definitions (12.3), (12.5), (12.7)-(12.9) provide other singular solutions. (Remember that in the isotropic case the complex traction operator is given by (11.13)). After elementary transformations we obtain the final results:

$$\begin{aligned} 4\pi\mu(\chi+1)U_i^1(z,\tau) &= -2\chi\ln(z-\tau) + \chi k_1 + k_2, \\ 4\pi\mu(\chi+1)U_i^2(z,\tau) &= i[-2\chi\ln(z-\tau) + \chi k_1 - k_2]; \end{aligned} \tag{12.12}$$

$$\begin{aligned} -2\pi(\chi+1)\exp(i\alpha_0)\sigma_S^1(z,\tau) &= -\exp(i\alpha_0)\left(\frac{\chi-1}{z-\tau} - \chi\frac{\partial k_1}{\partial z} - \frac{\partial k_2}{\partial z}\right), \\ -2\pi(\chi+1)\exp(i\alpha_0)\sigma_S^2(z,\tau) &= -i\exp(i\alpha_0)\left(\frac{\chi-1}{z-\tau} - \chi\frac{\partial k_1}{\partial z} + \frac{\partial k_2}{\partial z}\right); \end{aligned} \tag{12.13}$$

$$2\pi i(\chi+1)U_S^1(z,\tau)=-\frac{\chi-1}{z-\tau}\frac{\partial\tau}{\partial s}-\frac{\partial k_1}{\partial s}-\frac{\partial k_2}{\partial s},$$
$$2\pi i(\chi+1)U_S^2(z,\tau)=i\left(-\frac{\chi-1}{z-\tau}\frac{\partial\tau}{\partial s}-\frac{\partial k_1}{\partial s}+\frac{\partial k_2}{\partial s}\right); \tag{12.14}$$

$$2\pi i(\chi+1)\sigma_H^1(z,\tau)=-2\mu\left[\frac{2}{(z-\tau)^2}\frac{\partial\tau}{\partial s}-\frac{\partial^2 k_1}{\partial z\partial s}-\frac{\partial^2 k_2}{\partial z\partial s}\right],$$
$$2\pi i(\chi+1)\sigma_H^2(z,\tau)=-i2\mu\left[\frac{2}{(z-\tau)^2}\frac{\partial\tau}{\partial s}-\frac{\partial^2 k_1}{\partial z\partial s}+\frac{\partial^2 k_2}{\partial z\partial s}\right]; \tag{12.15}$$

$$2\pi(\chi+1)F_l^1(z,\tau)=(\chi-1)\ln(z-\tau)-\chi k_1-k_2,$$
$$2\pi(\chi+1)F_l^2(z,\tau)=i[(\chi-1)\ln(z-\tau)-\chi k_1+k_2]; \tag{12.16}$$

$$2\pi i(\chi+1)F_S^1(z,\tau)=-2\mu\left(\frac{2}{\tau-z}\frac{\partial\tau}{\partial s}-\frac{\partial k_1}{\partial s}-\frac{\partial k_2}{\partial s}\right)=-2\mu\frac{\partial}{\partial s}[2\ln(z-\tau)-k_1-k_2],$$
$$2\pi i(\chi+1)F_S^2(z,\tau)=-i2\mu\left(\frac{2}{\tau-z}\frac{\partial\tau}{\partial s}-\frac{\partial k_1}{\partial s}+\frac{\partial k_2}{\partial s}\right)=-i2\mu\frac{\partial}{\partial s}[2\ln(z-\tau)-k_1+k_2], \tag{12.17}$$

where

$$k_1(\tau,z)=\ln\frac{\tau-z}{\overline{\tau}-\overline{z}},\quad k_2(\tau,z)=\frac{\tau-z}{\overline{\tau}-\overline{z}}; \tag{12.18}$$

α_0 is an angle at a field point z; ds is the increment of the length of the complex differential $d\tau$ at a source point τ; $d\tau/ds=\exp(i\alpha)$; α is the angle at the point τ; the symbol ∂ of a partial derivative is used instead of d because the formulae include differentiation with respect to the co-ordinates of both the field point z and the source point τ; the derivative $\partial/\partial z$ with respect to the field point z is understood as a complete derivative of the form (11.18), which accounts for a dependence of $\overline{z}$ on z; similarly, the derivative $\partial/\partial s$ is also understood as a complete derivative, now with respect to the source point τ, accounting for the dependence of τ and $\overline{\tau}$ on s.

Note that the kernel $k_1(\tau, z)$ is a common kernel which appears when using complex variables for Laplace's and elasticity equations. It is not singular at the point $z=\tau$, it arises from the standard term $\ln r$ when the latter is transformed to a complex form as:

$$\ln r=\frac{1}{2}\ln[(z-\tau)(\overline{z}-\overline{\tau})]=\ln(z-\tau)-\frac{1}{2}k_1(\tau,z).$$

The kernel $k_2(\tau, z)$ is a kernel common in the complex integral equations of plane elasticity (see, e. g. Muskhelishvili [1,4,5], Sherman [1,2], Linkov [1-3,6,7],

Savruk [1-4]). Now we see that it is induced by the solution for a point force. Note also that in each pair of singular solutions, the brackets on r. h. s. differ only by the sign of the last term depending on k_2.

The complex kernels are more compact that their real counterparts which are represented by sums of terms in the components of real matrices. Eventually, this results in a significantly fewer number of integrals to evaluate when applying the complex BIE rather than their real counterparts.

Comment. The initial complex fields $U_l^1(z,,\tau)$ and $U_l^2(z,\tau)$ may be also easily obtained by substitution of the complex functions by Kolosov-Muskhelishvili (K-M) for point forces (Muskhelishvili [5, § 57]) into Muskhelishvili's equations for displacements (Muskhelishvili [5, § 32]). This was illustrated by Linkov and Mogilevskaya [4]. Here we have preferred to derive $U_l^1(z,\tau)$ and $U_l^2(z,\tau)$ directly from the *real* Kelvin's solution to follow the theory of real potentials without employing results based on other methods. Meanwhile, the functions and formulae by K-M will present the basis of the second part of the book. They will serve us for vast extensions, in particular, to obtain the initial complex fundamental solutions for periodic problems, for bonded half-planes, for a circular inclusion in a plane and also for bonded half-planes with a periodic system of point forces.

§ 13. POTENTIALS IN COMPLEX VARIABLE FORM

The equations of elasticity theory (1.4) are linear. Hence, a linear combination of real singular solutions also gives a solution for displacements or traction. For *complex* vectors of displacements and traction this means that multiplication of any of the complex singular solutions (12.3), (12.5) (or what is actually the same, (12.6)), (12.7), (12.8) by a *real* number leads to a solution of elasticity equations when $z \neq \tau$. (We emphasize that it is true only when multiplying by a real number!). Then by multiplying the complex singular solutions by *real* densities and integrating over a *real* length of an arbitrary sufficiently smooth contour L, we again obtain complex solutions for points z not on L. This implies that to obtain the complex analogies of well-known real potentials discussed in the previous chapter, we need to use a *real* density w^1 for the first equations in each pair (12.3), (12.5)-(12.10) and a real density w^2 for the second equations in these pairs. As a result, we arrive at the complex forms of the four common potentials (4.1)-(4.4):

$$u_l(z) = \int_L (U_l^1 w_l^1 + U_l^2 w_l^2) ds, \tag{13.1}$$

$$\sigma_S(z) = \int_L (\sigma_S^1 w_l^1 + \sigma_S^2 w_l^2) ds, \tag{13.2}$$

$$u_S(z) = \int_L (U_S^1 w_S^1 + U_S^2 w_S^2) ds, \tag{13.3}$$

$$\sigma_H(z) = \int_L (\sigma_H^1 w_S^1 + \sigma_H^2 w_S^2) ds, \tag{13.4}$$

and the two additional complex potentials corresponding to the resultant force:

$$f_l(z) = \int_L (F_l^1 w_l^1 + F_l^2 w_l^2) ds, \qquad (13.5)$$

$$f_S(z) = \int_L (F_S^1 w_S^1 + F_S^2 w_S^2) ds. \qquad (13.6)$$

The index l or S in the notation of densities marks the origin of the density: either from the displacement fields U_l^1, U_l^2 of the simple layer with the logarithmic singularity $\ln|\tau - z|$, or from the displacement fields U_S^1, U_S^2 of the double layer with the strong singularity $1/|\tau - z|$. In contrast with (4.2) and (4.4), the traction is given in the local co-ordinates at the point z; they coincide with the global co-ordinates when $\alpha = \pi/2$ (Figure 8).

We may combine the real densities w_l^1, w_l^2 and w_S^1, w_S^2 into the complex values:

$$w_l = w_l^1 + iw_l^2, \quad w_S = w_S^1 + iw_S^2.$$

Then

$$w_l^1 = \frac{w_l + \overline{w_l}}{2}, \quad w_l^2 = \frac{w_l - \overline{w_l}}{2i}, \quad w_S^1 = \frac{w_S + \overline{w_S}}{2}, \quad w_S^2 = \frac{w_S - \overline{w_S}}{2i}.$$

Substitution of these expressions into (13.1)-(13.6) yields the final *complex* form of the potentials:[1]

$$u_l(z) = \frac{1}{2}\int_L \left[(U_l^1 - iU_l^2) w_l + (U_l^1 + iU_l^2)\overline{w_l}\right] ds, \qquad (13.7)$$

$$\sigma_S(z) = \frac{1}{2}\int_L \left[(\sigma_S^1 - i\sigma_S^2) w_l + (\sigma_S^1 + i\sigma_S^2)\overline{w_l}\right] ds, \qquad (13.8)$$

$$u_S(z) = \frac{1}{2}\int_L \left[(U_S^1 - iU_S^2) w_S + (U_S^1 + iU_S^2)\overline{w_S}\right] ds, \qquad (13.9)$$

$$\sigma_H(z) = \frac{1}{2}\int_L \left[(\sigma_H^1 - i\sigma_H^2) w_S + (\sigma_H^1 + i\sigma_H^2)\overline{w_S}\right] ds, \qquad (13.10)$$

$$f_l(z) = \frac{1}{2}\int_L \left[(F_l^1 - iF_l^2) w_l + (F_l^1 + iF_l^2)\overline{w_l}\right] ds, \qquad (13.11)$$

$$f_S(z) = \frac{1}{2}\int_L \left[(F_S^1 - iF_S^2) w_S + (F_S^1 + iF_S^2)\overline{w_S}\right] ds. \qquad (13.12)$$

[1] Certainly, the length increment ds may also be written in a complex form by using (11.15). Below we will use this option.

The first four potentials (13.7)-(13.10) are common potentials of the simple layer, singular traction, double layer and hypersingular traction, but written in the complex form (the tractions are taken in the local co-ordinates). Hence, separation of the real and the imaginary parts in these potentials leads to their real counterparts discussed in the previous chapter. The potentials (13.11), (13.12) do not have real analogues in the classical theory. We may term the first of them the potential of the *simple layer*, while the second the potential of the *double layer*, this time, for the *resultant force*. Their structure and properties are similar to those of the potentials of the simple layer (13.7) and the double layer (13.9) for *displacements*.

Complex potentials for Kelvin's solution. For the simplest complex initial singular solutions U_l^1, U_l^2, given by Kelvin's solution, the kernels are defined by (12.12)-(12.16). Their substitution into the potentials (13.7)-(13.12) yields

$$4\pi\mu i(\chi+1)u_l(z) = i\int_L \left[-2\chi w_l \ln(z-\tau) + \chi w_l k_l + \overline{w_l k_2}\right]\exp(-ia)dr, \tag{13.13}$$

$$-2\pi(\chi+1)\exp(i\alpha_0)\sigma_S(z) = -\int_L \left[-(\chi-1)\frac{w_l}{\tau-z} - \chi w_l\frac{\partial k_1}{\partial z} - \overline{w_l}\frac{\partial k_2}{\partial z}\right]\exp[-i(\alpha-\alpha_0)]d\tau, \tag{13.14}$$

$$2\pi i(\chi+1)u_S(z) = \int_L \left[-\frac{(\chi-1)w_S d\tau}{\tau-z} - w_S dk_1 - \overline{w_S}dk_2\right], \tag{13.15}$$

$$2\pi i(\chi+1)\sigma_H(z) = -2\mu\int_L \left[\frac{2w_S d\tau}{(\tau-z)^2} - w_S\frac{\partial}{\partial z}dk_1 - \overline{w_S}\frac{\partial}{\partial z}dk_2\right], \tag{13.16}$$

$$2\pi(\chi+1)f_l(z) = \int_L \left[(\chi-1)w_l\ln(z-\tau) - \chi w_l k_1 - \overline{w_l k_2}\right]\exp(-i\alpha)d\tau, \tag{13.17}$$

$$2\pi i(\chi+1)f_S(z) = -2\mu\int_L \left[\frac{2w_S d\tau}{\tau-z} - w_S dk_1 - \overline{w_S}dk_2\right], \tag{13.18}$$

where $k_1(\tau, z)$, $k_2(\tau, z)$ are defined by (12.18). The differentials dk_1, dk_2 presume that the kernels $k_1(\tau, z)$ and $k_2(\tau, z)$ depend both on τ and $\bar{\tau}$ with $\bar{\tau}$ depending on τ along L. The potential (13.16) allows us integration by parts. When carrying out this operation, we can drop terms, which appear out of integrals, because in applied problems these terms are zero. We obtain

$$2\pi i(\chi+1)\sigma_H(z) = -2\mu\int_L \left(\frac{2w'_S d\tau}{\tau-z} + w'_S\frac{\partial k_1}{\partial z}d\tau + \overline{w'_S}\frac{\partial k_2}{\partial z}d\bar{\tau}\right), \tag{13.19}$$

where $w'(t) = dw/dt$. Now the kernel in (13.19) is not hypersingular: it has only a strong singularity. This property made such forms of the potentials quite popular before hypersingular kernels became commonplace. However, one needs to pay for this rather illusory advantage by using the *derivative* $w'(t)$ instead of the function itself. Eventually, we need to satisfy additional conditions when solving applied problems numerically. The hypersingular form, as we will see, does not have this shortcoming, and it is not more complicated for the evaluation of integrals. Note also that the potential (13.19), so easily obtained from the hypersingular potential (13.16), is far from being obvious in real variables. Thus the corresponding real equations, that are the equations containing *dislocation densities* as unknowns, usually appear without a connection with the classical theory of potentials.

§ 14. LIMIT VALUES OF COMPLEX POTENTIALS. PHYSICAL MEANING OF DENSITIES

The main terms in singular and hypersingular potentials are defined by integrals of Cauchy and Hadamard types. Thus, in accordance with the properties of such integrals discussed at the end of § 11, the limits of the potentials exist when the field point z tends to the contour L from the left or right. Just as for real counterparts of the potentials, this gives us the basis for the numerical procedure of coming to the limit *after* integration. It avoids evaluation of the direct values of integrals, divergent in a usual sense. On the other hand, we may use dependence of the limiting values of the integrals with their direct values evaluated on the basis of the definitions of principal value and finite-part integrals. Then for the limiting values of the potentials (13.7)-(13.10) and (13.12) we obtain:[1]

$$u_l^+(t) = u_l^-(t) = \frac{1}{2}\int_L \left[(U_l^1 - iU_l^2)w_l + (U_l^1 + iU_l^2)\overline{w_l}\right]ds, \tag{14.1}$$

$$\sigma_S^\pm(t) = \pm\frac{1}{2}i\exp(-i\alpha_0)w_l + \frac{1}{2}\int_L \left[(\sigma_S^1 - i\sigma_S^2)w_l + (\sigma_S^1 + i\sigma_S^2)\overline{w_l}\right]ds, \tag{14.2}$$

$$u_S^\pm(t) = \mp\frac{1}{2}w_S + \frac{1}{2}\int_L \left[(U_S^1 - iU_S^2)w_S + (U_S^1 + iU_S^2)\overline{w_S}\right]ds, \tag{14.3}$$

$$\sigma_H^+(t) = \sigma_H^-(t) = \frac{1}{2}\int_L \left[(\sigma_H^1 - i\sigma_H^2)w_S + (\sigma_H^1 + i\sigma_H^2)\overline{w_S}\right]ds, \tag{14.4}$$

$$f_S^+(t) = f_S^-(t) = \frac{1}{2}\int_L \left[(F_S^1 - iF_S^2)w_S + (F_S^1 + iF_S^2)\overline{w_S}\right]ds. \tag{14.5}$$

[1] We do not write down the limiting values of the potential of the simple layer for the resultant force to avoid lengthy discussion concerning with its discontinuity. This potential is not of interest by itself.

The divergent integrals on the r. h. s. of (14.2), (14.3), (14.5) are understood as complex principal value (Cauchy) integrals. Their kernels are marked with the index S. The divergent integral on the r. h. s. of (14.4) is understood as a complex finite-part (Hadamard) integral. Its kernel has the index H. These formulae serve us as the basis for the numerical procedure of coming to the limit *before* integration.

The simplest way to prove (14.1)-(14.5) is to take Kelvin's solution as an initial singular solution. Then, by using the potentials (13.7)-(13.12) and formulae (11.26), (11.27) for the limiting values of the integrals of Cauchy and Hadamard types, we immediately arrive at equations (14.1)-(14.5). After this, it is sufficient to note that the structure of the resulting equations does not depend on a particular choice of the initial singular solutions U_l^1, U_l^2. Equations (14.1)-(14.4) may be also obtained from their real analogies (6.1)-(6.4).

Physical meaning of densities. Equations (14.2) and (14.3) make evident the physical meaning of the complex densities. From them we have

$$w_l(t) = -i\exp(i\alpha_0)\Delta\sigma(t), \tag{14.6}$$

$$w_S(t) = -\Delta u(t), \tag{14.7}$$

where $\Delta\sigma = \sigma_S^+ - \sigma_S^-$ is the *complex traction discontinuity* on the contour; it is defined in the local co-ordinates while the r. h. s. of (14.6), as follows from (11.11), gives this discontinuity in the global co-ordinates; $\Delta u = u^+ - u^-$ is the *complex displacement discontinuity* on the contour (in the global co-ordinates). We could also write (14.6), (14.7) by using their real analogies (5.1), (5.2).

In addition to the density w_S consider the density $b = -i\,w_S'$, that is the derivative of w_S with respect to t taken with the multiplier $-i$. From (14.7) we have:

$$b = -iw_S' = i\Delta u'. \tag{14.8}$$

Note that (11.15) implies $\Delta u' = \exp(-i\alpha_0)d\Delta u/ds$. Write also the derivative $d\Delta u/ds$ of the displacement discontinuity in the local co-ordinates (n, t) by using (11.9): $d\Delta u/ds = -i\exp(i\alpha_0)d\Delta v/ds$. Recollect that α_0 is the angle between the tangent to the contour and the x-axis. Then from (14.8) we have

$$b = -iw_S' = \frac{d\Delta v}{ds}, \tag{14.9}$$

where $\Delta v = \Delta v_n + i\Delta v_t$ is the complex displacement discontinuity vector in the local co-ordinates. In dislocation theory, the derivatives $d\Delta v_n/ds$ and $d\Delta v_t/ds$ are respectively the normal and shear density of dislocations, while the differentials $d\Delta v_n$ and $d\Delta v_t$ represent components of Burger's vector on a contour element ds. Hence, (14.9) means that b is the *complex density of dislocations*: $b = d\Delta v_n/ds + i(d\Delta v_t/ds)$, while $d\Delta v$ is the complex Burger's vector on the element ds. As the derivative w_S' differs from b only by the multiplier i, it also may be called a density of dislocations. We see that the density in the potential (13.19) is a density of dislocations.[1]

1 Strictly speaking, the physical meaning of a dislocation refers only to plane strain. In a computational context, one may use this convenient term for plane stress as well.

Chapter 3

COMPLEX INTEGRAL EQUATIONS OF THE INDIRECT APPROACH

The potentials (13.7) and (13.9) satisfy plane elasticity equations for displacements. Hence, they may serve to represent displacement fields. The potentials (13.8) and (13.10) satisfy elasticity equations for stresses; they serve to find stresses corresponding to displacements (13.7) or (13.9). The potentials (13.11) and (13.12) satisfy equations for the resultant force. They may serve to find the resultant force corresponding to (13.7) or (13.9). However, before using these formulae for points within a considered region, we need to choose the density in the potential so that its limiting values satisfy prescribed boundary conditions. These conditions, given in real variables by (1.17)-(1.19), have the same form in the complex variables: it is sufficient to refer displacements, traction and their discontinuities to complex vectors introduced by formulae (11.6), (11.10). When using the resultant force, we obtain it by integration of traction in accordance with (11.17). In this case, there appear constants, sometimes unknown, which express the value of the resultant force at the initial point of each isolated part of the contour.

§ 15. CLOSED CONTOURS

Suppose that the contour L consists of m closed parts L_k, which we consider to be the external boundaries of a body. For instance, it may be a finite body with holes (Figure 11, *a*). For these parts, the limiting values are given only from the positive side of L. (Recall that we have agreed to consider the normal to be outward and to mark the limits from the body, that is in the direction of the normal, with the index "+"). Then the discontinuities $\Delta\sigma$ and Δu, representing as follows from (14.6), (14.7) the densities of potentials, do not have a *direct* physical meaning in this particular case. This explains the origin of the term "indirect" for this approach because primarily the approach was applied to closed contours.[1]

Under prescribed *displacements* $u(t)$, we may try to satisfy the boundary condition by using the limiting values of the potentials (13.7) or (13.9):

$$\lim_{z\to t}\frac{1}{2}\int_L\left[\left(U_l^1-iU_l^2\right)w_l+\left(U_l^1+iU_l^2\right)\overline{w_l}\right]ds=u(t), \qquad (15.1)$$

[1] Certainly, as was explained for real equations, we may interpret the discontinuities in terms of *two* problems: 1) that for the region D, and 2) that for the region $D^{\cdot}$, the complement of D in the whole plane (see comments in § 7 and 10).

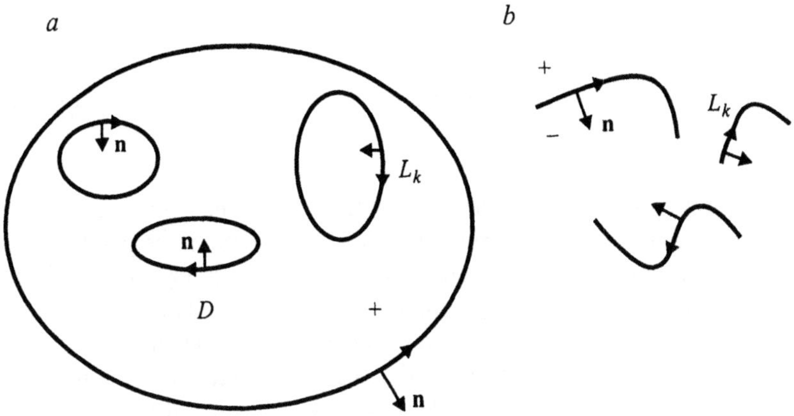

Fig. 11.

$$\lim_{z\to t^+}\frac{1}{2}\int_L\left[\left(U_S^1-iU_S^2\right)w_S+\left(U_S^1+iU_S^2\right)\overline{w_S}\right]ds=u(t). \tag{15.2}$$

Recall that the limits in (15.1) from both sides coincide. The potential (13.7) is continuous through L and has a weak (logarithmic) singularity. This implies that (15.1) is a Fredholm equation of the *first* kind, which is inconvenient for finding the solution under displacements prescribed on L. Note, however, that if a dependence between the displacements and traction is prescribed on L, equation (15.1) may be useful.

Equation (15.2) does not have this drawback. Its real analogies commonly serve to find the solution of the problems with displacements prescribed on a closed contour L. Note that the form (15.2) does not contain the direct value of a divergent integral; it serves as the basis for the numerical procedure of coming to the limit *after* integration.

For prescribed traction $\sigma(t)$, we may use the potentials (13.8), (13.10) or (13.12). If Kelvin's solution is taken as an initial solution, we may also apply (13.19). By taking the limits and satisfying the boundary condition, we obtain:

$$\lim_{z\to t^+}\frac{1}{2}\int_L\left[\left(\sigma_S^1-i\sigma_S^2\right)w_l+\left(\sigma_S^1+i\sigma_S^2\right)\overline{w_l}\right]ds=\sigma(t), \tag{15.3}$$

$$\lim_{z\to t^+}\frac{1}{2}\int_L\left[\left(\sigma_H^1-i\sigma_H^2\right)w_S+\left(\sigma_H^1+i\sigma_H^2\right)\overline{w_S}\right]ds=\sigma(t), \tag{15.4}$$

$$\lim_{z\to t^+}\frac{1}{2}\int_L\left[\left(F_S^1-iF_S^2\right)w_S+\left(F_S^1+iF_S^2\right)\overline{w_S}\right]ds=f(t)+C(t), \tag{15.5}$$

$$\lim_{z\to t^+}\frac{-2\mu}{2\pi i(\chi+1)}\int_L\left(\frac{2w_S'd\tau}{\tau-z}+w'\frac{\partial k_1}{\partial z}d\tau+\overline{w'}\frac{\partial k_2}{\partial z}d\overline{\tau}\right)=\sigma(t). \tag{15.6}$$

In (15.5) in accordance with (11.17) we assume

$$f(t) = \int_{t_k}^{t} \sigma(\tau)d\tau, \tag{15.7}$$

t_k is an arbitrary point on the isolated part L_k ($k = 1,\ldots, m$) of the contour L; $C(t) = C_k$ on L_k; C_k is a constant equal to the value of the resultant force at the point t_k. We may take one of these constants to be zero. To find the remaining constants, we must employ additional conditions of continuity of the resultant force when traveling from one contour to another along an arbitrary arc connecting them. This is a common inconvenience when one uses the resultant force for a number of isolated contours both closed and open. Besides, to have the integral (15.7) single-valued when traveling along L_k, the total resultant force, applied to the whole contour L_k, should be zero. If we do not assume this condition, we need to introduce additional terms, which serve to suppress the multi-valued nature of the integral (15.7). These terms are easily found on the basis of the Kolosov-Muskhelishvili theory. Meanwhile, it is not an easy task to find them by following the way used in this section.

Using traction instead of the resultant force avoids these shortcomings. In this respect, the equations (15.3), (15.4) and (15.6) have a significant advantage. The real analogue of (15.3) was commonly used for problems with prescribed traction before hypersingular equations became popular.

The equation with the hypersingular kernel (15.4) is in essence the complex form of real equations used in the displacement discontinuity method. As noted in § 7, Crouch [1] suggested this method for the case when (i) Kelvin's solution serves as an initial singular solution, (ii) the contour is represented by straight segments and (iii) the density is piece-wise constant along the segments. The method got its title because the density in (15.4), as follows from (14.7), represents the displacement discontinuity.

Equation (15.6) with the strong singularity contains the complex density of dislocations as the density. As mentioned in § 13, its derivation using the real variables is not so clear as it is for complex variables.

The forms (15.1)-(15.6) do not employ the direct values of divergent integrals. They serve for the numerical procedure of coming to the limit *after* integration. Alternative forms, corresponding to coming to the limit *before* integration and employing the direct values, follow when we use (14.1)-(14.5) in (15.1)-(15.5) and an analogous formula in (15.6). The result is

$$\frac{1}{2}\int_L \left[\left(U_l^1 - iU_l^2\right)w_l + \left(U_l^1 + iU_l^2\right)\overline{w_l}\right]ds = u(t), \tag{15.8}$$

$$-\frac{1}{2}w_S + \frac{1}{2}\int_L \left[\left(U_S^1 - iU_S^2\right)w_S + \left(U_S^1 + iU_S^2\right)\overline{w_S}\right]ds = u(t). \tag{15.9}$$

$$\frac{i}{2}\exp(-i\alpha_0)w_l+\frac{1}{2}\int_L\left[\left(\sigma_S^1-i\sigma_S^2\right)w_l+\left(\sigma_S^1+i\sigma_S^2\right)\overline{w_l}\right]ds=\sigma(t), \quad (15.10)$$

$$\frac{1}{2}\int_L\left[\left(\sigma_H^1-i\sigma_H^2\right)w_S+\left(\sigma_H^1+i\sigma_H^2\right)\overline{w_S}\right]ds=\sigma(t), \quad (15.11)$$

$$\frac{1}{2}\int_L\left[\left(F_S^1-iF_S^2\right)w_S+\left(F_S^1+iF_S^2\right)\overline{w_S}\right]ds=f(t)+C(t), \quad (15.12)$$

$$-\frac{2\mu}{2\pi i(\chi+1)}\int_L\left(\frac{2w_S'd\tau}{\tau-t}+w'\frac{\partial k_1}{\partial t}d\tau+\overline{w'}\frac{\partial k_1}{\partial t}d\overline{\tau}\right)=\sigma(t). \quad (15.13)$$

The integrals in (15.9), (15.10), (15.12) and (15.13) are understood as principal value integrals. The integral in (15.11) is a finite-part integral.

§ 16. OPEN CONTOURS

For open contours, the conditions (1.17)-(1.19) involve both sides of L (Figure 11, *b*). All the complex BIE of the previous section keep their form, but now the densities have a physical meaning in the terms of the problem itself: $w_l = -\exp(i\alpha)\Delta\sigma$ is the traction discontinuity in the global coordinates, $w_S = -\Delta u$ is the displacement discontinuity taken with the minus sign. We shall use this interpretation when writing BIE.

Equation (15.1) practically keeps its form. It states that the limiting values of the displacements from the left and right of L coincide. It may serve as a CV-BIE in problems when the displacements are continuous, and the traction has discontinuity through open arcs. Such are, for instance, the problems for plates with stringers. Note that in this case the resulting equations are not Fredholm equations of the first kind due to a connection between the displacements and the traction discontinuities along the stringers.

Equations (15.2) and (15.3) are now taken for the limits from both sides ($z \to \pm t$). They are not of interest for open arcs because they just give the limiting values of displacements or traction by using their discontinuity.

Equations (15.4)-(15.6) are of greater interest for applications. The limits in them do not depend on the side. As $w_S = -\Delta u$, these equations serve for problems with displacement discontinuities, such as problems for cracks.

Comparing the conclusions of this paragraph for open contours with those made in the previous paragraph for closed contours, we see that only equations (15.4)-(15.6) are equally applicable in both cases. As equation (15.4) with a *hypersingular* kernel, in contrast with (15.5) and (15.6), contains neither unknown constants nor the derivative of the density, it is the most convenient for numerical application. Besides, it contains the very values which characterize contact interaction on the surfaces of cracks: the traction and the displacement discontinuity. These virtues distinguish the hypersingular equations from other BIE and make them very attractive for solving applied problems.

As usual, the forms of equations with the symbol of the limit serve for the procedure of coming to the limit *after* integration. They do not involve the direct values of divergent integrals.

Alternative forms employ such values. We will write them only for equations (15.4) and (15.5):

$$\frac{1}{2}\int_L\left[\left(\sigma_H^1 - i\sigma_H^2\right)\left(-\Delta u\right)+\left(\sigma_H^1 + i\sigma_H^2\right)\left(-\overline{\Delta u}\right)\right]ds = \sigma(t), \tag{16.1}$$

$$\frac{1}{2}\int_L\left[\left(F_S^1 - iF_S^2\right)\left(-\Delta u\right)+\left(F_S^1 + iF_S^2\right)\left(-\overline{\Delta u}\right)\right]ds = f(t)+C(t). \tag{16.2}$$

The integral in (16.1) is understood as a complex finite-part (Hadamard) integral. The integral in (16.2) is taken as a complex principal value (Cauchy) integral. We repeat that these forms serve for the numerical procedure of coming to the limit *before* integration.

Concluding this paragraph we emphasize that, as follows from the structure of the potentials used in the indirect approach, the elastic properties of a medium are the same on both sides of L. To some extent, this restricts applications of the indirect approach. In the next chapter, we will see that the direct approach, when used properly, avoids this restriction.

§ 17. EQUATIONS OF THE INDIRECT APPROACH FOR KELVIN'S SOLUTION

In the simplest case, when we take Kelvin's solution as an initial singular solution, the potentials are given by (13.13)-(13.19). Take into account the physical meaning of the densities expressed by (14.6), (14.7) and (14.9). For brevity, write only the equations corresponding to the procedure of coming to the limit after integration. For Kelvin's solution, (15.8)-(15.13) become

$$\frac{1}{2\pi i}\int_L\left[-2\chi\,\Delta\sigma\ln(t-\tau)d\tau+\chi\,\Delta\sigma k_1 d\tau-\overline{\Delta\sigma k_2 d\tau}\right]=2\mu(\chi+1)u(t). \tag{17.1}$$

$$\frac{\chi+1}{2}\Delta u(t)+\frac{1}{2\pi i}\int_L\left[\frac{(\chi-1)\Delta u d\tau}{\tau-t}+\Delta u dk_1+\overline{\Delta u dk_2}\right]=(\chi+1)u(t), \tag{17.2}$$

$$\frac{\chi+1}{2}\Delta\sigma(t)+\frac{1}{2\pi i}\int_L\left[-\frac{(\chi-1)\Delta\sigma d\tau}{\tau-t}-\chi\Delta\sigma\frac{\partial k_1}{\partial t}d\tau+\overline{\Delta\sigma}\frac{\partial k_2}{\partial t}d\overline{\tau}\right]=$$
$$=(\chi+1)\sigma(t), \tag{17.3}$$

$$\frac{1}{2\pi i}\int_L \left[\frac{2\Delta u d\tau}{(\tau - t)^2} - \Delta u \frac{\partial}{\partial t} dk_1 - \overline{\Delta u} \frac{\partial}{\partial t} dk_2\right] = \frac{\chi + 1}{2\mu}\sigma(t), \tag{17.4}$$

$$\frac{1}{2\pi i}\int_L \left(\frac{2\Delta u d\tau}{\tau - t} - \Delta u dk_1 - \overline{\Delta u} dk_2\right) = \frac{\chi + 1}{2\mu}[f(t) + C(t)], \tag{17.5}$$

$$-\frac{1}{2\pi}\int_L \left(\frac{2b d\tau}{\tau - t} + b\frac{\partial k_1}{\partial t} d\tau - \bar{b}\frac{\partial k_2}{\partial t} d\bar{\tau}\right) = \frac{\chi + 1}{2\mu}\sigma(t). \tag{17.6}$$

Recall that (14.8) connects the density b of dislocations with the derivative of the displacement discontinuity Δu:

$$b = i\Delta u'. \tag{17.7}$$

We comment on these equations.

Equations (17.2), (17.3) are the complex analogues of well-known real singular equations used for closed contours (see, e.g. Ugodchikov and Khutorianski [1]). In the complex form, they are presented here for the first time.

Hypersingular equation (17.4) presents, in complex variables, the basis of the displacement discontinuity method suggested by Crouch [1] in real variables. It allows us to use advantages of complex variables and complex finite-part integrals. First, this equation was obtained by direct differentiation of equation (17.5) (Linkov [7], see also Linkov, Mogilevskaya [1]).

Equation (17.5) is a particular case of a more general equation including the case when traction may be different on opposite surfaces of cracks (see § 29). In this more general case, equation (17.5) was first derived in another way (Linkov [1]). It has been studied both for open and closed contours (Linkov [1, 6]).

Equation (17.6) corresponds to real equations involving the density of dislocations (Goldshtein, Salganik [1]). For systems of cracks it was first derived in the general case by differentiation of (17.5) and integration by parts (Linkov [1]). Later on, other authors (see, e. g. Savruk [1], Ioakimidis [1], Parton, Perlin [1]) repeatedly derived it.

The derived equations hold for both open and closed contours. Meanwhile, the elastic properties of a medium have to be the same on both sides of an arc. In the following chapter we shall see that all these equations may be considered as particular cases of the equations for blocky systems (§ 23). This allows us to remove this restriction, wholly or partially.

We could see that all the CV-BIE of the *indirect* approach, except (17.5), are complex variable forms of the well-known real equations. But being equivalent to real equations in mathematical sense, they have significant computational advantages over their real counterparts. These advantages are explained in Chapter 12.

Chapter 4

COMPLEX INTEGRAL EQUATIONS OF THE DIRECT APPROACH

The common procedure for deriving the equations of the direct approach is as follows: reciprocal (Betti's) formula – Somigliana's identities – integral equations. Certainly, we may try to shorten this way by starting directly from the real Somigliana's formulae for displacements (8.6) and substituting the complex expressions for kernels and densities into it. In practice this is longer than following the whole chain. Thus we choose to take all the steps. In addition this will provide us with the complex variable form of Betti's formula, which is important in itself.

§ 18. BETTI'S FORMULA IN COMPLEX VARIABLE FORM

Consider a region D, finite or infinite, inside a closed contour L. As usual, the normal **n** at contour points is taken outward to D (Figure12, *a*). Rewrite Betti's reciprocal formula (8.1) in the notation of § 8:

$$\int_L \left[\mathbf{t}_n^*(\mathbf{x})\right]^T \mathbf{u}(\mathbf{x}) ds_x = \int_L \left[\mathbf{u}^*(\mathbf{x})\right]^T \mathbf{t}_n(\mathbf{x}) ds_x, \tag{18.1}$$

where **u(x)** and **u*(x)** are two arbitrary sufficiently smooth displacement fields in the region D; $\mathbf{t}_n(\mathbf{x})$ and $\mathbf{t}_n^*(\mathbf{x})$ are the corresponding traction on the contour.

Write first scalar products under the integral sign in complex variable form. We have:

$$\left[\mathbf{t}_n^*(\mathbf{x})\right]^T \mathbf{u}(\mathbf{x}) = \left[\mathbf{u}(\mathbf{x})\right]^T \mathbf{t}_n^*(\mathbf{x}) = \sigma_{nx}^* u_x + \sigma_{nx}^* u_y .$$

In the complex form this equation becomes

$$\left[\mathbf{t}_n^*(\mathbf{x})\right]^T \mathbf{u}(\mathbf{x}) = \mathrm{Re}\left[\left(\overline{\sigma_{nx}^* + i\sigma_{ny}^*}\right)(u_x + iu_y)\right]$$

In accordance with (11.6), (11.10) put $u = u_x + iu_y$, $\sigma = \sigma_{nn} + i\sigma_{nt}$. Then accounting for (11.11) we have:

$$\left[\mathbf{t}_n^*(\mathbf{x})\right]^T \mathbf{u}(\mathbf{x}) = \sigma_{nx}^* u_x + \sigma_{ny}^* u_y = \mathrm{Re}\left[i\overline{\sigma}\exp(-i\alpha)u^*\right].$$

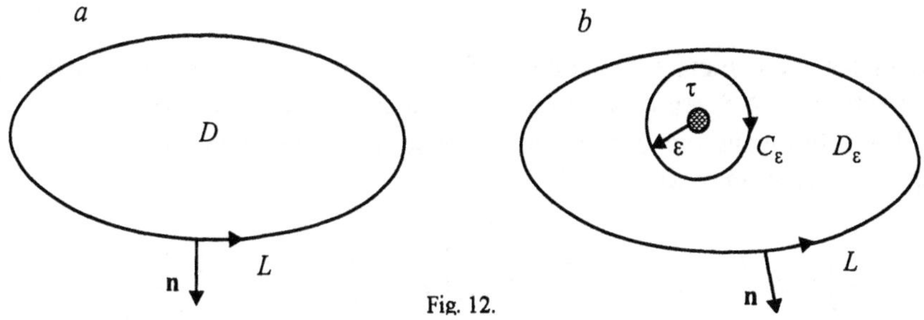

Fig. 12.

Analogously

$$\left[\mathbf{u}^*(\mathbf{x})\right]^{\mathrm{T}} \mathbf{t}_{\mathbf{n}}(\mathbf{x}) = \sigma_{nx}u_x^* + \sigma_{ny}u_y^* = \mathrm{Re}[i\overline{\sigma}\exp(-i\alpha)u^*].$$

Hence, the real Betti's formula (18.1) takes its complex form

$$\mathrm{Re}\int_L i\left(\overline{\sigma}^* u - \overline{\sigma}u^*\right)d\bar{t} = 0, \tag{18.2}$$

where we took into account that $d\bar{t} = \exp(-i\alpha)ds$. Recall that α is the angle between the x-axis and the traveling direction leaving the region D on the left. This formula in its slightly changed form served as the starting point in the papers by Kovneristov (Kovneristov [1], Kovneristov et al. [1]).

§ 19. SOMIGLIANA'S IDENTITIES IN COMPLEX VARIABLE FORM

Take as a displacement field $u(z)$ some solution of an elasticity problem for the region D; $\sigma(z)$ is the corresponding traction field. As a field $u^*(z)$ we take the initial singular solution $U_l^1(z,\tau)$. This gives the displacement field induced by a point force applied at the point τ in the direction of the x-axis. The singular solution $\sigma_l^1(z,\tau)$ provides the corresponding traction.

We may immediately apply Betti's formula (18.2) to these two fields when the singular source point τ is *outside* D. But if it is *within* D, we need to exclude a small vicinity of the point $z = \tau$ with the contour C_ε from D (Figure 12, b). Then we obtain the region D_ε with the contour $L + C_\varepsilon$, which does not have singular points; we may apply Betti's formula (18.2) to it:

$$\mathrm{Re}\int_L i\left[\overline{\sigma_S^1(t,\tau)}u(t) - \overline{\sigma(t)}U_l^1(t,\tau)\right]d\bar{t} =$$

$$= \begin{cases} -\mathrm{Re}\displaystyle\int_{C_\varepsilon} i\left[\overline{\sigma_S^1(t,\tau)}u(t) - \overline{\sigma(t)}U_l^1(t,\tau)\right]d\bar{t}, & \tau \in D, \\ 0, & \tau \notin D + L. \end{cases} \tag{19.1}$$

By taking the contour C_ε as a circumference with a small radius ε and letting ε tend to zero as in § 8, we conclude that the integral on the r. h. s. of (19.1) tends to $u_x(z)$. Then, following our convention to denote the integration point by τ, and the field point by z, we may write (19.1) as

$$\operatorname{Re}\int_L i\left[\overline{\sigma_S^1(\tau,z)}u(\tau)-\overline{\sigma(\tau)}U_l^1(\tau,z)\right]d\bar{\tau}=\begin{cases}-u_x(z), & z\in D,\\ 0, & z\notin D+L.\end{cases} \tag{19.2}$$

Analogously, by taking $U_l^2(z,\tau)$ as a field of displacements, we obtain:

$$\operatorname{Re}\int_L i\left[\overline{\sigma_S^2(\tau,z)}u(\tau)-\overline{\sigma(\tau)}U_l^2(\tau,z)\right]d\bar{\tau}=\begin{cases}-u_y(z), & z\in D,\\ 0, & z\notin D+L.\end{cases} \tag{19.3}$$

Multiply (19.3) by i and add to (19.2), using $u = u_x + iu_y$. Then after changing the sign of the result we have:

$$\begin{aligned}&\operatorname{Re}\int_L i\left[U_l^1(\tau,z)\overline{\sigma(\tau)}-\overline{\sigma_S^1(\tau,z)}u(\tau)\right]d\bar{\tau}+\\ &+i\operatorname{Re}\int_L i\left[U_l^2(\tau,z)\overline{\sigma(\tau)}-\overline{\sigma_S^2(\tau,z)}u(\tau)\right]d\bar{\tau}=\begin{cases}u(z), & z\in D,\\ 0, & z\notin D+L.\end{cases}\end{aligned} \tag{19.4}$$

These are Somigliana's identities for displacements. As $\operatorname{Re}(ia) = \operatorname{Im}(\bar{a} - a)/2$ for any complex vector a, we may write (19.4) as

$$\begin{aligned}&-\frac{i}{2}\int_L\Big\{\left[\overline{U_l^1(\tau,z)}+i\,\overline{U_l^2(\tau,z)}\right]\sigma(\tau)d\tau-\left[U_l^1(\tau,z)+i\,U_l^2(\tau,z)\right]\overline{\sigma(\tau)}d\bar{\tau}+\\ &+\left[\overline{\sigma_S^1(\tau,z)}+i\,\overline{\sigma_S^2(\tau,z)}\right]u(\tau)d\tau-\left[\sigma_S^1(\tau,z)+i\,\sigma_S^2(\tau,z)\right]\overline{u(\tau)}d\bar{\tau}\Big\}=\\ &=\begin{cases}u(z), & z\in D,\\ 0, & z\notin D+L.\end{cases}\end{aligned} \tag{19.5}$$

Now we need to transform the result to the form analogous to the real identities (8.16). To this end, note that

$$\begin{aligned}&i\left[\overline{\sigma_S^1(\tau,z)}+i\,\overline{\sigma_S^2(\tau,z)}\right]d\bar{\tau}=\left\{\left[\overline{-i\exp(i\alpha)\,\sigma_S^1(\tau,z)}\right]+i\left[\overline{-i\exp(i\alpha)\,\sigma_S^2(\tau,z)}\right]\right\}ds=\\ &=\left\{\operatorname{Re}\left[-i\exp(i\alpha)\,\sigma_S^1(\tau,z)\right]-i\operatorname{Im}\left[-i\exp(i\alpha)\,\sigma_S^1(\tau,z)\right]\right\}ds+\\ &+i\left\{\operatorname{Re}\left[-i\exp(i\alpha)\,\sigma_S^2(\tau,z)\right]-i\operatorname{Im}\left[-i\exp(i\alpha)\,\sigma_S^2(\tau,z)\right]\right\}ds.\end{aligned}$$

From this, by using (12.7), we have:

$$i\left[\overline{\sigma_S^1(\tau,z)}+i\,\overline{\sigma_S^2(\tau,z)}\right]d\bar{\tau}=\left[U_S^1(z,\tau)-iU_S^2(z,\tau)\right]ds. \tag{19.6}$$

Analogously

$$i\left[\sigma_S^1(\tau,z)+i\,\sigma_S^2(\tau,z)\right]d\tau=-\left[U_S^1(z,\tau)+iU_S^2(z,\tau)\right]ds. \tag{19.7}$$

Besides, equations (12.4) give:

$$\begin{aligned}\overline{U_l^1(\tau,z)}+i\,\overline{U_l^2(\tau,z)}&=U_l^1(z,\tau)-i\,U_l^2(z,\tau),\\ U_l^1(\tau,z)+i\,U_l^2(\tau,z)&=U_l^1(z,\tau)+i\,U_l^2(z,\tau).\end{aligned} \tag{19.8}$$

By using (19.6)-(19.8) in (19.5), we obtain complex Somigliana's identities for *displacements* in the final form:

$$\frac{1}{2}\int_L\left\{-i\exp(i\alpha)\sigma(\tau)\left[U_l^1(z,\tau)-iU_l^2(z,\tau)\right]+i\exp(-i\alpha)\overline{\sigma(\tau)}\left[U_l^1(z,\tau)+i\,U_l^2(z,\tau)\right]\right\}ds-$$

$$-\frac{1}{2}\int_L\left\{\left[U_S^1(z,\tau)-i\,U_S^2(z,\tau)\right]u(\tau)+\left[U_S^1(z,\tau)+i\,U_S^2(z,\tau)\right]\overline{u(\tau)}\right\}ds=$$

$$=\begin{cases}u(z), & z\in D,\\ 0, & z\notin D+L.\end{cases} \tag{19.9}$$

Comparing the first integral on the l. h. s. of (19.9) with the potential (13.7) we see that it represents the potential of a simple layer with the density $-i\exp(i\alpha)\sigma(\tau)$, which, as is clear from (11.11), is the complex traction in global coordinates. The second integral in (19.9), as is evident from (13.9), is the potential of a double layer with the density $u(\tau)$. Hence, the l. h. s. of Somigliana's identity is a linear combination of the potentials of the simple and double layers with the densities equal to actual traction and displacements. This combination is such that the obtained displacements are equal to the actual displacements within the region D, and equal to zero outside D. Certainly, this conclusion is nothing more than a restatement of the analogous conclusion for the real Somigliana's identity (8.6) for displacements (see discussion of equation (8.6) in § 8).

Complex Somigliana's identities for *traction* immediately follow from (19.9) after applying the complex traction operator. Accounting for (12.5), (12.8) we obtain:

$$\frac{1}{2}\int_L\left\{-i\exp(i\alpha)\sigma(\tau)\left[\sigma_S^1(z,\tau)-i\sigma_S^2(z,\tau)\right]+i\exp(-i\alpha)\overline{\sigma(\tau)}\left[\sigma_S^1(z,\tau)+i\sigma_S^2(z,\tau)\right]\right\}ds-$$

$$-\frac{1}{2}\int_L \left\{\left[\sigma_H^1(z,\tau)-i\sigma_H^2(z,\tau)\right]u(\tau)+\left[\sigma_H^1(z,\tau)+i\sigma_H^2(z,\tau)\right]\overline{u(\tau)}\right\}ds=\begin{cases}\sigma(z), & z\in D,\\ 0, & z\notin D+L.\end{cases} \quad (19.10)$$

Comment. One could write complex Somigliana's identities for displacements and for traction directly by analogy with their real counterparts (8.6), (8.7). In particular, it is sufficient to make the following changes in (8.6). The real potential of the simple layer is changed to its *complex* form (14.1), the real double layer potential to its *complex* form (14.3), the real vector of displacements is changed to the *complex* vector of displacements, and the real vector of traction in the global coordinates to the *complex* vector, also in global coordinates. Similarly, equation (19.10) is the complex analogue of Somigliana's real identities for traction (8.7). Their structure, and the meaning of the kernels and densities are similar.

§ 20. INTEGRAL EQUATIONS OF THE DIRECT APPROACH

Somigliana's identities for displacements (19.9) and for traction (19.10) turn into integral equations in the limit when the field point z tends to the contour L from either side. As we mentioned when discussing the potentials, these limits always exist for sufficiently smooth contours and densities. By taking the limits from inside and outside the region D, we obtain:

$$\begin{aligned}&\lim_{z\to t}\frac{1}{2}\int_L\Big\{-i\exp(i\alpha)\sigma(\tau)\left[U_l^1(z,\tau)-iU_l^2(z,\tau)\right]+\\&\quad+i\exp(-i\alpha)\overline{\sigma(\tau)}\left[U_l^1(z,\tau)+iU_l^2(z,\tau)\right]\Big\}ds-\\&-\lim_{z\to t}\frac{1}{2}\int_L\left\{\left[U_S^1(z,\tau)-iU_S^2(z,\tau)\right]u(\tau)+\left[U_S^1(z,\tau)+iU_S^2(z,\tau)\right]\overline{u(\tau)}\right\}ds\,\}=\\&=\begin{cases}u(t), & z\to t^+,\\ 0, & z\to t^-,\end{cases}\quad t\in L.\end{aligned} \quad (20.1)$$

$$\begin{aligned}&\lim_{z\to t}\frac{1}{2}\int_L\Big\{-i\exp(i\alpha)\sigma(\tau)\left[\sigma_S^1(z,\tau)-i\sigma_S^2(z,\tau)\right]+\\&\quad+i\exp(-i\alpha)\overline{\sigma(\tau)}\left[\sigma_S^1(z,\tau)+i\sigma_S^2(z,\tau)\right]\Big\}ds-\\&-\lim_{z\to t}\frac{1}{2}\int_L\left\{\left[\sigma_H^1(z,\tau)-i\sigma_H^2(z,\tau)\right]u(\tau)+\left[\sigma_H^1(z,\tau)+i\sigma_H^2(z,\tau)\right]\overline{u(\tau)}\right\}ds=\\&=\begin{cases}\sigma(t), & z\to t^+,\\ 0, & z\to t^-,\end{cases}\quad t\in L.\end{aligned} \quad (20.2)$$

In numerical procedures of coming to the limit *after* integration, one may use any of the two limits $z\to\pm t$. We may also use common forms of these equations by expressing the limiting values of the integrals through their direct values. The resulting equations do not depend on the side from which we come to the limit. They are

$$\begin{aligned}&\frac{1}{2}\int_L \left\{-i\exp(i\alpha)\sigma(\tau)\left[U_l^1(t,\tau)-iU_l^2(t,\tau)\right]+i\exp(-i\alpha)\overline{\sigma(\tau)}\left[U_l^1(t,\tau)+iU_l^2(t,\tau)\right]\right\}ds-\\&\quad-\frac{1}{2}\int_L \left\{\left[U_S^1(t,\tau)-iU_S^2(t,\tau)\right]u(\tau)+\left[U_S^1(t,\tau)+iU_S^2(t,\tau)\right]\overline{u(\tau)}\right\}ds=\frac{1}{2}u(t),\end{aligned}\tag{20.3}$$

$$\begin{aligned}&\frac{1}{2}\int_L \left\{-i\exp(i\alpha)\sigma(\tau)\left[\sigma_S^1(t,\tau)-i\sigma_S^2(t,\tau)\right]+i\exp(-i\alpha)\overline{\sigma(\tau)}\left[\sigma_S^1(t,\tau)+i\sigma_S^2(t,\tau)\right]\right\}ds-\\&\quad-\frac{1}{2}\int_L \left\{\left[\sigma_H^1(t,\tau)-i\sigma_H^2(t,\tau)\right]u(\tau)+\left[\sigma_H^1(t,\tau)+i\sigma_H^2(t,\tau)\right]\overline{u(\tau)}\right\}ds=\frac{1}{2}\sigma(t).\end{aligned}\tag{20.4}$$

Now the second integral in (20.3) and the first integral in (20.4) are understood as principal value integrals, while the second integral in (20.4) is understood as a finite-part integral. These forms of complex boundary equations serve for the numerical procedure of coming to the limit *before* integration.

Comment 1. In all the previous formulae, one may exclude the multiplier $\exp(i\alpha)$ and use the differential $d\tau$. Indeed, from (11.15), (11.9) we have

$$\exp(i\alpha)ds=d\tau,\quad u(\tau)ds=-i\exp(i\alpha)v\,ds=-iv\,d\tau,$$

where $v = v_n + i\, v_t$ is the vector of complex displacements in the local coordinates at a point on the contour. Then, for instance, hypersingular equation (20.4) takes the form

$$\begin{aligned}&\frac{1}{2i}\int_L \left\{\left[\sigma_S^1(t,\tau)-i\sigma_S^2(t,\tau)\right]\sigma(\tau)d\tau-\left[\sigma_S^1(t,\tau)+i\sigma_{Sl}^2(t,\tau)\right]\overline{\sigma(\tau)}d\overline{\tau}\right\}+\\&+\frac{1}{2i}\int_L \left\{\left[\sigma_H^1(t,\tau)-i\sigma_H^2(t,\tau)\right]v(\tau)d\tau-\left[\sigma_H^1(t,\tau)+i\sigma_H^2(t,\tau)\right]\overline{v(\tau)}d\overline{\tau}\right\}=\frac{1}{2}\sigma(t).\end{aligned}\tag{20.5}$$

In equation (20.5), both the traction and displacements are used in local coordinates.

Comment 2. We emphasize the fact, already noted in § 9, that the equations of the direct approach in their usual form refer only to *closed* contours. Meanwhile, by properly employing this approach we may extend it to *open* contours. For real variables, this was done in § 9 for the particular case of Kelvin's solution. Below in § 23 we shall do it in the complex form.

§ 21. COMPLEX EQUATIONS OF THE DIRECT APPROACH FOR KELVIN'S SOLUTION

Formulae (12.12)-(12.15) give the explicit forms of the kernels entering Somigliana's identities (19.9), (19.10) and equations (20.1)-(20.5). Their substitution into (19.9), (19.10) and elementary transformation yields:

$$\frac{1}{2\pi i}\int_L \left[(\chi-1)u\frac{d\tau}{\tau-z}+u\,dk_1(\tau,z)+\bar{u}\,dk_2(\tau,z)-2\chi\frac{\sigma}{2\mu}\ln(\tau-z)\,d\tau+\right.$$

$$\left.+\chi\frac{\sigma}{2\mu}k_1(\tau,z)d\tau-\frac{\bar{\sigma}}{2\mu}k_2(\tau,z)d\bar{\tau}\right]=\begin{cases}(\chi+1)u(z), & z\in D,\\ 0, & z\notin D+L,\end{cases} \tag{21.1}$$

$$\frac{1}{2\pi i}\int_L \left[2u\frac{d\tau}{(\tau-z)^2}-u\frac{\partial}{\partial z}dk_1(\tau,z)-\bar{u}\frac{\partial}{\partial z}dk_2(\tau,z)-(\chi-1)\frac{\sigma}{2\mu}\frac{d\tau}{\tau-z}-\right.$$

$$\left.-\chi\frac{\sigma}{2\mu}\frac{\partial k_1(\tau,z)}{\partial z}d\tau+\frac{\bar{\sigma}}{2\mu}\frac{\partial k_2(\tau,z)}{\partial z}d\bar{\tau}\right]=\begin{cases}(\chi+1)\dfrac{\sigma(z)}{2\mu}, & z\in D,\\ 0, & z\notin D+L,\end{cases} \tag{21.2}$$

where $k_1(\tau,z)$ and $k_2(\tau,z)$ are standard kernels defined by (12.18). For brevity, we do not write arguments of the functions under the integral sign when it is clear from the context. Henceforth, we always integrate with respect to the variable τ.

Note that in the identities (21.2), the terms containing the displacement vector, *do not depend on elastic constants*. This fact, already emphasized for real variables in § 9, plays a key role in extending the equations of the direct approach to systems of elastic blocks (see § 23).

Each of Somigliana's identities (21.1), (21.2) generates three additional equations, which are not known in real variables. They are obtained by integration by parts employing the equations $\dfrac{d\tau}{\tau-t}=d\ln(\tau-t)$, $\dfrac{d\tau}{(\tau-t)^2}=-d\dfrac{1}{\tau-t}$ to replace the displacement $u(t)$ by its derivative $u'(t)$; and/or $\sigma d\tau=df$ to replace the traction $\sigma(t)$ with its integral, that is the resultant force $f(t)$.

In this way one obtains equations containing either (i) $u'(t)$, $\sigma(t)$, or (ii) $u(t)$, $f(t)$, or (iii) $u'(t)$, $f(t)$. These additional equations, in their turn, generate new equations obtained by their differentiation or integration with respect to the field point t. There is no sense in writing down these multiple equations, many of which have mostly historical interest (some of them are given in the Russian edition of this book). We will present here only one option, which gives a singular CV-BIE often used to solve problems involving cracks. It follows from (21.2) after integrating the terms containing displacements by parts. As a result, we obtain Somigliana's identities for traction in the form containing traction and the derivative of the displacements under the integral sign:

$$\frac{1}{2\pi i}\int_L\left[2u'\frac{d\tau}{\tau-z}+u'\frac{\partial k_1}{\partial z}d\tau+\overline{u'}\frac{\partial k_2}{\partial z}d\bar{\tau}-(\chi-1)\frac{\sigma}{2\mu}\frac{d\tau}{\tau-z}-\right.$$

$$\left.-\chi\frac{\sigma}{2\mu}dk_1(\tau,z)+\frac{\bar{\sigma}}{2\mu}dk_2(\tau,z)\right]=\begin{cases}(\chi+1)\dfrac{\sigma(z)}{2\mu}, & z\in D,\\ 0, & z\notin D+L.\end{cases}\qquad(21.3)$$

One may change the derivative $u'(t)$ of the displacements to the "density of dislocations" $b(t)$ by using the definition (14.8).

Take now the limit from within the region D in the first lines of (21.1)-(21.3) and keep the second lines. Then, by using the direct values of integrals to express their limiting values, we obtain:

$$\frac{1}{2\pi i}\int_L\left[(\chi-1)u\,\frac{d\tau}{\tau-t}+u\,dk_1(\tau,t)+\bar{u}\,dk_2(\tau,t)-2\chi\frac{\sigma}{2\mu}\ln(\tau-t)d\tau+\right.$$

$$\left.+\chi\frac{\sigma}{2\mu}k_1(\tau,t)d\tau-\frac{\bar{\sigma}}{2\mu}k_2(\tau,t)d\bar{\tau}\right]=\begin{cases}\dfrac{1}{2}(\chi+1)u(t), & t\in D,\\ 0, & t\notin D+L,\end{cases}\qquad(21.4)$$

$$\frac{1}{2\pi i}\int_L\left\{2u\,\frac{d\tau}{(\tau-t)^2}-u\frac{\partial}{\partial t}dk_1(\tau,t)-\bar{u}\frac{\partial}{\partial t}dk_2(\tau,t)-(\chi-1)\frac{\sigma}{2\mu}\frac{d\tau}{\tau-t}-\right.$$

$$\left.-\chi\frac{\sigma}{2\mu}\frac{\partial k_1}{\partial t}d\tau+\frac{\bar{\sigma}}{2\mu}\frac{\partial k_2}{\partial t}d\bar{\tau}\right\}=\begin{cases}\dfrac{1}{2}(\chi+1)\dfrac{\sigma(t)}{2\mu}, & t\in L,\\ 0, & t\notin D+L,\end{cases}\qquad(21.5)$$

$$\frac{1}{2\pi i}\int_L\left[2u'\frac{d\tau}{\tau-t}+u'\frac{\partial k_1}{\partial t}d\tau+\overline{u'}\frac{\partial k_2}{\partial t}d\bar{\tau}-(\chi-1)\frac{\sigma}{2\mu}\frac{d\tau}{\tau-t}-\right.$$

$$\left.-\chi\frac{\sigma}{2\mu}\frac{\partial k_1}{\partial t}d\tau+\frac{\bar{\sigma}}{2\mu}\frac{\partial k_2}{\partial t}d\bar{\tau}\right]=\begin{cases}\dfrac{1}{2}(\chi+1)\dfrac{\sigma(t)}{2\mu}, & t\in L,\\ 0, & t\notin D+L,\end{cases}\qquad(21.6)$$

The first lines in (21.4), (21.5) are the complex forms of the equations of the direct approach known in real variables (see, e.g. Hartmann [1], Ugodchikov and Khutorianski [1], Krishnasamy, Schmerr et al. [1]). But being equivalent in a mathematical sense, they present computational advantages over their real counterparts, particularly, easy evaluation of singular and hypersingular integrals over curvilinear elements.

The equations (21.4)-(21.6) will also serve us to derive the most general equations for blocky systems including all other equations as particular cases.

Before deriving them, for completeness, we reveal the connection between the equations of the direct approach with the classical CV-BIE by Muskhelishvili, and the equations by the author.

§ 22. THE CONNECTION TO MUSKHELISHVILI'S EQUATIONS

As mentioned, there are numerous other equations containing $u'(t)$ and/or $f(t)$ by integration (21.4), (21.5) by parts. In particular, by replacing the traction $\sigma(t)$ in (21.4), (21.5) by the resultant force $f(t)$ we obtain:

$$\frac{1}{2\pi i}\int_L \Big\{(\chi-1)u\frac{d\tau}{\tau-t}+u\,dk_1(\tau,t)+\bar{u}\,dk_2(\tau,t)+2\chi\frac{f}{2\mu}\frac{d\tau}{\tau-t}-$$

$$-\chi\frac{f}{2\mu}dk_1(\tau,t)+\frac{\bar{f}}{2\mu}dk_2(\tau,t)\Big\}=\begin{cases}\frac{1}{2}(\chi+1)u(t), & t\in L,\\ 0, & t\notin D+L,\end{cases} \tag{22.1}$$

$$\frac{1}{2\pi i}\int_L \Big[2u\frac{d\tau}{\tau-t}-u\,dk_1(\tau,t)-\bar{u}\,dk_2(\tau,t)-(\chi-1)\frac{f}{2\mu}\frac{d\tau}{\tau-t}+$$

$$+\chi\frac{f}{2\mu}dk_1(\tau,t)-\frac{\bar{f}}{2\mu}dk_2(\tau,t)\Big]=\begin{cases}\frac{1}{2}(\chi+1)\frac{f(t)}{2\mu}, & t\in L,\\ 0, & t\notin D+L.\end{cases} \tag{22.2}$$

By summing these equations we have:

$$\frac{1}{\pi i}\int_L \frac{(2\mu u+f)d\tau}{\tau-t}=\begin{cases}2\mu u(t)+f(t), & t\in L,\\ 0, & t\notin D+L.\end{cases}$$

From the holomorphicity theorem (see equation (11.30)) this means that the function under the integral sign represents the limiting value of the function $2\mu u(z) + f(z)$, holomorphic within the region D. By dividing this function by $\chi + 1$, we obtain the function

$$\varphi(z)=\frac{2\mu u(z)+f(z)}{\chi+1}, \tag{22.3}$$

that is also holomorphic in D and has the limiting values $\varphi(t) = [2\mu u(t) + f(t)]/(\chi + 1)$ on the contour L. By using (22.3) and simple transformations we may write (22.1) and (22.2) as (Linkov and Mogilevskaya [5], Linkov [6])

$$\frac{1}{\pi i}\int_L \varphi\frac{d\tau}{\tau-t}-\frac{1}{2\pi i}\int_L \varphi dk_1-\frac{1}{2\pi i}\int_L \overline{\varphi}dk_2=\frac{1}{2}f(t)+\frac{1}{2\pi i}\int_L f\frac{d\overline{\tau}}{\overline{\tau}-\overline{t}},\quad t\in L, \quad (22.4)$$

$$\frac{\chi}{\pi i}\int_L \varphi\frac{d\tau}{\tau-t}-\frac{\chi}{2\pi i}\int_L \varphi dk_1+\frac{1}{2\pi i}\int_L \overline{\varphi}dk_2=\mu u(t)+\frac{1}{\pi i}\int_L \mu u\frac{d\overline{\tau}}{\overline{\tau}-\overline{t}},\quad t\in L. \quad (22.5)$$

We may exclude the singular integrals from (22.4), (22.5) by employing the holomorphicity of the function $\varphi(z)$. Indeed, applying (11.30), we write these equations as

$$\varphi(t)-\frac{1}{2\pi i}\int_L \varphi dk_1-\frac{1}{2\pi i}\int_L \overline{\varphi}dk_2=\frac{1}{2}f(t)+\frac{1}{2\pi i}\int_L f\frac{d\overline{\tau}}{\overline{\tau}-\overline{t}},\quad t\in L, \qquad (22.6)$$

$$\chi\varphi(t)-\frac{\chi}{2\pi i}\int_L \varphi dk_1+\frac{1}{2\pi i}\int_L \overline{\varphi}dk_2=\mu u(t)+\frac{1}{\pi i}\int_L \mu u\, d\frac{\overline{\tau}}{\overline{\tau}-\overline{t}},\quad t\in L. \qquad (22.7)$$

Equations (22.6), (22.7) are Muskhelishvili's equations (Muskhelishvili [1]) for the prescribed displacement and traction respectively (see also Muskhelishvili [5, § 98]). These are regular Fredholm equations of the second kind. However, they contain as the unknown the function $\varphi(t)$ instead of physically significant values that are the resultant force and the displacements. Substitution (22.3) into (22.6), (22.7) leads to the equations with respect to these physical values (Linkov [2]). We see that the equations by Muskhelishvili [1, 5] and the author [2, 6] for closed contours may be obtained from the complex equations of the direct approach. *Vice versa*, all the equations obtained by the direct approach may be derived from those of this section.

§ 23. COMPLEX EQUATIONS FOR BLOCKY SYSTEMS WITH STRINGERS OR CRACKS

Now we will use the CV-BIE of the direct approach to derive general equations for blocky systems with stringers or/and cracks. These equations include the previous CV-BIE of the direct and indirect approaches as particular cases. Actually, they reproduce the real equations of § 9 in the complex variable form. This form being

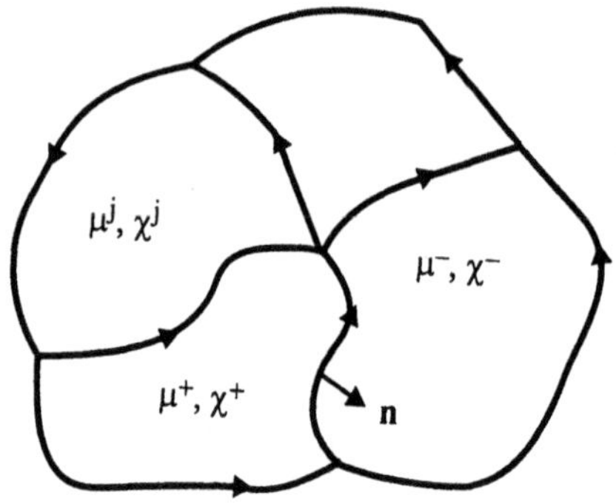

Fig. 13.

computationally advantageous, it presents an efficient new tool for computational mechanics.

Consider a piece-wise homogeneous body comprised of a finite number of blocks which interact on their boundaries (Figure 13). The blocks may be inserted into an infinite matrix. In this case the matrix is also considered as a block. Elastic constants of the j-th block are μ^j and χ^j. For each bock we may write equations (21.4)-(21.6) and similar equations obtained by integration by parts. As all other equations are just modifications of the first two, which directly yield from Somigliana's identities, we will dwell only on (21.4) and (21.5).

Equations (21.4), (21.5) for the j-th block are

$$\frac{1}{2\pi i}\int_{L^j}\Bigg[\left(\chi^j-1\right)u^j\frac{d\tau}{\tau-t}+u^j dk_1+\overline{u^j}dk_2-2\frac{\chi^j}{2\mu^j}\sigma^j\ln(\tau-t)d\tau+$$

$$+\frac{\chi^j}{2\mu^j}\sigma^j k_1 d\tau-\frac{1}{2\mu^j}\overline{\sigma}^j k_2 d\overline{\tau}\Bigg]=\begin{cases}\frac{1}{2}\left(\chi^j+1\right)u^j(t), & t\in L^j,\\ 0, & t\notin D^j+L^j,\end{cases}\tag{23.1}$$

$$\frac{1}{2\pi i}\int_{L^j}\Bigg[2u^j\frac{d\tau}{(\tau-t)^2}-u^j\frac{\partial}{\partial t}dk_1-\overline{u^j}\frac{\partial}{\partial t}dk_2-\frac{\chi^j-1}{2\mu^j}\sigma^j\frac{d\tau}{\tau-t}-$$

$$-\frac{\chi^j}{2\mu^j}\sigma^j\frac{\partial k_1}{\partial t}d\tau+\frac{1}{2\mu^j}\overline{\sigma^j}\frac{\partial k_2}{\partial t}d\overline{\tau}\Bigg]=\begin{cases}\frac{1}{2}\frac{\chi^j+1}{2\mu^j}\sigma^j(t), & t\in L^j,\\ 0, & t\notin D^j+L^j,\end{cases}\tag{23.2}$$

Sum each of (23.1), (23.2) over all the blocks, recalling that the second lines of (23.1), (23.2) hold when the point t is outside a block. We obtain:

$$\frac{1}{2\pi i}\int_L\Bigg\{[\Delta\chi u+(\chi-1)\Delta u]\frac{d\tau}{\tau-t}+\Delta u dk_1+\overline{\Delta u}dk_2-[(a_3-a_1)\sigma+$$

$$+\frac{1}{2}(a_2 - a_4)\Delta\sigma][2\ln(\tau - t) - k_1]d\tau - (a_1\bar{\sigma} + \frac{1}{2}a_4\Delta\bar{\sigma})k_2 d\bar{\tau}\Big\} =$$

$$= (\chi + 1)u(t) + \frac{1}{4}\Delta\chi\,\Delta u(t), \qquad t \in L, \tag{23.3}$$

$$\frac{1}{2\pi i}\int_{L^j}\Big\{2\Delta u\frac{d\tau}{(\tau - t)^2} - \Delta u\frac{\partial}{\partial t}dk_1 - \overline{\Delta u}\frac{\partial}{\partial t}dk_2 + [(2a_1 - a_3)\sigma + (a_4 - \frac{1}{2}a_2)\Delta\sigma]\frac{d\tau}{\tau - t} +$$

$$+[(a_1 - a_3)\sigma + \frac{1}{2}(a_4 - a_2)\Delta\sigma]\frac{\partial k_1}{\partial t}d\tau + (a_1\bar{\sigma} + \frac{1}{2}a_4\Delta\bar{\sigma})\frac{\partial k_2}{\partial t}d\bar{\tau}\Big\} =$$

$$= \frac{1}{2}a_2\,\sigma + \frac{1}{4}a_3\Delta\sigma, \quad t \in L, \tag{23.4}$$

where, as usual, the symbol Δ denotes discontinuity: $\Delta\chi = \chi^+ - \chi^-$, $\Delta u = u^+ - u^-$, $\Delta\sigma = \sigma^+ - \sigma^-$; the symbols χ, u and σ denote the mean values:

$$\chi = \frac{\chi^+ + \chi^-}{2}, \; u = \frac{u^+ + u^-}{2}, \; \sigma = \frac{\sigma^+ + \sigma^-}{2}; \tag{23.5}$$

a_1, a_2, a_3, a_4, are piece-wise constant characteristics of the properties of adjacent blocks:

$$a_1 = \frac{1}{2\mu^+} - \frac{1}{2\mu^-}, \quad a_2 = \frac{\chi^+ + 1}{2\mu^+} + \frac{\chi^- + 1}{2\mu^-},$$
$$a_3 = \frac{\chi^+ + 1}{2\mu^+} - \frac{\chi^- + 1}{2\mu^-}, \quad a_4 = \frac{1}{2\mu^+} + \frac{1}{2\mu^-}, \tag{23.6}$$

now L is the total boundary of the system of blocks, and we take the contact between surfaces of blocks as a single line at which physical properties may experience discontinuities; $\sigma = \sigma_{nn} + i\sigma_{nt}$ is the traction vector at the point t with the normal **n** to the contour directed to the right of the traveling direction; this direction is taken arbitrarily on the contacts but for the external boundaries, when they are present, we, as usual, consider the normal to be outward (the traveling path leaves the blocks on the left). Also as usual, the index "plus" ("minus") refers to a value from the left (right) to the traveling path. If there is no embedding matrix, we assume that on the external boundary of the blocks $\chi^- = 0$, $1/\mu^- = 0$, $u^- = 0$, $\sigma^- = 0$.

On the parts of L where both displacements and traction are continuous, while the elastic properties are the same on the both sides, all the integrals on the l. h. s. of (23.3), (23.4) vanish. Hence, these parts may be excluded from the contour L. This means that the equations are applicable to open arcs (Figure14, *a*) and also to arcs which terminate on the boundary of a block (Figure14, *b*).

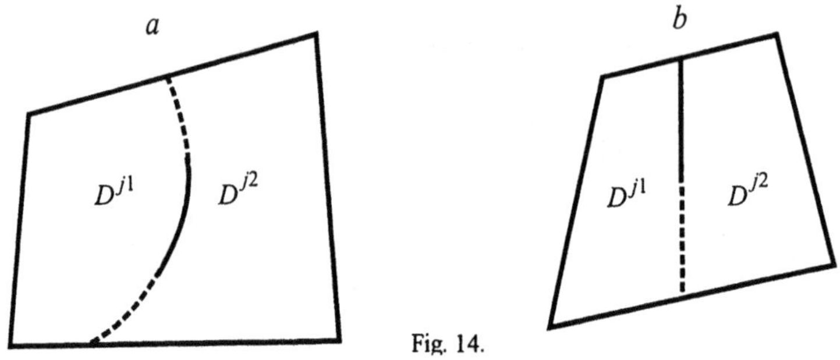

Fig. 14.

In general, the system (23.3), (23.4) serves for problems involving the discontinuities of both the displacements and tractions. In such problems, we have two contact conditions imposed on the four values σ, u, $\Delta\sigma$ and Δu. The conditions when used in (23.3), (23.4), make the system complete. We may efficiently solve it by the complex variable boundary element method (see Chapter 12).

In a number of applications one of the equations may be solved separately from the other. For instance, this is the case in problems for stringers welded to homogeneous plates. Then according (23.6), $a_1 = 0$, $a_3 = 0$ and since the displacements are continuous through the stringers, we have $\Delta u = 0$. Therefore, (23.3) contains only the values $\Delta\sigma$ and u connected by a law of stringer deformation. Being used in (23.3), the law allows us to solve the equation separately from (23.4). (The latter equation may serve to find the mean traction σ, defined by the second of (23.5), after u and consequently $\Delta\sigma$ are found from (23.3)).

The other case refers to problems of micromechanics. Commonly in these problems, the traction is continuous through the contacts of blocks ($\sigma^+ = \sigma^- = \sigma$; $\Delta\sigma = 0$), while the displacement discontinuity Δu is either independent or a prescribed function of the traction σ. Then (23.4) takes the form (Linkov [7], Linkov and Mogilevskaya [1,2]):

$$\frac{1}{2\pi i}\int_L \left[2\Delta u \frac{d\tau}{(\tau - t)^2} - \Delta u \frac{\partial}{\partial t} dk_1 - \overline{\Delta u}\frac{\partial}{\partial t} dk_2 + (2a_1 - a_3)\sigma - \right.$$

$$\left. +(a_1 - a_3)\sigma \frac{\partial k_1}{\partial t} d\tau + a_1\overline{\sigma}\frac{\partial k_2}{\partial t}d\overline{\tau} \right] = \frac{1}{2}a_2\sigma(t), \tag{23.7}$$

where now the symbol σ denotes the actual traction both on contacts and on the external boundary.

Comment 1. As usual, one can transform the complex variable equations for the tractions and the displacement discontinuities into equations containing the *resultant forces* and/or the *derivative* of the displacements. Thus, each of (23.3), (23.4), (23.7) generates at least three new equations. Together with (23.3), (23.4), they include all the CV-BIE of the indirect and direct approaches derived above as the simplest particular cases. We will not write down any of the additional equations: containing either the resultant forces and/or the derivative of the displacements, they are less convenient for satisfying contact conditions than (23.3), (23.4), (23.7).

Comment 2. Equations (23.3), (23.4), being derived from *complex* variable Somigliana's identities (23.1), (23.2) for displacements and traction, reproduce analogous equations (9.2), (9.5) obtained in a similar way from *real* variables Somigliana's identities (9.1), (9.4). Respectively, (23.7) is the complex variable form of equation (9.6) written for Kelvin's fundamental solution.

We emphasize again that re-writing of real equations in the complex variable form is not just a formal operation; it greatly eases the numerical solution of plane problems, now for blocky systems. The hypersingular equation (23.7) is the most attractive for numerical modeling of a medium with internal structure. First, it is rather general: the elastic properties of the blocks (grains) may be different; the blocks may contain cracks, pores and inclusions; the cracks may be inter-, intro- and/or infra-granular. Inclusions represent a particular case of internal blocks. Pores are accounted for by including their surfaces into the contour L. For irreversible deformations on contacts, the traction and the displacement discontinuities in (23.7) are changed to their increments. This makes (23.7) applicable to many problems of micromechanics (see, e.g. Dobroskok et al. [1], Mogilevskaya et al. [1] and the examples in § 32, 57). We will present further enhancement of this important equation, in § 30 of the next chapter.

Secondly, equation (23.7) has significant computational advantages additional to those common to all CV-BIE. Normally, it has a *zero* index; in practice this means that there is no need to satisfy additional conditions, as is the case when one uses singular equations for cracks. As mentioned, it contains the values, displacement discontinuities and traction (or their increments), which characterize the contact interaction. Finally, it is easily solved by the CV-BEM, as explained and illustrated in Chapters 12, 13.

§ 24. CALCULATION OF STRESSES, RESULTANT FORCE AND DISPLACEMENTS AT POINTS WITHIN THE BLOCKS

Assume that we have solved the hypersingular equation (23.7) for a blocky system with cracks. Now we know both the displacement discontinuity and the traction along the contour L. Then we can immediately find stresses at points within the blocks. Indeed, applying (21.2) to each of the blocks and summing we have:

$$(\chi^j+1)\frac{\sigma(z)}{2\mu^j}=\frac{1}{2\pi i}\int_L\left[2\Delta u\frac{d\tau}{(\tau-z)^2}-\Delta u\frac{\partial}{\partial z}dk_1(\tau,z)-\overline{\Delta u}\frac{\partial}{\partial z}dk_2(\tau,z)+\right.$$

$$\left.+(2a_1-a_3)\sigma\frac{d\tau}{\tau-z}+(a_1-a_3)\sigma\frac{\partial k_1}{\partial z}d\tau+a_1\overline{\sigma}\frac{\partial k_2}{\partial z}d\overline{\tau}\right],\qquad z\in D^j. \quad (24.1)$$

The point z in (24.1) belongs to the block with the number j, the index j in χ^j, $2\mu^j$ means that these elastic constants refer to this block. The constants a_1, a_2, a_3 are defined by (23.6). When calculating the vector σ, we take the direction of the normal to an area at the point z arbitrarily. A particular choice influences only the angle α defining its orientation and entering the derivative $d\bar{z}/dz = \exp(-2i\alpha)$. Hence, as mentioned in § 11, we have the traction on an arbitrary oriented area; actually, (24.1) gives the tensor of stresses.

Unfortunately, it is impossible to obtain such simple results for the displacements within the blocks. The reason is that Somigliana's identities for *displacements* (21.1) contain displacements under the integral sign with the multiplier $\chi - 1$. It is different for blocks with different Poisson's ratio. Hence, after summing equations (21.1) over the blocks we have:

$$u(z) = \frac{1}{2\pi i\left(\chi^j + 1\right)} \int_L \Big\{ [\Delta\chi u + (\chi - 1)\Delta u] \frac{d\tau}{\tau - z} + \Delta u \, dk_1(\tau, z) + \\ + \overline{\Delta u} \, dk_2(\tau, z) - 2(a_3 - a_1)\sigma \ln(\tau - z) d\tau + \\ + (a_3 - a_1)\sigma k_1(\tau, z) d\tau - a_1 \bar{\sigma} k_2(\tau, z) d\bar{\tau} \Big\}, \quad z \in D^j, \tag{24.2}$$

where, as in (24.1), we assume that the traction is continuous through contacts and the point z is within the j-th bock.

From (24.2) we see that the displacements within the blocks are expressed through the displacement discontinuities only when the Poisson's coefficients of all the blocks are the same (in this case, $\chi^+ = \chi^- = \chi$, $\Delta\chi = 0$). In the general case, when $\chi^+ \neq \chi^-$ at least on some part of L, we need to substitute Δu into (23.3), to solve the latter equation with respect to the mean displacement u, and only after this, use (24.2). Another, more convenient way to find the displacements within blocks when we know the displacement discontinuity on their boundaries, is based on the Kolosov-Muskhelishvili theory; we do not have to solve an additional equation. We will present this method at the end of § 30.

To conclude this part, we point out that by writing real BIE in the complex variable form we obtain a powerful new means for solving boundary value problems. Besides, we have additional equations, containing the resultant force. They are not known in real variables and may be useful for coupling the method of boundary elements (BEM) with the method of finite elements (FEM).

PART II

METHODS BASED ON THE THEORY BY KOLOSOV-MUSKHELISHVILI

In part I, following the usual procedure of the real potentials, we simply reproduced the two-dimensional real equations in complex form. This opened an opportunity to employ computational advantages of the CV-BIE. Some additional results were obtained in passing. So far, we have not employed the Kolosov-Muskhelishvili (K-M) theory. This theory is specially tailored to derive the advantages of using complex variables in plane elasticity to the fullest extent; it provides additional important ways to solve plane problems. In this part we present its applications to our theme, aiming, as noted, to enhance and extend the most general and useful equations derived for blocky systems. We will start from holomorphicity theorems. We will show how to employ them to obtain the extensions. A few examples from micromechanics will illustrate the computational advantages of the derived CV-BIE.

Chapter 5

FUNCTIONS OF KOLOSOV-MUSKHELISHVILI AND HOLOMORPHICITY THEOREMS

§ 25. FUNCTIONS AND REPRESENTATIONS OF KOLOSOV-MUSKHELISHVILI

It is well known (see, e. g. Muskhelishvili [5]) that the equations of plane elasticity can be reduced to the bi-harmonic equation for Airy's stress function. Hence, two *harmonic* functions are sufficient to represent a general solution. This means that the general solution in complex variables is represented by two holomorphic functions of the complex argument $z = x + iy$. The classical Goursat formula gives the connection between the real Airy function and these two holomorphic functions. By taking displacements and stresses expressed in terms of Airy's function and using Goursat's formula, we obtain the displacements and stresses expressed in terms of these two complex functions. This idea was first employed by Kolosov [1] and comprehensively exploited by Muskhelishvili [5].

The final formulae by K-M are (see Muskhelishvili [5, § 31, 32]):

$$f(z)=\varphi(z)+z\overline{\varphi'(z)}+\overline{\psi(z)}, \tag{25.1}$$

$$2\mu u(z)=\chi\varphi(z)-z\overline{\varphi'(z)}-\overline{\psi(z)}, \tag{25.2}$$

where, as previously, $f(z)$ is a complex vector defined by (11.16); it is equal to i times the vector of the resultant force $h_x + i\,h_y$ in global coordinates:

$$f(z)=i(h_x+ih_y)=\int_{z_0}^{z}\sigma\,dz \tag{25.3}$$

(recall that in § 11 we agreed for brevity to call $f(z)$ the resultant force); $u = u_x + iu_y$ is the complex vector of displacements in the global coordinates (x, y); χ is Muskhelishvili's parameter.

In (25.3), $\sigma = \sigma_{nn} + i\sigma_{nt}$ is the complex vector of traction on an area with normal $\mathbf{n}$ in the local coordinates $(\mathbf{n},\mathbf{t})$ (see Figure 8 in § 11); it is expressed through the total derivative of $f(z)$ in accordance with (11.18). The latter presumes that $f(z)$ is a composite function of z and $\bar{z}$ (z), so that

$$\sigma=\frac{df}{dz}=\frac{\partial f}{\partial z}+\frac{\partial f}{\partial \bar{z}}\frac{\partial \bar{z}}{\partial z}. \tag{25.4}$$

The functions $\varphi(z)$ and $\psi(z)$ are assumed to be holomorphic in any simply connected part of a considered region; they are termed the Kolosov-Muskhelishvili (K-M) functions (potentials). For a multi-connected region these functions may be multi-valued and we need special terms that allow us to have single-valued displacements. Thus, side by side with the functions $\varphi(z)$ and $\psi(z)$, we sometimes use their derivatives

$$\Phi(z)=\frac{d\varphi}{dz}, \qquad \Psi(z)=\frac{d\psi}{dz}, \tag{25.5}$$

that are more convenient because they are always single-valued, which implies that they are holomorphic in any considered region. By taking the total derivatives from (25.1), (25.2) and accounting for (25.4), (25.5) we obtain equations for the complex traction vector in the local coordinates:

$$\sigma(z)=\Phi(z)+\overline{\Phi(z)}+\left[z\overline{\Phi'(z)}+\overline{\Psi(z)}\right]\frac{d\bar{z}}{dz} \tag{25.6}$$

and for the derivative of the complex displacement vector in the direction dz:

$$2\mu u'(z)=\chi\Phi(z)-\overline{\Phi(z)}-\left[z\overline{\Phi'(z)}+\overline{\Psi(z)}\right]\frac{d\bar{z}}{dz}. \tag{25.7}$$

The functions $\Phi(z)$ and $\Psi(z)$ are also called K-M functions.

We may express the function $\varphi(z)$ in terms of the physically significant values $u(z)$ and $f(z)$. Indeed, summing (25.1) and (25.2) and dividing by $\chi + 1$ we have:

$$\varphi(z)=\frac{2\mu\, u(z)+f(z)}{\chi+1}. \tag{25.8}$$

Analogously from (25.6), (25.7) it follows:

$$\Phi(z)=\varphi'(z)=\frac{2\mu\, u'(z)+\sigma(z)}{\chi+1}. \tag{25.9}$$

Unfortunately, there are no such simple expressions in terms of physical values for the functions $\psi(z)$ and $\Psi(z)$.

Suppose a region D has a closed contour L, and that the point z tends from within the region to a point t on its contour, we have from (25.1), (25.2), (25.6), (25.7):

$$\varphi(t)+t\overline{\varphi'(t)}+\overline{\psi(t)}=f(t), \tag{25.10}$$

$$\chi\varphi(t)-t\overline{\varphi'(t)}-\overline{\psi(t)}=2\mu u(t), \tag{25.11}$$

$$\Phi(t)+\overline{\Phi(t)}+\left[t\overline{\Phi'(t)}+\overline{\Psi(t)}\right]\frac{d\bar{t}}{dt}=\sigma(t), \tag{25.12}$$

$$\chi\,\Phi(t)-\overline{\Phi(t)}-\left[t\overline{\Phi'(t)}+\overline{\Psi(t)}\right]\frac{d\bar{t}}{dt}=2\mu u'(t). \tag{25.13}$$

In accordance with the conventions of § 11, we assume that the traction vector in (25.12) is defined on the elements of L with the normal **n** outward to D. The derivative of displacements in (25.13) is taken in the direction that leaves the region on the left.

For open arcs, the limit depends on the side from which the point tends from a region to a contour. Denoting, as usual, the limits from the left and the right of the traveling path on L by plus and minus, we have:

$$\varphi^{\pm}(t)+t\overline{\varphi'^{\pm}(t)}+\overline{\psi^{\pm}(t)}=f^{\pm}(t), \tag{25.14}$$

$$\chi\varphi^{\pm}(t)-t\overline{\varphi'^{\pm}(t)}-\overline{\psi^{\pm}(t)}=2\mu u^{\pm}(t), \tag{25.15}$$

$$\Phi^{\pm}(t)+\overline{\Phi^{\pm}(t)}+\left[t\overline{\Phi'^{\pm}(t)}+\overline{\Psi^{\pm}(t)}\right]\frac{d\bar{t}}{dt}=\sigma^{\pm}(t) \tag{25.16}$$

$$\chi\Phi^{\pm}(t)-\overline{\Phi^{\pm}(t)}-\left[t\overline{\Phi'^{\pm}(t)}+\overline{\Psi^{\pm}(t)}\right]\frac{d\bar{t}}{dt}=2\mu u'^{\pm}(t). \tag{25.17}$$

The direction of travel along L is taken arbitrarily; it may be opposite on adjacent parts of a same arc. However, the normal **n** in (25.16) is always directed to the *right* of the chosen travel direction. Thus, the pair (**n**,**t**) always represents the unit vectors of the *right* handed Cartesian system (see Figure 8 in § 11).

§ 26. HOLOMORPHICITY THEOREMS

The classical way of obtaining the complex variable boundary integral equations (CV-BIE) of plane elasticity was to employ the holomorphicity of the functions $\varphi(z)$ and $\psi(z)$ in the region D with contour L (Muskhelishvili [1], see also [5, § 98]). In fact, it employs the necessary and sufficient condition for a function defined on L to represent the limiting values of a function holomorphic in D. This interpretation gives us a key to derive the CV-BIE in a general case when the region contains arbitrary holes and cuts, and also for periodic and doubly periodic problems. Thus, in this paragraph we present the necessary holomorphicity theorems. As usual, we assume the contour and the functions to be sufficiently smooth to justify the operations of coming to the limit, integration by parts and evaluation of the direct values of integrals.

Finite simply connected region. Consider a finite simply connected region D with contour L (Figure 15, *a*). Let a function $\varphi(z)$ be holomorphic within D, continuous up to its contour and have the limiting values $\varphi(t)$ on L. Then Cauchy's theorem (see, e. g. Churchill [1]) reads

$$\frac{1}{2\pi i}\int_L \frac{\varphi(\tau)d\tau}{\tau - z}=\varphi(z), \quad z\in D. \tag{26.1}$$

For points z outside D the integrand taken as a function of τ does not have poles within D. Hence,

$$\frac{1}{2\pi i}\int_L \frac{\varphi(\tau)d\tau}{\tau - z}=0, \quad z\notin D+L. \tag{26.2}$$

We can consider equation (26.2) as a necessary and sufficient condition for the function $\varphi(t)$ to represent the limiting values of a function $\varphi(z)$ holomorphic in D (see, e. g. Muskhelishvili [5, § 73]). By applying Sokhotski-Plemelj formula (11.26)

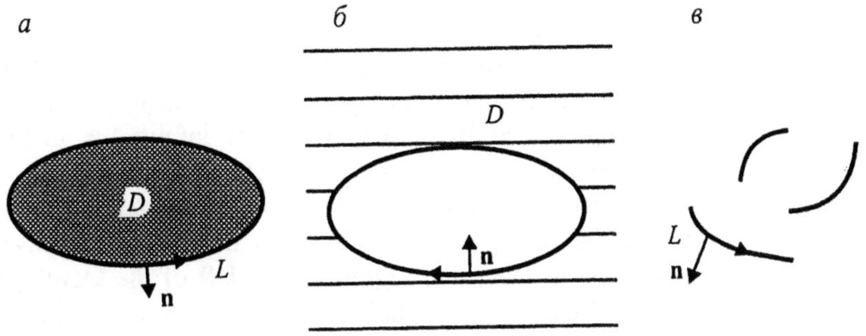

Fig. 15.

to (26.1) when z tends to L from within D, or to (26.2) when z tends from outside D, we obtain the same result:

$$\varphi(t)=\frac{1}{\pi i}\int_L \frac{\varphi(\tau)d\tau}{\tau - t}, \quad t \in L. \tag{26.3}$$

The integral on the r. h. s. is understood as a principal value (Cauchy) integral. This equation is a convenient form of the necessary and sufficient condition of holomorphicity. We will formulate this condition as a theorem.

Theorem (Plemelj). *A function $\varphi(t)$ of the Holder class represents the limiting values of a function $\varphi(z)$ holomorphic in D if and only if the condition (26.3) is satisfied for all points t on L.*

The *necessity* of this condition follows from the derivation of (26.3) given above. To prove the *sufficiency* of (26.3), consider the function

$$F(z)=\frac{1}{2\pi i}\int_L \frac{\varphi(\tau)d\tau}{\tau - z},$$

where $\varphi(t)$ is a Holder class function prescribed *only* on L. We want to show that if $\varphi(t)$ satisfies (26.3) then there exists a function $\varphi(z)$ holomorphic in D such that its limiting values on L are equal to $\varphi(t)$.

The function $F(z)$ is holomorphic within D. From the Sokhotski-Plemelj formula (11.26) it follows that its limiting value on the contour is

$$F^+(t)=\tfrac{1}{2}\varphi(t)+\frac{1}{2\pi i}\int_L \frac{\varphi(\tau)d\tau}{\tau - t}.$$

Then employing (26.3) we have $F^+(t) = \varphi(t)$, which implies that the function $\varphi(t)$ really represents the limiting values of a function holomorphic within D: this is the function $\varphi(z) = F(z)$. Equation (26.1) serves to define it within D.

Infinite simply connected region. Consider now an infinite region D (Figure 15, *b*), outside a closed contour L. Assume that $\varphi(z)$ is holomorphic within D, continuous up to its contour and has the limiting values $\varphi(t)$ on L. For further applications and for simplicity, we will assume, unless otherwise stated, that a function holomorphic in an infinite region is *equal to zero at infinity.* (The general case, that the function is represented by a polynomial at infinity, is included immediately, see, e.g. Muskhelishvili [5, § 70]). Under these conditions, the argument used for a finite region holds for an infinite region. In particular, formulae (26.1)-(26.3) hold, as does the formulation of Plemelj's theorem. Naturally, in accordance with our convention to leave the region on the left of the travel path, now we travel L *clock-wise*, (Figure 15, *b*), while for a finite region the direction was *counter clock-wise* (Figure 15, *a*).

Comment on multi-connected regions. All the formulae and the theorem hold also for multi-connected regions. This is due to the fact that the function $\varphi(z)$, being holomorphic, is single-valued in such regions (see Muskhelishvili [5, §74]).

Assume now that a function $\varphi(t)$ given on the contour L has the derivative $\varphi'(t)$ of the Holder class. Then we may use the following theorem involving a complex finite-part integral.

Theorem. *A function $\varphi(t)$ with the derivative $\varphi'(t)$ of the Holder class represents the limiting values of a function $\varphi(z)$ holomorphic in D if and only if the condition*

$$\varphi'(t)=\frac{1}{\pi i}\int_L \frac{\varphi(\tau)d\tau}{(\tau-t)^2} \tag{26.4}$$

is satisfied at any point t on L. The integral is understood as a finite-part integral. *The function $\varphi(z)$ is defined by (26.1).*

We may see the *necessity* of this condition by noting that (26.4) is the result of formal differentiation of (26.3). *Sufficiency* follows from the inverse procedure, - integration (26.4) which leads to the sufficient condition (26.3). The procedures of differentiation and integration are justified in part III (see § 47, 48 in chapter 10).

Open arcs in an infinite plane. Consider a set of p non-intersecting arcs L_j (j = 1,..., p) in an infinite plane (Figure 15, *c*). Arbitrarily chose the direction of travel along each arc and denote, as usual, the left and the right vicinity by + and - respectively. Let a_j, b_j denote the beginning and end of the j-th arc, respectively. The total contour is denoted $L = L_1 \ldots + L_p$.

The following theorem is stated on the basis of a general theory (Muskhelishvili [4]).

Theorem. *Functions $\varphi^+(t)$, $\varphi^-(t)$ of the H^* class on L (including beginning and end points of the arcs) represent the limiting values of a function $\varphi(z)$, piece-wise holomorphic in D and equal to zero at infinity if and only if the condition*

$$\varphi^+(t)+\varphi^-(t)=\frac{1}{\pi i}\int_L \frac{\varphi^+(\tau)-\varphi^-(\tau)}{\tau-t}d\tau \tag{26.5}$$

is satisfied at all points t on L. The function $\varphi(z)$ is defined by equation

$$\varphi(z)=\frac{1}{2\pi i}\int_L \frac{\varphi^+(\tau)-\varphi^-(\tau)}{\tau - z}d\tau. \qquad (26.6)$$

Besides, if the functions $\varphi^+(t)$, $\varphi^-(t)$ *have the derivatives* $\varphi^{+'}(t)$, $\varphi^{-'}(t)$ *of the H* class and* $\varphi^+(a_j) - \varphi^-(a_j) = \varphi^+(b_j) - \varphi^-(b_j) = 0$, *then*

$$\varphi'^{\pm}(t)=\varphi^{\pm'}(t),$$

$$\varphi^{+'}(t)+\varphi^{-'}(t)=\frac{1}{\pi i}\int_L \frac{\varphi^{+'}(\tau)-\varphi^{-'}(\tau)}{\tau - t}d\tau. \qquad (26.7)$$

The first of these equations states that one can interchange the order of the operations of taking the limits and differentiation.

Necessity. Assume that $\varphi^+(t)$ and $\varphi^-(t)$ are the limiting values of a function $\varphi(z)$ piece-wise holomorphic outside L. Consider the function

$$F(z)=\frac{1}{2\pi i}\int_L \frac{\varphi^+ - \varphi^-}{\tau - z}d\tau \qquad (26.8)$$

and show that it coincides with $\varphi(z)$. From Sokhotski-Plemelj formulae (11.26) we have

$$F^+ - F^- = \varphi^+ - \varphi^- \qquad (26.9)$$

This implies that the function $F(z) - \varphi(z)$, being piece-wise holomorphic outside L, satisfies the equation

$$(F-\varphi)^+ - (F-\varphi)^- = 0.$$

As it is zero at infinity, it equals to zero in the whole plane. Hence, $F(z) = \varphi(z)$, and for the limiting values we have $F^+(t) = \varphi^+(t)$, $F^-(t) = \varphi^-(t)$. Again applying the Sokhotski-Plemelj formulae (11.26), we obtain:

$$F^+ + F^- = \frac{1}{\pi i}\int_L \frac{\varphi^+ - \varphi^-}{\tau - t}d\tau. \qquad (26.10)$$

As we have stated that $F^+(t) = \varphi^+(t)$, $F^-(t) = \varphi^-(t)$, this coincides with (26.5).

Sufficiency of (26.5) follows immediately if we consider again the function $F(z)$ defined by (26.8). Now (26.5) is satisfied by assumption. Then from (26.10) and

(26.5) we have: $F^+(t) + F^-(t) = \varphi^+(t) + \varphi^-(t)$. From this, accounting for (26.9) we obtain $\varphi^+(t) = F^+(t)$, $\varphi^-(t) = F^-(t)$, that is the functions $\varphi^+(t)$ and $\varphi^-(t)$ really represent the limiting values of the piece-wise holomorphic function $\varphi(z)$ defined by (26.6). The proof of the first of equations (26.7) needs more sophisticated discussion. Actually, Muskhelishvili gave it in § 115 of his monograph [4] (see also Gahov [1]).

The condition (26.5) contains a principal value integral. We can also formulae the holomorphicity theorem in terms of a finite-part integral.

Theorem. *Functions $\varphi^+(t)$, $\varphi^-(t)$ with the derivatives $\varphi^{+'}(t)$, $\varphi^{-'}(t)$ of the H^* class on L represent the limiting values of a function $\varphi(z)$ piece-wise holomorphic in D and equal to zero at infinity if and only if the condition*

$$\varphi^{+'} + \varphi^{-'} = \frac{1}{\pi i} \int_L \frac{\varphi^+ - \varphi^-}{(\tau - t)^2} d\tau$$

is satisfied at any point t on L. Besides, $\varphi^{\pm'}(t) = \varphi'^{\pm}(t)$ and the functions $\varphi(z)$, $\varphi'(z)$ are defined by equations

$$\varphi(z) = \frac{1}{2\pi i} \int_L \frac{\Delta\varphi d\tau}{\tau - z}, \quad \varphi'(z) = \frac{1}{2\pi i} \int_L \frac{\Delta\varphi d\tau}{(\tau - z)^2},$$

where $\Delta\varphi(t) = \varphi^+(t) - \varphi^-(t)$.

Comment on sets of closed and open contours. Note that we may consider the formulae for closed contours as particular cases of those for open arcs. Indeed, in accordance with our convention, we may write $\varphi^+(t) = \varphi(t)$, $\varphi^{+'}(t) = \varphi'(t)$ in (26.1)-(26.4) and take into account that due to (26.2) we may consider $\varphi^-(t) = 0$. Then the holomorphicity condition (26.3) for closed contours takes the form of the condition (26.5) for open arcs. Hence, the same arguments, which served for closed and open contours separately, when joined, serve to show that the condition (26.5) is the necessary and sufficient condition in a general case. Now the contour of a considered region (finite or infinite, simply- or multi-connected) may contain both closed and open non-intersecting parts. Below we will use the condition (26.5) in the general case. This will allow us to obtain the CV-BIE for finite or infinite regions with holes and cracks.

§ 27. HOLOMORPHICITY THEOREMS FOR PERIODIC PROBLEMS

The holomorphicity theorems for periodic problems are similar to those for non-periodic problems. Their proofs are close to those of the previous paragraph. The

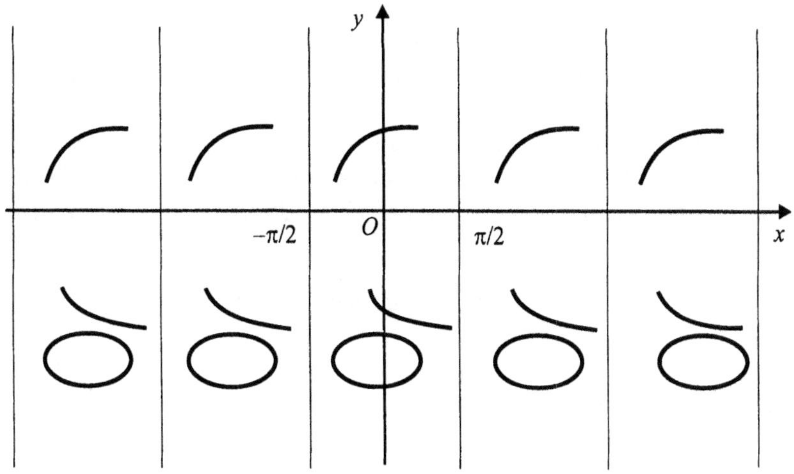

Fig. 16.

differences mostly concern the behavior of the periodic functions at infinity, and the cyclic constants in doubly periodic problems.

A complex function $\varphi(z)$ is termed a *periodic* function with the period $2\omega = 2\omega_x + i2\omega_y$, if $\varphi(z + k2\omega) = \varphi(z)$ for any integer k. The linear transformation $Z = \pi z/(2\omega)$ makes the problem periodic with the period π along the real axis x (Figure 16). Hence, without loss of generality, we may consider a periodic function $\varphi(z)$ with the period π along the real axis: $\varphi(z + k\pi) = \varphi(z)$ $(k = \ldots, -2, -1, 0, 1, 2, \ldots)$.

For non-periodic problems, the kernel with the Cauchy singularity $1/z$ was of special importance. For periodic problems with period π, the corresponding singularity is $\cot z$; the latter is defined by the series

$$\cot z = \frac{1}{z} + \sum_k{}' \left(\frac{1}{z - w} + \frac{1}{w} \right), \tag{27.1}$$

where $w = k\pi$. Summation in (27.1) is over all integers $k \neq 0$. The term corresponding $k = 0$ is written separately. We see that this periodic function has a singularity of $1/z$ type in each period: for periodic problems, the kernel $\cot z$ replaces $1/z$. The second addend under the summation sign in (27.1) does not depend on z; it serves to make the series convergent. We will see that in doubly periodic problems, Weierstrass' zeta-function plays the same role; this function results from $1/z$ in a way analogous to summing in (27.1).

Take as a main strip the strip $-\pi/2 < x < \pi/2$. The geometry of a periodic problem is prescribed by a contour L in the main strip and by congruent contours in each other strip (Figure 16). We assume that the contour L consists of a number of sufficiently smooth, non-intersecting closed and open parts.

Consider functions $\varphi^+(t)$, $\varphi^-(t)$ of the H^* class on L. In accordance with the comment at the end of the previous paragraph, for closed contours we assume $\varphi^+(t) = \varphi(t)$, $\varphi^-(t) = 0$. The direction of travel is as before.

The question is this: when do these functions represent the limiting values of a piece-wise holomorphic function $\varphi(z)$ periodic with period π? The answer is given by the following theorem (Linkov [4]).

Theorem. *Functions $\varphi^+(t)$, $\varphi^-(t)$ of the H^* class on L represent the limiting values of a periodic function $\varphi(z)$ with period π, which is piece-wise holomorphic outside L and contours congruent to L, and tends to $\pm B$ when $z \to \pm i\infty$, if and only if the condition*

$$\varphi^+ + \varphi^- = \frac{1}{\pi i}\int_L (\varphi^+ - \varphi^-)\cot(\tau - t)\,d\tau \qquad (27.2)$$

is satisfied under the additional equation

$$\frac{1}{2\pi}\int_L (\varphi^+ - \varphi^-)\,d\tau = B. \qquad (27.3)$$

Necessity. Let $\varphi^+(t)$, $\varphi^-(t)$ be the limiting values of a function $\varphi(z)$ which satisfies the conditions of the theorem. Consider a function:

$$F(z) = \frac{1}{2\pi i}\int_L (\varphi^+ - \varphi^-)\cot(\tau - z)\,d\tau. \qquad (27.4)$$

It is a periodic function with period π, piece-wise holomorphic outside L and the congruent contours. To obtain its limiting values when $y \to \pm\infty$, we note that the formula $\cot(\tau - z) = i\dfrac{\exp[2i(\tau - z)] + 1}{\exp[2i(\tau - z)] - 1}$ implies:

$$F(\pm i\infty) = \pm\frac{1}{2\pi}\int_L (\varphi^+ - \varphi^-)\,d\tau = \pm B_1.$$

Besides, from (27.1) we have $\cot(\tau - z) = \dfrac{1}{\tau - z} + g(z)$ where $g(z)$ is holomorphic in the main strip when τ belongs this strip. Then applying Sokhotski-Plemelj formulae (11.26) to (27.4) we have

$$F^+(t) - F^-(t) = \varphi^+(t) - \varphi^-(t), \qquad (27.5)$$

that is $(F - \varphi)^+ - (F - \varphi)^- = 0$. The latter implies that the difference $F(z) - \varphi(z)$ is an *entire* function periodic with period π, and equal to $\pm(B_1 - B)$ when $z = \pm\infty$. Then from the general theory (see, e. g. Carrier et al. [1]) it follows that this function is zero. Hence, $F(z) = \varphi(z)$, $F^{\pm}(t) = \varphi^{\pm}(t)$, $B_1 = B$. Now again applying Sokhotski-Plemelj formulae to (27.4), we come to (27.2).

Sufficiency. Let the functions $\varphi^{+}(t)$, $\varphi^{-}(t)$ satisfy (27.2), (27.3). Again use (27.4) to introduce the piece-wise holomorphic function $F(z)$, periodic with period π. As above, the Sokhotski-Plemelj formulae (11.26) yield (27.5) and

$$F^{+}(t)+F^{-}(t)=\frac{1}{\pi i}\int_{L}\left(\varphi^{+}-\varphi^{-}\right)\cot(\tau-t)d\tau.$$

Accounting for (27.2), we have from this equation $F^{\pm}(t) = \varphi^{\pm}(t)$. Hence, $\varphi^{+}(t)$, $\varphi^{-}(t)$ are the limiting values of the piece-wise holomorphic function $F(z)$. Besides, $F(\pm i\infty) = \pm B$, which completes the proof.

This formulation of the theorem uses a principal value integral. We may also formulate a holomorphicity theorem in terms of a finite-part integral.

Theorem. *Functions $\varphi^{+}(t)$, $\varphi^{-}(t)$ with the derivatives $\varphi^{+'}(t)$, $\varphi^{-'}(t)$ of the H^* class on L represent the limiting values of a periodic function $\varphi(z)$ with period π, which is piece-wise holomorphic outside L and the contours congruent to L, and tends to $\pm B$ when $z\to\pm i\infty$, if and only if the equation*

$$\varphi^{+'}+\varphi^{-'}=\frac{1}{\pi i}\int_{L}\left(\varphi^{+}-\varphi^{-}\right)\operatorname{cosec}^{2}(\tau-t)d\tau \tag{27.6}$$

and the condition (27.3) are satisfied.

In the considered case we have:

$$\varphi(z)=\frac{1}{2\pi i}\int_{L}\left(\varphi^{+}-\varphi^{-}\right)\cot(\tau-z)d\tau,\quad \varphi'(z)=\frac{1}{2\pi i}\int_{L}\left(\varphi^{+}-\varphi^{-}\right)\operatorname{cosec}^{2}(\tau-z)d\tau.$$

We comment on the hypersingular kernel in the last integral. From $\frac{d}{dz}\cot z=-\operatorname{cosec}^{2}z$, we see that, as could be expected, it is obtained from the function $\cot z$ in the same way as the kernel $\frac{1}{z^{2}}$ is obtained from $\frac{1}{z}$ in non-periodic problems: $\frac{d}{dz}\left(\frac{1}{z}\right)=-\frac{1}{z^{2}}$.

This theorem immediately follows from the previous one if we take into consideration that under the conventions accepted for functions $\varphi^{+}(t)$, $\varphi^{-}(t)$, equation (27.6) is equivalent to (27.2). (The equivalence is stated on the basis of the results presented in § 47, 48 of chapter 10).

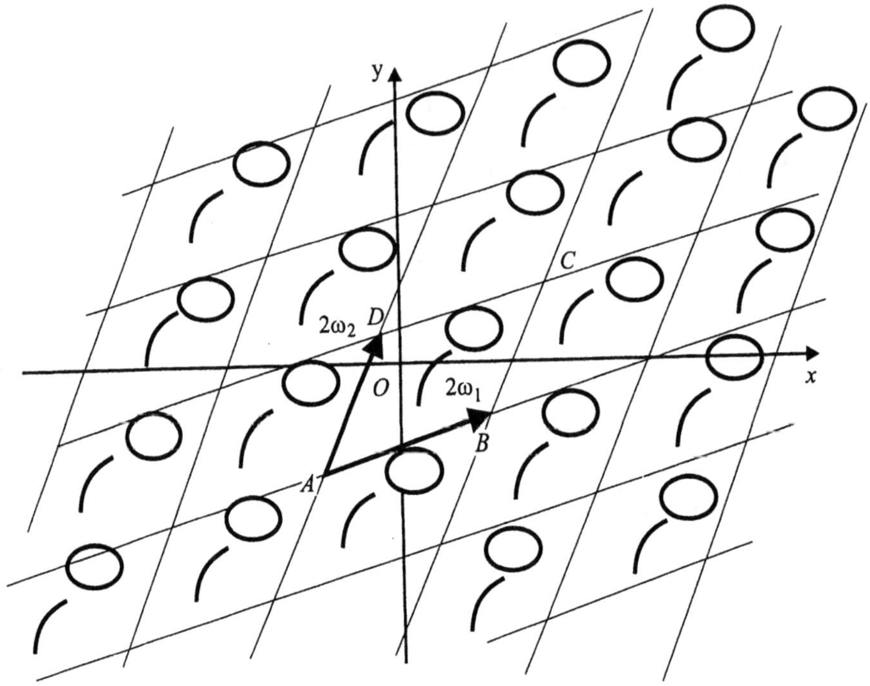

Fig. 17.

§ 28. HOLOMORPHICITY THEOREMS FOR DOUBLY PERIODIC PROBLEMS

Consider doubly periodic problems (Figure 17). A function $\varphi(z)$ is termed *doubly periodic* with the (complex) periods $2\omega_1$ and $2\omega_2$ if for any integers k_1 and k_2 we have $\varphi(z + k_1 2\omega_1 + k_2 2\omega_2) = \varphi(z)$. The function $\varphi(z)$ is termed *quasi-periodic* with the (complex) periods $2\omega_1$ and $2\omega_2$ and the *cyclic constants* $2\alpha_1$ and $2\alpha_2$ if for any integers k_1 and k_2 we have $\varphi(z + k_1 2\omega_1 + k_2 2\omega_2) = \varphi(z) + k_1 2\alpha_1 + k_2 2\alpha_2$. We will be interested in quasi-periodic functions because these functions may serve to represent fields of displacements. Doubly periodic functions are a particular case of quasi-periodic ones when the cyclic constants are zero.

The periods $2\omega_1$ and $2\omega_2$ are always assumed to be non-collinear, i.e. ω_1/ω_2 is not real. This is equivalent to

$$\operatorname{Im}\left(\overline{\omega_1}\omega_2\right) \neq 0. \tag{28.1}$$

For further discussion, particularly in Chapter 8, we will also need Weierstrass' zeta-function $\zeta(z)$, Legendre's identity for it, and Natanzon's function $Q(z)$ which is closely related to $\zeta(z)$.

Weierstrass' zeta-function with the periods $2\omega_1$ and $2\omega_2$ is defined as

$$\zeta(z)=\frac{1}{z}+\sum{}'\left(\frac{1}{z-w}+\frac{1}{w}+\frac{z}{w^2}\right),\tag{28.2}$$

where $w = k_1 2\omega_1 + k_2 2\omega_2$. Summation in (28.2) is assumed over all integer pairs k_1 and k_2 positive, negative and zero except $k_1 = k_2 = 0$. As for the complex cotangent in the periodic problem, the zeta-function reproduces the function $1/z$ in each cell (cf. formula (27.1)). The second term in the sum in (28.2) does not depend on z; the third term, being linear in z, is quasi-periodic for any period. These terms serve to provide convergence of the series.

It is easy to see that the function $\zeta(z)$ is quasi-periodic. Its cyclic constants $2\eta_1$ and $2\eta_2$ are completely defined by the periods $2\omega_1$ and $2\omega_2$. In particular, we may find them by using the formula

$$2\eta_k = 2\zeta(\omega_k),\quad k=1,2.\tag{28.3}$$

The periods and cyclic constants of the zeta-function are connected by *Legendre's identity*

$$\eta_1\omega_2-\eta_2\omega_1=\frac{\pi i}{2}.\tag{28.4}$$

Finally, Natanzon's function $Q(z)$ (Natanzon [1]) is introduced by the equation

$$Q(z)=\sum{}'\left[\frac{\overline{w}}{(z-w)^2}-\frac{\overline{w}}{w^2}-\frac{2z\overline{w}}{w^3}\right],\tag{28.5}$$

where w and the summation have the same meaning as in the definition (28.2) of the zeta-function. This function is connected with the zeta-function by an important equation, which makes $Q(z)$ useful for our applications:

$$Q(z+2\omega_k)=Q(z)-2\overline{\omega}_k\zeta'(z)+2\lambda_k,\quad k=1,2.\tag{28.6}$$

Herein, $2\lambda_k$ (k = 1, 2) are Natanzon's constants. They may be found from the formula

$$2\lambda_k=2\overline{\omega}_k\zeta'(\omega_k)+2Q(\omega_k),\quad k=1,2,$$

where $\zeta'(\omega_k)$ is the value of the derivative of the zeta-function at the point $z = \omega_k$. Natanzon's constants are connected with the cyclic constants of the zeta-function by the relation $\lambda_2\omega_1 - \lambda_1\omega_2 = \eta_1\overline{\omega}_2 - \eta_2\overline{\omega}_1$.

The periods allow us to divide the plane into congruent parallelograms (Figure 17). The main parallelogram is that which contains the origin as an internal point. In Figure 17 this is the parallelogram *ABCD*. In a doubly periodic problem, the geometry of holes and cuts is defined by the contour L in the main parallelogram. The holes and cuts in other parallelograms are congruent to L. As usual, we consider the contour L to be sufficiently smooth. We fix the travel direction on each part in accordance with the usual convention. If there are closed contours (holes), we locate the origin outside them.

Consider a function $\varphi(z)$, piece-wise holomorphic outside L and congruent contours, quasi-periodic with the periods $2\omega_1$ and $2\omega_2$ with the cyclic constants $2\alpha_1$ and $2\alpha_2$. Its limiting values on L are $\varphi^+(t)$, $\varphi^-(t)$ and, as usual, we assume $\varphi^+(t) = \varphi(t)$, $\varphi^-(t) = 0$ for closed contours (holes). Let us integrate $\varphi(z)$ over the parallelogram *ABCD* and over any other contour including L within the main parallelogram. Due to the holomorphicity of $\varphi(z)$ in the region between the integration contours, this integral is zero. Then accounting for the quasi-periodicity of $\varphi(z)$ and shrinking the internal contour to L, we obtain:

$$\frac{1}{2\pi i}\int_L \left(\varphi^+ - \varphi^-\right)d\tau = -\frac{2}{\pi i}\left(\alpha_1\omega_2 - \alpha_2\omega_1\right). \tag{28.7}$$

We will use this equation when proving the following holomorphicity theorem (Linkov [5]).

T h e o r e m . *Functions* $\varphi^+(t)$, $\varphi^-(t)$ *of the* H^* *class on L represent the limiting values of a function* $\varphi(z)$, *quasi-periodic with the periods* $2\omega_1$, $2\omega_2$ *and the cyclic constants* $2\alpha_1$, $2\alpha_2$, *which is piece-wise holomorphic outside L and contours congruent to L, and zero at* $z = 0$, *if and only if the conditions*

$$\varphi^+ + \varphi^- = \frac{1}{\pi i}\int_L \left(\varphi^+ - \varphi^-\right)\left[\zeta(\tau - t) - \zeta(\tau)\right]d\tau - \frac{4}{\pi i}\left(\alpha_1\eta_2 - \alpha_2\eta_1\right)t, \tag{28.8}$$

$$\frac{1}{2\pi i}\int_L \left(\varphi^+ - \varphi^-\right)d\tau = -\frac{2}{\pi i}\left(\alpha_1\omega_2 - \alpha_2\omega_1\right) \tag{28.9}$$

are satisfied. In this case

$$\varphi(z) = \frac{1}{2\pi i}\int_L \left(\varphi^+ - \varphi^-\right)\left[\zeta(\tau - z) - \zeta(\tau)\right]d\tau - \frac{2}{\pi i}\left(\alpha_1\eta_2 - \alpha_2\eta_1\right)z. \tag{28.10}$$

If, in addition, the functions $\varphi^+(t)$, $\varphi^-(t)$ have the derivatives $\varphi^{-'}(t)$, $\varphi^{-'}(t)$ of the H^* class and the conditions $\varphi^+(a_j) - \varphi^-(a_j) = \varphi^+(b_j) - \varphi^-(b_j) = 0$ are satisfied at the ends a_j, b_j of cuts, then from the general theory (Gahov [1], Muskhelishvili [4]) applied to the main parallelogram, it follows that $\varphi^{\pm'}(t) = \varphi'^{\pm}(t)$ and

$$\varphi^{+\prime}+\varphi^{-\prime}=\frac{1}{\pi i}\int_L\left(\varphi^{+\prime}-\varphi^{-\prime}\right)\zeta(\tau-t)d\tau-\frac{4}{\pi i}\left(\alpha_1\eta_2-\alpha_2\eta_1\right),$$

$$\varphi'(z)=-\frac{1}{2\pi i}\int_L\left(\varphi^{+}-\varphi^{-}\right)\zeta'(\tau-z)d\tau-\frac{2}{\pi i}\left(\alpha_1\eta_2-\alpha_2\eta_1\right)=$$

$$=\frac{1}{2\pi i}\int_L\left(\varphi^{+\prime}-\varphi^{-\prime}\right)\zeta(\tau-z)d\tau-\frac{2}{\pi i}\left(\alpha_1\eta_2-\alpha_2\eta_1\right).$$

Necessity. Let $\varphi^+(t)$, $\varphi^-(t)$ be the limiting values of a function $\varphi(z)$ which satisfies the conditions of the theorem. For $\varphi(z)$, equation (28.7) is met, hence the condition (28.9) is fulfilled. Consider another function:

$$F(z)=\frac{1}{2\pi i}\int_L\left(\varphi^{+}-\varphi^{-}\right)\left[\zeta(\tau-z)-\zeta(\tau)\right]d\tau-\frac{2}{\pi i}\left(\alpha_1\eta_2-\alpha_2\eta_1\right)z \quad (28.11)$$

and show that it coincides with $\varphi(z)$. From the definition (28.11) it follows that $F(z)$ is a quasi-periodic function with the cyclic constants equal to

$$\frac{\eta_k}{\pi i}\int_L\left(\varphi^{+}-\varphi^{-}\right)d\tau-4\frac{\omega_k}{\pi i}\left(\alpha_1\eta_2-\alpha_2\eta_1\right). \quad (28.12)$$

By employing (28.9) and Legendre's identity (28.4), we see that the expression (28.12) is equal to $2\alpha_k$, ($k = 1, 2$). Hence, $F(z)$ has the same cyclic constants as $\varphi(z)$. From (28.11), it also follows that $F(0) = 0$. Besides, accounting for the definition of Weierstrass' zeta function (28.2), we may apply the formulae of Sokhotski-Plemelj (11.26). From them we obtain:

$$F^+(t)-F^-(t)=\varphi^+(t)-\varphi^-(t), \quad (28.13)$$

$$F^+(t)+F^-(t)=\frac{1}{\pi i}\int_L\left(\varphi^{+}-\varphi^{-}\right)\left[\zeta(\tau-t)-\zeta(\tau)\right]d\tau-\frac{4}{\pi i}\left(\alpha_1\eta_2-\alpha_2\eta_1\right)t. \quad (28.14)$$

Hence, $F(z)$ - $\varphi(z)$ is a function without discontinuities. In accordance with the definition (28.11), it may have a pole of the first order at infinity. This implies that $F(z) - \varphi(z) = C + C_1 z$. As $F(0) = \varphi(0) = 0$, we have $C = 0$. As $F(z)$ and $\varphi(z)$ have the same cyclic constants, $C_1 = 0$. Thus $F(z) = \varphi(z)$, and now (28.14) implies that (28.8) is satisfied.

Sufficiency. Let the functions $\varphi^+(t)$, $\varphi^-(t)$ satisfy (28.8), (28.9). Again use (28.11) to introduce the function $F(z)$. The Sokhotski-Plemelj formulae (11.26) imply (28.13), (28.14), which being combined with (28.8) yield $F^{\pm}(t) = \varphi^{\pm}(t)$. Hence,

$\varphi^+(t)$, $\varphi^-(t)$ are the limiting values of the piece-wise holomorphic function $F(z)$. As shown above, its cyclic constants are $2\alpha_k$ $(k = 1, 2)$ and $F(0) = 0$.

This formulation of the theorem uses a principal value integral. We may also formulate a holomorphicity theorem in terms of a finite-part integral.

Theorem. *Functions $\varphi^+(t)$, $\varphi^-(t)$ with the derivatives $\varphi^{+\prime}(t)$, $\varphi^{-\prime}(t)$ of the H^* class on L represent the limiting values of a function $\varphi(z)$, quasi-periodic with the periods $2\omega_1$, $2\omega_2$ and the cyclic constants $2\alpha_1$, $2\alpha_2$, which is piece-wise holomorphic outside L and contours congruent to L, and zero at z = 0, if and only if the condition*

$$\varphi^{+\prime}+\varphi^{-\prime} = \frac{1}{\pi i}\int_L \left(\varphi^+ - \varphi^-\right)\wp(\tau - t)d\tau - \frac{4}{\pi i}\left(\alpha_1\eta_2 - \alpha_2\eta_1\right), \qquad (28.15)$$

is satisfied. Herein, $\wp(z) = - d\zeta/dz$ is a hypersingular kernel. This is the *gamma-function* of Weierstrass; it is obtained from $\zeta(z)$ in the same way as the kernel $1/z^2$ is obtained from $1/z$ in non-periodic problems, and $\operatorname{cosec}^2 z$ is obtained from $\cot z$ in periodic problems with period π.

In this case we have:

$$\varphi^{\pm\prime}(t) = \varphi'^{\pm}(t),$$

$$\varphi'(z) = \frac{1}{2\pi i}\int_L \left(\varphi^{+\prime} - \varphi^{-\prime}\right)\wp(\tau - z)d\tau - \frac{2}{\pi i}\left(\alpha_1\eta_2 - \alpha_2\eta_1\right).$$

The function $\varphi(z)$ may contain an arbitrary constant which we will fix by the condition $\varphi(0) = 0$. Then the function $\varphi(z)$ is given by (28.10).

This theorem immediately follows from the previous one if we take into consideration that under the requirements imposed on $\varphi^+(t)$, $\varphi^-(t)$, equation (28.15) is equivalent to (28.8) when $\varphi(0) = 0$.

Chapter 6

COMPLEX VARIABLE INTEGRAL EQUATIONS

§ 29. GENERAL APPROACH

The Kolosov-Muskhelishvili formulae (25.1), (25.2) show that two functions $\varphi(z)$ and $\psi(z)$, holomorphic in a simply connected region, define the complex displacements and the resultant force. On the other hand, the holomorphicity theorems are formulated in terms of the *limiting values* of such functions. Thus we may obtain CV-BIE by substituting the limiting values of $\varphi(z)$ and $\psi(z)$ into the holomorphicity conditions and using the connection between these limiting values and the physical values expressed by the boundary conditions (25.10), (25.11) (or, in more general form, by (25.14), (25.15)). Alternatively, we may start from equations (25.6), (25.7), employing the functions $\Phi(z)$ and $\Psi(z)$, and use the boundary conditions (25.12), (25.13) (or, in more general case (25.16), (25.17)). The latter choice leads to equations that follow from the equations of the former after differentiation with respect to the free variable. It involves lengthier derivation; we prefer the first choice.

First, we will illustrate the approach by considering the simplest case of a homogeneous medium with holes and cracks. Our aim is to prepare starting equations, which will serve to obtain general equations for blocky systems. We will also reveal the links with the results of the previous part.

Problem formulation. Consider a plane region D, finite or infinite, with p cuts L_k ($k = 1, \ldots, p$) and m holes L_j ($j = p+1, \ldots, p+m$). For a finite region we have the additional contour: the external boundary L_0 of the region (Figure 18). Denote the total contour by L. As usual, we assume that the travel path along the closed contours leaves the region D on the left. The travel direction on an open contour is fixed arbitrarily; we denote the start point a_k, and end point b_k ($k = 1,\ldots, p$). For definiteness, we also fix start and end points on a closed contour: $a_j = b_j$ ($j = 0, p+1, \ldots, p+m$). The normal $\mathbf{n}$ is always directed to the right of the travel direction. To shorten derivation we assume stresses at infinity to be zero. (The case of nonzero stresses is included by additional terms below).

We use the boundary conditions on L of the (25.14) form:

$$\varphi^{\pm}(t) + t\overline{\varphi'^{\pm}(t)} + \overline{\psi^{\pm}(t)} = f^{\pm}(t), \tag{29.1}$$

where $f^{\pm} = f_c^{\pm} + C^{\pm}(t)$,

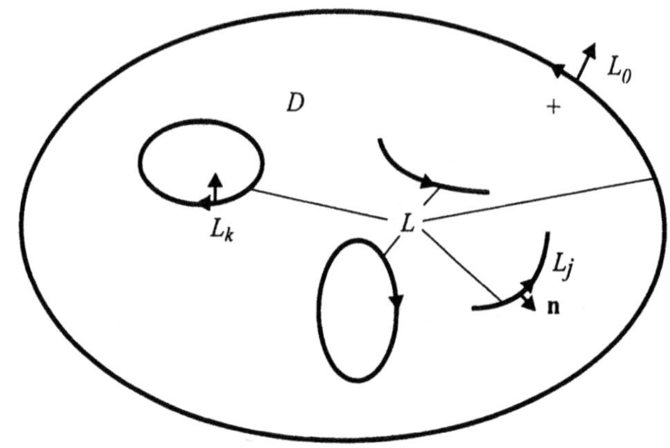

Fig. 18.

$$f_c^{\pm}(t) = \int_{a_j}^{t} \sigma^{\pm}(\tau)d\tau, \quad t \in L_j, j = 0,\ldots,m+p,$$

$C^{\pm}(t) = C_k^{\pm}$ on L_k, $C_k^{\pm}$ are constants ($k = 0,1,\ldots, p + m$). For *open* contours we assume $C_k^{+} = C_k^{-}$ ($k = 1,\ldots,p$). For *closed* contours we assume $\sigma^{+} = \sigma$, $\sigma^{-} = 0$, $C_j^{+} = C_j$, $C_j^{-} = 0$ ($j = 0, p+1,\ldots, p + m$). We may prescribe one of the constants arbitrarily. For definiteness, we will take it zero on the external contour. For an infinite region, this implies that $f(\infty) = 0$. The remaining $p + m$ constants are to be found when solving the problem.

Suppose that a set of contours is loaded with the self-equilibrated traction.[1] This means that

$$\int_{L} \Delta\sigma \; d\tau = 0, \quad k = 0,\ldots,m+p, \tag{29.2}$$

where $\Delta\sigma = \sigma^{+} - \sigma^{-}$; as usual, the symbol "plus" ("minus") denotes the limiting value from the left ("right") of the travel path. For closed contours we assume $\sigma^{+} = \sigma$, $\sigma^{-} = 0$. For an infinite region, we change the condition (29.2) for $j = 0$ by the condition of zero stresses at infinity.

We are looking for a solution that has no displacement discontinuities at the tips of cracks (cuts), that is

[1] Note that here and in our previous publications (Linkov [1,5]) the term "self-equilibrated load" is used when the *total* resultant force applied to a whole open or closed contour is equal to zero. Naturally, for open contours representing cracks, this does not exclude traction discontinuity through the contour: we may have $\sigma^{+} \neq \sigma^{-}$ ($f^{+} \neq f^{-}$). This comment is necessary because Savruk, for instance, in his papers (see Savruk [1, 4], Savruk, Osiv, Prokopchuk [1]) uses the term "self-equilibrated load" in a different sense. He refers it only to a particular case of (29.1) when the tractions on the opposite sides of a crack are equal and opposite; this gives $\sigma^{+} = \sigma^{-}$. In such cases, our terminology presumes *continuity of traction* through the contour (there is no traction discontinuity, $\Delta\sigma = 0$).

$$\Delta u(a_k) = u^+(a_k) - u^-(a_k) = 0, \quad \Delta u(b_k) = u^+(b_k) - u^-(b_k) = 0, \quad k = 1,\ldots,p.$$

Then from (29.2) and (25.8) it follows that that $\Delta\varphi(a_k) = \Delta\varphi(b_k) = 0$ ($k = 1,\ldots, p$). These conditions imply that the function $\Delta\varphi(t) = \varphi^+(t) - \varphi^-(t)$ has to belong to Muskhelishvili's class h_{2p} on the contour L (see brief comments in § 11).

General procedure. Let us use a standard procedure leading to CV-BIE. From (29.1) we have the expressions for the limiting values of the piece-wise holomorphic function $\psi^{\pm}(t)$:

$$\psi^{\pm}(t) = \overline{f^{\pm}(t)} - \overline{\varphi^{\pm}(t)} - \bar{t}\,\varphi'^{\pm}(t). \tag{29.3}$$

By applying the holomorphicity condition (26.5) to the function $\psi(t)$ and using (29.3) we obtain:

$$\overline{f^+} + \overline{f^-} - \left(\overline{\varphi^+} + \overline{\varphi^-}\right) - \bar{t}\left(\varphi^{+'} + \varphi^{-'}\right) = \\ = \frac{1}{2\pi i}\int_L \left[\overline{f^+} - \overline{f^-} - \left(\overline{\varphi^+} - \overline{\varphi^-}\right) - \bar{t}\left(\varphi^{+'} - \varphi^{-'}\right)\right]\frac{d\tau}{\tau - t}.$$

To exclude the derivatives, apply the second of the formulae (26.7) multiplied by $-\bar{t}$:

$$-\bar{t}\left(\varphi^{+'} + \varphi^{-'}\right) = -\frac{\bar{t}}{\pi i}\int_L \left(\varphi^{+'} - \varphi^{-'}\right)\frac{d\tau}{\tau - t}.$$

Besides, take into account that from the holomorphicity of $\varphi(z)$ it follows that

$$\overline{\varphi^+} + \overline{\varphi^-} = \overline{\frac{1}{\pi i}\int_L \left(\varphi^+ - \varphi^-\right)\frac{d\tau}{\tau - t}}.$$

After subtracting these equations and integration by parts we arrive at an equation without derivatives:

$$\overline{f^+} + \overline{f^-} - 2\left(\overline{\varphi^+} + \overline{\varphi^-}\right) = \frac{1}{\pi i}\int_L \overline{\Delta f}\,\frac{d\tau}{\tau - t} - \overline{\frac{1}{\pi i}\int_L \Delta\varphi\, dk_1} + \frac{1}{\pi i}\int_L \Delta\varphi\, d\overline{k_2}, \tag{29.4}$$

where $\Delta\varphi = \varphi^+(t) - \varphi^-(t)$; the kernels $k_1(\tau,t)$ and $k_2(\tau,t)$ are given by (12.18). As usual, for closed contours we assume $\varphi^+(t) = \varphi(t)$, $\varphi^-(t) = 0$, $f^+ = f$, $f^- = 0$. If all the contours are open, then (29.4) turns into Muskhelishvili's equation (22.6). For a single finite block with a contour L_j, accounting for the holomorphicity condition (26.3), we can write (29.4) as an equation

$$\frac{1}{2\pi i}\int_{L_j}\left[\frac{2\varphi_j\, d\tau}{\tau-t}-\varphi_j\, dk_1(\tau,t)-\overline{\varphi_j}\, dk_2(\tau,t)\right]=\frac{1}{2}f_j+\frac{1}{2\pi i}\int_{L_j}f_j\frac{d\bar\tau}{\bar\tau-\bar t}, \quad (29.5)$$

which will serve us to derive equations for blocky systems.

In general, equation (29.4) contains the sum $\varphi^+(t)+\varphi^-(t)$ which should be expressed through the difference $\Delta\varphi=\varphi^+(t)-\varphi^-(t)$. We will take this additional step by employing the holomorphicity condition (26.5). As a result, after taking the complex conjugation and dividing by two, we arrive at the singular CV-BIE:

$$\frac{1}{2\pi i}\int_L\left(2\Delta\varphi\frac{d\tau}{\tau-t}-\Delta\varphi dk_1-\overline{\Delta\varphi}dk_2\right)=\frac{f^++f^-}{2}+\frac{1}{2\pi i}\int_L\Delta f\frac{d\bar\tau}{\bar\tau-\bar t}. \quad (29.6)$$

Differentiation (29.6) with respect to t, accompanied by integration by parts, yields an equation for the derivative $\Delta\varphi'(t)$:

$$\frac{1}{2\pi i}\int_L\left[\frac{2\Delta\varphi' d\tau}{\tau-t}+\Delta\varphi'\left(\frac{\partial k_1}{\partial t}\right)d\tau+\overline{\Delta\varphi'}\left(\frac{\partial k_2}{\partial t}\right)d\bar\tau\right]=$$

$$=\frac{\sigma^++\sigma^-}{2}+\frac{1}{2\pi i}\int_L\Delta\sigma\left(\frac{1}{\tau-t}+\frac{\partial k_1}{\partial t}\right)d\tau. \quad (29.7)$$

From (25.8) we have

$$\Delta\varphi=\varphi^+-\varphi^-=\frac{2\mu\,\Delta u+\Delta f}{\chi+1},$$

and substitution $\Delta\varphi(t)$ into (29.6), (29.7) transforms these equations into equations for physical values, the displacement discontinuities or their derivatives:

$$\frac{1}{2\pi i}\int_L\left[\frac{2\Delta u\, d\tau}{\tau-t}-\Delta u\, dk_1-\overline{\Delta u}\, dk_2-\right.$$

$$\left.-(\chi-1)\frac{\Delta f}{2\mu}\frac{d\tau}{\tau-t}+\chi\frac{\Delta f}{2\mu}dk_1-\frac{\overline{\Delta f}}{2\mu}dk_2\right]=\frac{\chi+1}{2\mu}\frac{f^++f^-}{2}, \quad (29.8)$$

$$\frac{1}{2\pi i}\int_L\left[2\Delta u'\frac{d\tau}{\tau-t}+\Delta u'\frac{\partial k_1}{\partial t}+\overline{\Delta u'}\frac{\partial k_2}{\partial t}d\bar\tau-(\chi-1)\frac{\Delta\sigma}{2\mu}\frac{d\tau}{\tau-t}-\right.$$

$$-\chi\frac{\Delta\sigma}{2\mu}\frac{\partial k_1}{\partial t}d\tau+\frac{\overline{\Delta\sigma}}{2\mu}\frac{\partial k_2}{\partial t}d\bar{\tau}\Bigg]=\frac{\chi+1}{2\mu}\frac{\sigma^+ +\sigma^-}{2}. \qquad (29.9)$$

Additional equations for solving (29.9) express the conditions that there is no displacement discontinuity at a tip of a crack:

$$\int_{L_j}\Delta u'(t)d\tau=0, \quad j=1,\dots,p.$$

At first, equations (29.6)-(29.9) were obtained by the method used above for the particular case of a plane with arbitrary cracks (Linkov [1]). Afterwards, for the same particular case, equation (29.9) was independently derived and systematically employed by Ioakimidis (Ioakimidis [1], Theocaris and Ioakimidis [7]), and Savruk (Savruk [1, 4], Savruk, Osiv and Prokopchuk [1]). We see that equations (29.6)-(29.10) are true also in the general case of a region, finite or infinite, containing holes and cracks.

Comparing the derived equations with the equations of the direct approach given in Part I, we see that for closed contours, equations (29.8), (29.9) reproduce the equations of the direct approach when Kelvin's solution serves as an initial singular solution. Indeed, equation (29.8) coincides with the first equation in (22.2), equation (29.9) – with the first in (21.6). Thus, the equations obtained by using the boundary conditions in terms of the *traction*, are the same as those obtained by the direct approach from Somigliana's identity for the *traction*. We see that the two branches of the theory are not isolated: they lead to the same CV-BIE. This refers also to the equations derived starting from the boundary condition in terms of the displacements: they reproduce those obtained by the direct approach from Somigliana's identity for the *displacements*. We will not write down these equations explicitly again.

Hypersingular equation. A *hypersingular* equation follows from (29.9) if we formally integrate by parts the terms containing the derivative of the displacement discontinuity (Linkov [1], Linkov and Mogilevskaya [1-3]). This operation is justified under the conditions accepted for the densities in the integrals (see § 47). As a result, we have the displacements instead of their derivative and the singular integral becomes hypersingular; the latter is understood as a finite-part integral. We obtain:

$$\frac{1}{2\pi i}\int_L\Bigg[2\Delta u\frac{d\tau}{(\tau-t)^2}-\Delta u\frac{\partial}{\partial t}dk_1-\overline{\Delta u}\frac{\partial}{\partial t}dk_2-$$

$$-(\chi-1)\frac{\Delta\sigma}{2\mu}\frac{d\tau}{\tau-t}-\chi\frac{\Delta\sigma}{2\mu}\frac{\partial k_1}{\partial t}d\tau+\frac{\overline{\Delta\sigma}}{2\mu}\frac{\partial k_2}{\partial t}d\bar{\tau}\Bigg]=\frac{\chi+1}{2\mu}\frac{\sigma^+ +\sigma^-}{2}. \qquad (29.10)$$

The hypersingular equation (29.10) has an important advantage over all the preceding equations of this section: it contains displacements and traction that are the values that are commonly prescribed or sought on the boundary. In contrast, the singular equations either contain the resultant force, that is the *integral of the traction*, or, alternatively, the *derivative of the displacements*. This complicates using the singular equations in mixed and contact problems. Besides, as mentioned, for problems involving open contours (in particular, cracks) we are forced to use additional equations. The hypersingular equation (29.10) is free of these shortcomings.

Equation (29.10) coincides with equation in (21.5) of the direct approach. Thus, *all* the equations of the direct approach derived from the first and the second Somigliana's identities are now obtained from Muskhelishvili's integral equations on the basis of K-M theory. Naturally, the equations containing the traction (or the resultant force) as a free term are obtained from the boundary condition for the traction (or the resulting force), and correspond to the equations resulting from Somigliana's identity for traction. The equations containing the displacements as a free term are obtained from the boundary condition for displacements and correspond to the equations resulting from Somigliana's identity for displacements.

Holomorphicity representations for prescribed traction. After the derived CV-BIE is solved, we may find the stresses and displacements within the region by employing K-M formulae (25.1), (25.2) or (25.6), (25.7). In accordance with the holomorphicity theorem, the functions $\varphi(z)$, $\psi(z)$, $\Phi(z)$ and $\Psi(z)$ are given by formulae

$$\varphi(z)=\frac{1}{2\pi i}\int_L \Delta\varphi\frac{d\tau}{\tau-z},\quad \psi(z)=\frac{1}{2\pi i}\int_L\left(\frac{\overline{\Delta f}-\overline{\Delta\varphi}}{\tau-z}d\tau+\Delta\varphi d\frac{\bar{\tau}}{\tau-z}\right),\qquad (29.11)$$

$$\varphi'(z)=\Phi(z)=\frac{1}{2\pi i}\int_L \Delta\varphi'\frac{d\tau}{\tau-z},$$

$$\psi'(z)=\Psi(z)=\frac{1}{2\pi i}\int_L\left[\frac{\left(\overline{\Delta\sigma}-\overline{\Delta\varphi'}\right)d\bar{\tau}}{\tau-z}-\frac{\bar{\tau}\Delta\varphi' d\tau}{(\tau-z)^2}\right].\qquad (29.12)$$

The discontinuities are understood as explained in the beginning of the paragraph. In particular, for closed contours, we assume that the discontinuity simply coincides with the function itself. We call the integral representations of K-M functions, based on densities satisfying the holomorphicity theorem, *holomorphicity representations*.

In (29.11) and we may express $\Delta\varphi$ through physical values by employing the equation used above: $\Delta\varphi = (2\mu\Delta u + \Delta f)/(\chi + 1)$. In (29.12) we may express $\Delta\varphi'$ through physical values by employing the immediate corollary: $\Delta\varphi' = (2\mu\Delta u' + \Delta\sigma)/(\chi + 1)$.

Procedure of integration before coming to limit. The singular and hypersingular equations of this paragraph contain *direct* values of integrals.

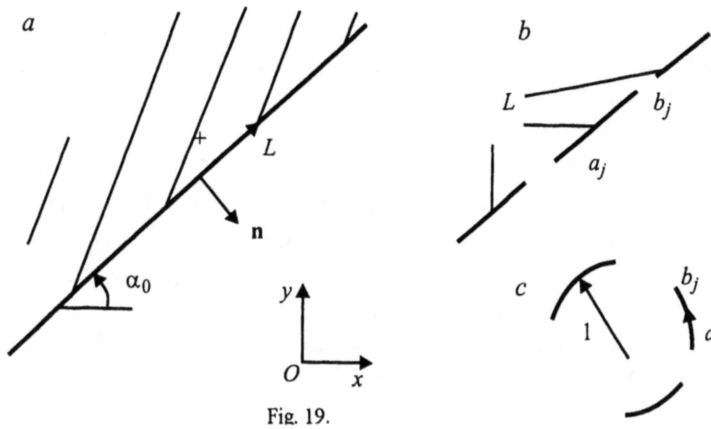

Fig. 19.

Namely, singular integrals are understood as principal value integrals, hypersingular as finite-part integrals. Thus, in our classification, any numerical procedure used to solve the derived equations corresponds to coming to the limit *before* integration. The integral representations (29.11)-(29.12) give an opportunity to use the alternative procedure many times mentioned in the Part I, of coming to the limit *after* integration.

The simplest particular cases. We mentioned that all the equations obtained in this chapter correspond to the simplest initial singular solution of the direct approach of the theory of potentials, that is Kelvin's solution. Their general feature is that in addition to the standard singular or hypersingular integral they contain regular integrals with only *two* standard expressions: $k_1(\tau,t) = \ln\dfrac{\tau - t}{\bar{\tau} - \bar{t}}$, $k_2(\tau,t) = \dfrac{\tau - t}{\bar{\tau} - \bar{t}}$.

It is clear that simplifications of the CV-BIE are possible when these expressions are simple. There are two such cases:

1) when the points τ and t belong to a same straight line;

2) when the points τ and t belong to a same circumference.

It is easy to see that there are no other simple cases. Indeed, a straight line is the simplest contour, while a circumference can be transformed into it by the simplest linear-fractional transformation. Any other contour, elliptic for instance, is transformed into a straight line (or circumference) by more complicated formula. There may be simplifications for some closed contours, but they are not as significant as those in these two cases.

Contours along a same straight line. Let the straight line subtend the angle α_0 with the x-axis (Figure 19, *a*, *b*). For any two points τ and t on it, we have $\tau - t = |\tau - t|\exp(i\alpha_0)$, $\bar{\tau} - \bar{t} = |\tau - t|\exp(-i\alpha_0)$. Then $k_1(\tau,t) = 2i\alpha_0$, $k_2(\tau,t) = \exp(2i\alpha_0)$; they are constant on the line. Hence, $dk_1 = dk_2 = 0$, $\partial k_1/\partial t = \partial k_2/\partial t = 0$. As a result, all the integrals except the standard singular or hypersingular vanish in the equations of the previous paragraphs. This allows us to obtain solutions in analytical form. They include solutions for a half-plane with prescribed traction and/or displacements on its boundary (Figure 19, *a*) and for cracks along a straight line (Figure 19, *b*). We will not reproduce these well-known solutions: they are given in Muskhelishvili [5].

Contours along a circumference. Let the points τ and t belong to a circumference with the center at the origin and with unit radius (Figure 19, *c*). The general case of an arbitrary circumference can be reduced to this by an elementary linear transformation.

Introduce a polar co-ordinate system with the polar angle counted from the x-axis. Then $\overline{\tau}=1/\tau$, $\overline{t}=1/t$ and the kernels become $k_1(\tau,t) = \ln(-\tau t)$, $k_2(\tau,t) = -\tau t$. Hence, in equations (29.6)-(29.10), the integrals containing $k_1(\tau,t)$, $k_2(\tau,t)$ either become constants multiplied by prescribed functions, or vanish. Finally, the problems for a circle or a plane with a circular hole or for circular-arc cracks along a circumference have simple solutions in quadratures, which often may be carried out analytically. The monograph by Muskhelishvili [5] contains these solutions as well.

§ 30. EQUATIONS FOR BLOCKY SYSTEMS WITH DISPLACEMENT AND/OR TRACTION DISCONTINUITIES

For finite systems of blocks and for systems with holes (voids, pores), equations of § 23 have eigenfunctions, which cause difficulties in numerical implementation. The K-M theory allows us to remove the obstacle.

Initial identities. Suppose we have a system, finite or infinite, of n plane blocks (Figure 20). The matrix, which embeds the blocks, is treated as an infinite block. A block within another block is an inclusion. The values referring to the j-th block are indicated by the index j.

Consider for simplicity a *finite* block j. Denote D_j its internal region, L_j its contour and D_j^- the region complementing D_j in the whole pane. We show that for field points $t = z$ outside the region D_j occupied by the j-th block, the r. h. s. in equation (29.5) is zero. The analogous property will be preserved for all other equations, which are corollaries of the equation for prescribed resultant force (29.5).

Suppose, the limiting values $\varphi_j(t)$ of K-M function $\varphi_j(z)$ for the j-th block satisfy equation (29.5) for this block:

$$\frac{1}{2\pi i}\int_{L_j}\left[\frac{2\varphi_j\,d\tau}{\tau-t}-\varphi_j\,dk_1(\tau,t)-\overline{\varphi_j}\,dk_2(\tau,t)\right]=\frac{1}{2}f_j+\frac{1}{2\pi i}\int_{L_j} f_j\frac{d\overline{\tau}}{\overline{\tau}-\overline{t}}. \quad (30.1)$$

Employ holomorphicity representations (29.11) and consider the functions

$$\varphi_j^*(z)=\frac{1}{2\pi i}\int_{L_j}\varphi_j\frac{d\tau}{\tau-z},\quad \psi_j^*(z)=\frac{1}{2\pi i}\int_{L_j}\left(\overline{f_j}-\overline{\varphi_j}-\overline{\tau}\varphi_j'\right)\frac{d\tau}{\tau-z} \quad (30.2)$$

for z outside D_j. Equations (30.2) show that these functions are holomorphic in the region D_j^- complementing D_j in the whole pane. By coming in (30.2) to the limit outside D_j (from D_j^-) and by using the Sokhotski-Plemelj formulae we may write (30.1) as

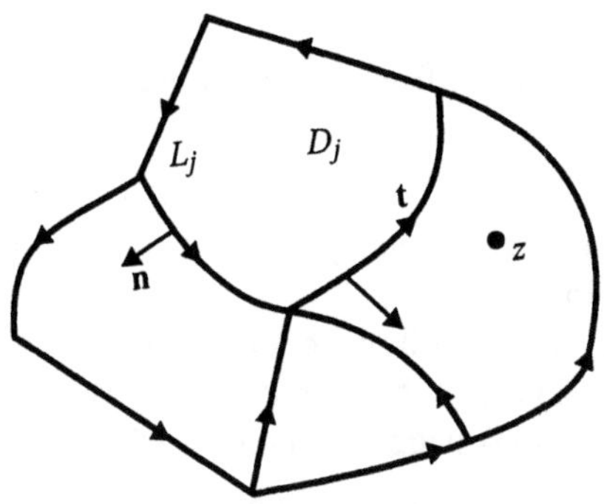

Fig. 20.

$$\left[\varphi_j^*(z)+z\overline{\varphi_j^{*\prime}(z)}+\overline{\psi_j^*(z)}\right]_e=0. \tag{30.3}$$

The symbol "e" in (30.4) denotes the limit from the side *external* to D_j. Comparison of (30.3) with the boundary condition (25.10) shows that the functions $\varphi_j^*(z)$ and $\psi_j^*(z)$, holomorphic in the region D_j^-, give the solution of the elasticity problem under zero traction on the boundary of D_j^-. This implies (Muskhelishvili [5, § 34]) that $\varphi_j^*(z) = C_1 + iC_2z$, $\psi_j^*(z) = C_3$, where C_2 is a real, while C_1 and C_3 are complex constants. From (30.2) it follows that $\varphi_j^*(z)$ and $\psi_j^*(z)$ tend to zero when z tends to infinity. This implies that $C_1 = C_2 = C_3 = 0$. Hence the functions $\varphi_j^*(z)$ and $\psi_j^*(z)$ are identically zero outside the j-th block. Then

$$\frac{1}{2\pi i}\int_{L_j}\varphi_j\frac{d\tau}{\tau-t}=0,\quad \frac{1}{2\pi i}\int_{L_j}\left[2\varphi_j\frac{d\tau}{\tau-t}-\varphi_j\,dk_1-\overline{\varphi_j}\,dk_2\right]=0,\quad t\in D_j^-. \tag{30.4}$$

The last equation expresses the fact that $\varphi_j^*(z)+z\overline{\varphi_j^{*\prime}(z)}+\overline{\psi_j^*(z)}=0$ in D_j^- because $\varphi_j^*(z) = 0$ and $\psi_j^*(z) = 0$ in D_j^-. Comparison of the second in (30.4) with (30.1) shows that equation (30.1) is satisfied not only at the points on the contour L_j but also in the external region D_j^- if we assume that the r. h. s. of (30.1) is zero for points outside the j-th block. Thus, we have

$$2S_j\varphi_j-K_{1j}\varphi_j-K_{2j}\overline{\varphi_j}=\begin{cases}\frac{1}{2}f_j(t)+S_jf_j-K_{1j}f_j, & t\in L_j\\ 0, & t\in D_j^-\end{cases}, \tag{30.5}$$

where S_j is a singular, while K_{1j}, K_{2j} are regular operators defined as

$$S_jg=\frac{1}{2\pi i}\int_{L_j}\frac{g\,d\tau}{\tau-t},\quad K_{1j}g=\frac{1}{2\pi i}\int_{L_j}g\,dk_1,\quad K_{2j}\overline{g}=\frac{1}{2\pi i}\int_{L_j}\overline{g}\,dk_2,$$

we also used the identity following from the definition (12.18) of the kernel $k_1(\tau,t)$:
$$\frac{d\bar{\tau}}{\bar{\tau}-\bar{t}}=\frac{d\tau}{\tau-t}-dk_1(\tau,t).$$
Equations (30.5) for a single block present the initial identities for deriving equations for a system of blocks.

Fixing body motions. In the following discussion we will need to fix body motions of some blocks. Consider such a block k. We can eliminate its body motions (translation and rotation) under the prescribed resultant force by taking the values of $\varphi_j(z_{k1})$ and $\operatorname{Im}\varphi_j'(z_{k2})$ at arbitrary points z_{k1} and z_{k2} within the block (the point z_{k1} may coincide with the point z_{k2}). The additional conditions serving for the elimination, being written for the k-th block, are

$$S_k\varphi_k\Big|_{t=z_{k1}}=0, \qquad \operatorname{Im}(H_k\varphi_k)\Big|_{t=z_{k2}}=0, \tag{30.6}$$

where the hypersingular operator H_k is defined as

$$H_k g=\frac{1}{2\pi i}\int_{L_k}\frac{g\,d\tau}{(\tau-t)^2}.$$

The first of (30.6) excludes translation, the second rotation.

Differentiation of the first of (30.4) with respect to t gives after taking the imaginary part:

$$\operatorname{Im}(H_k\varphi_k)\Big|_{t\in D_k^-}=0. \tag{30.7}$$

Joining the first of (30.6) with the first of (30.4) and the second of (30.6) with (30.7) we can write:

$$S_k\varphi_k=0, \quad t=z_{k1} \quad \text{or} \quad \in D_k^- \tag{30.8}$$
$$\operatorname{Im}(H_k\varphi_k)=0, \quad t=z_{k2} \quad \text{or} \quad \in D_k^- \tag{30.9}$$

Equations for a system of blocks. We use the expression (25.8). Then equation (30.5) takes the form

$$2S_j\left(u_j+\frac{f_j}{2\mu_j}\right)-K_{1j}\left(u_j+\frac{f_j}{2\mu_j}\right)-K_{2j}\left(u_j+\frac{f_j}{2\mu_j}\right)=$$

$$
=\begin{cases} \dfrac{\chi_j+1}{2\mu_j}\left[\dfrac{1}{2}f_j(t)+S_jf_j-K_{1j}f_j\right], & t\in L_j, \\ 0, & t\in D_j^-. \end{cases} \tag{30.10}
$$

Now we may sum the l. h. s. of (30.10) over all the blocks (j = 1,..., n). When summing, we should take into account that for a fixed t the r. h. s. of (30.10) is non zero only for those two blocks which have the point t on their boundary. As each contact is traveled twice in opposite directions, after some algebra we arrive at the equation

$$
\frac{1}{2\pi i}\int_L\left\{\left[2\Delta u+(2a_1-a_3)f+\left(a_4-\frac{1}{2}a_2\right)\Delta f\right]\frac{d\tau}{\tau-t}-[\Delta u+(a_1-a_3)f+\right.
$$
$$
\left.+\frac{1}{2}(a_4-a_2)\Delta f\right]dk_1-\left(\Delta\bar{u}+a_1\bar{f}+\frac{1}{2}a_4\Delta\bar{f}\right)dk_2\Bigg\}=\frac{1}{2}a_2f+\frac{1}{4}a_3\Delta f, \tag{30.11}
$$

where $f=\frac{1}{2}\left(f^++f^-\right)$ is the mean value of the resultant force, $\Delta f=f^+-f^-$ is the discontinuity of the resultant force; the coefficients a_1, a_2, a_3, a_4 are the same as in (23.6):

$$
a_1=\frac{1}{2\mu^+}-\frac{1}{2\mu^-},\quad a_2=\frac{\chi^++1}{2\mu^+}+\frac{\chi^-+1}{2\mu^-},
$$
$$
a_3=\frac{\chi^++1}{2\mu^+}-\frac{\chi^-+1}{2\mu^-},\quad a_4=\frac{1}{2\mu^+}+\frac{1}{2\mu^-}. \tag{30.12}
$$

L is the total contour of blocks. Now we travel along each contact only once and the normal to it is fixed. The direction of travel along the external boundary, and the meaning of the symbols "plus" and "minus" are taken as usual in accordance with the conventions of § 25. Also as usual, $\Delta u = u^+ - u^-$. On the contours of the holes and on the external contour (for a finite region) we assume $u^- = 0$, $1/\mu^- = 0$ (below we will change the latter definition to a more convenient one to eliminate the influence of eigenfunctions).

Henceforth, we will presume that the discontinuity of the traction $\Delta\sigma$ and therefore the discontinuity of the resultant force Δf are prescribed functions. In particular, they are zero for continuous traction through contacts. Then for given dependence between the displacement discontinuity Δu and the mean traction $\sigma = (\sigma^+ + \sigma^-)/2$ on contacts, the equation (30.11) contains as unknown *only one* value at each point of L: either the displacement discontinuity Δu or the mean value of the resultant force f.

Actually, (30.11) is equation (23.4) integrated with respect to t. It may be obtained by summing (22.2) over the blocks. Thus, so far we have not obtained a new useful equation for blocky systems. To do this we need to make one more step.

New interpretation of (30.11). For a finite system with prescribed traction on its external boundary and/or for blocks encircling holes (Figure 21, *a*), the equation (30.11) has eigenfunctions. This presents difficulties when solving it. To avoid them in a physically clear way, we will modify the interpretation of this equation.

Consider a system of blocks, which encircle m holes with the contours L_k. We will term these blocks *real* blocks. Place an additional *fictitious* block in each hole; their contours, congruent to the contours of the surrounding real blocks, are shown in Figure 21, *b* by the dotted lines. For a finite system we introduce an additional *fictitious* infinite block, represented by a surrounding matrix D_0^- with the contour L_0. For certainty, assume that the traveling direction along L_k (k = 0,..., m) leaves the real blocks on the left.

Elastic constants χ_k, μ_k of the fictitious blocks may be taken arbitrarily. In particular, when the properties of the real blocks surrounding the k-th hole are the same, it is convenient to use their elastic constants as χ_k, μ_k.

We assume that the fictitious blocks are not in contact with the real blocks, but along their contours they have the same traction (and the resultant force) as that at the adjacent points of the real blocks. Therefore, the traction and the resultant force are continuous through the contours L_k (k = 0,..., m). Obviously, not being connected with the real blocks, the fictitious blocks may have arbitrary body motions with respect to the surrounding real blocks.

Apply the summation procedure used to derive (30.11) to both real and fictitious blocks. This leads to (30.11) with one difference: along the boundary of the k-th fictitious block we have $\chi^- = \chi_k$, $\mu^- = \mu_k$, instead of previously used definitions $u^- = 0$, $1/\mu^- = 0$. Henceforth, we will use the new interpretation.

Removing body motions. Certainly, with the new interpretation, equation (30.11) still has eigenfunctions. Now they have a clear physical meaning: they correspond to rigid translation and rotation of the fictitious blocks relative to the real blocks. This allows us to easily eliminate their unfavorable effect.

Indeed, we may eliminate the body motion of each of the fictitious blocks, as explained above, by employing (30.8), (30.9) with the points z_{k1}, z_{k2} within the k-th

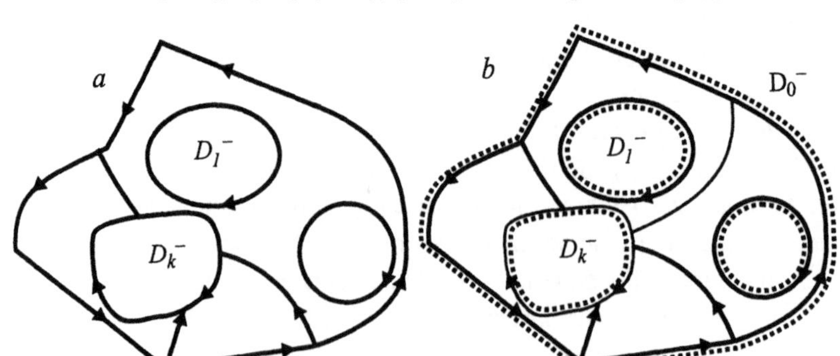

Fig. 21.

hole and for a finite system z_{01}, z_{02} within an arbitrary chosen real block. Then by using (25.8) and summing (30.8), (30.9) over both real and fictitious blocks we obtain additional conditions, which eliminate the unfavorable effect of eigenfunctions of (30.11) corresponding to body motions:

$$\frac{1}{2\pi i}\int_L (\Delta u + a_1 f + \frac{1}{2}a_4 \Delta f)\frac{d\tau}{\tau - z_{k1}} = 0, \quad k = 0,...,m, \tag{30.13}$$

$$\mathrm{Im}\left[\frac{1}{2\pi i}\int_L (\Delta u + a_1 f + \frac{1}{2}a_4 \Delta f)\frac{d\tau}{(\tau - z_{k2})^2}\right] = 0, \quad k = 0,...,m, \tag{30.14}$$

The equation ((30.11) together with (30.13), (30.14) comprise an overdetermined system. It has a unique solution given the corresponding elasticity problem has a unique solution. (For a finite system of blocks with prescribed traction on its external boundary, the solution of a physical problem is naturally defined with body motion, which is removed by (30.13), (30.14) for $k = 0$).

Modified equations for blocky systems. Equation (30.11) under the additional conditions (30.13), (30.14) is equivalent to the following equation *without* additional conditions:

$$\frac{1}{2\pi i}\int_L \left\{\left[2\Delta u + (2a_1 - a_3)f + \left(a_4 - \frac{1}{2}a_2\right)\Delta f\right]\frac{d\tau}{\tau - t} - \right.$$

$$\left. - \left[\Delta u + (a_1 - a_3)f + \frac{1}{2}(a_4 - a_2)\Delta f\right]dk_1 - \left(\Delta\bar{u} + a_1\bar{f} + \frac{1}{2}a_4\Delta\bar{f}\right)dk_2\right\} - C(t) -$$

$$- i\sum_{k=0}^{m}\frac{A_k}{\bar{t} - \bar{z}_{k2}}\mathrm{Im}\frac{1}{2\pi i}\int_L\left(\Delta u + a_1 f + \frac{1}{2}a_4\Delta f\right)\frac{d\tau}{(\tau - z_{k2})^2} = \frac{1}{2}a_2 f + \frac{1}{4}a_3\Delta f, \tag{30.15}$$

where $C(t) = C_k$ on the contours of holes and on the external boundary of a finite system of blocks ($k = 0,\dots, m$); the constants C_k are defined by the l. h. s. of (30.13):

$$C_k = \frac{1}{2\pi i}\int_L (\Delta u + a_1 f + \frac{1}{2}a_4\Delta f)\frac{d\tau}{\tau - z_{k1}},$$

A_k are arbitrary non-zero *real* constants with the physical dimension of area; they serve to make the dimension of the sum on the l. h. s. the same as that of other terms. One may take them equal to characteristic areas of holes. The equivalence of the overdetermined system (30.11), (30.13), (30.14) to the equation (30.15) may be proved following the line presented in Linkov [6].

Differentiation of (30.15), accompanied by integration by parts, yields the equation for the traction and the derivative of the displacement discontinuity:

$$\frac{1}{2\pi i}\int_L\left\{\left[2\Delta u'+(2a_1-a_3)\sigma+\left(a_4-\frac{1}{2}a_2\right)\Delta\sigma\right]\frac{d\tau}{\tau-t}+\right. \tag{30.16}$$

$$+\left[\Delta u'+(a_1-a_3)\sigma+\frac{1}{2}(a_4-a_2)\Delta\sigma\right]\frac{\partial k_1}{\partial t}d\tau+\left(\overline{\Delta u'}+a_1\overline{\sigma}+\frac{1}{2}a_4\overline{\Delta\sigma}\right)\frac{\partial k_2}{\partial t}\overline{d\tau}\left.\right\}+$$

$$+i\sum_{k=0}^{m}\frac{A_k\exp(-2i\alpha_0)}{(\overline{t}-\overline{z}_{k2})^2}\mathrm{Im}\frac{1}{2\pi i}\int_L\left(\Delta u'+a_1\sigma+\frac{1}{2}a_4\Delta\sigma\right)\frac{d\tau}{\tau-z_{k2}}=\frac{1}{2}a_2\sigma+\frac{1}{4}a_3\Delta\sigma,$$

where α_0 is the angle between the direction of traveling and the x-axis, and, as in (23.4),

$$\sigma=\frac{1}{2}(\sigma^+ +\sigma^-),\qquad \Delta\sigma=\sigma^+ -\sigma^-.$$

In particular cases of a *homogeneous* region with cracks and/or inclusions, equation (30.16) has been often used to solve problems by the method of mechanical quadratures (see, e.g. Savruk and Timoshuk [1], Savruk et al [1]) discussed in Chapter 14. The method is quite efficient when the contour L consists of a number of isolated closed and open arcs. But for an arbitrary blocky system with a branched contour L and with cracks terminating at other contours, it becomes inconvenient. The equation (30.16) becomes inconvenient itself when there is linear or non-linear interaction on the boundaries of blocks because a law of interaction involves displacement discontinuities themselves rather than their derivative.

Hypersingular equation. In such problems, as noted earlier, a hypersingular formulation, which involves the displacement discontinuities and the traction, is preferable. We obtain a hypersingular equation by differentiating (30.15), and then integrating by parts those terms containing the resultant force:

$$\frac{1}{2\pi i}\int_L\left\{\frac{2\Delta u d\tau}{(\tau-t)^2}-\Delta u\frac{\partial}{\partial t}dk_1-\overline{\Delta u}\frac{\partial}{\partial t}dk_2+\left[(2a_1-a_3)\sigma+\left(a_4-\frac{1}{2}a_2\right)\Delta\sigma\right]\frac{d\tau}{\tau-t}+\right.$$

$$+\left[(a_1-a_3)\sigma+\frac{1}{2}(a_4-a_2)\Delta\sigma\right]\frac{\partial k_1}{\partial t}d\tau+\left(a_1\overline{\sigma}+\frac{1}{2}a_4\overline{\Delta\sigma}\right)\frac{\partial k_2}{\partial t}\overline{d\tau}\left.\right\}+ \tag{30.17}$$

$$+i\sum_{k=0}^{m}\frac{A_k\exp(-2i\alpha_0)}{(\overline{t}-\overline{z}_{k2})^2}\mathrm{Im}\frac{1}{2\pi i}\int_L\left[\frac{\Delta u d\tau}{(\tau-z_{k2})^2}-(a_1\sigma+\frac{1}{2}a_4\Delta\sigma)\frac{d\tau}{\tau-z_{k2}}\right]=\frac{1}{2}a_2\sigma+\frac{1}{4}a_3\Delta\sigma,$$

where the coefficients a_1, a_2, a_3, a_4 are given by (30.12).

For a region with holes and/or for a finite system of blocks with prescribed traction on the entire external boundary, the equation (30.17) has eigenvalues corresponding to translations of the fictitious blocks, while their rotations are excluded by the sum on the l. h. s. To avoid the influence of the translations, one may solve (30.17) under the additional conditions (30.13). Alternatively, one may add the term $-C(t)$ entering the l. h. s. of (30.15) into the l. h. s. of (30.17). Then if a unique solution of (30.15) exists, it gives Δu and $\sigma = df/dt$ satisfying (30.17) while $C(t) = 0$.

To account for the stresses s_{xx}, s_{yy}, s_{xy} at infinity for a system of blocks embedded into a matrix with the elastic constants χ_∞, μ_∞, it is sufficient to add the term

$$F(t) = -\frac{\chi_\infty + 1}{2\mu_\infty}\sigma_\infty(t), \tag{30.18}$$

to the r. h. s. of (30.17); herein, $\sigma_\infty(t)$ is the traction corresponding to the stresses at infinity

$$\sigma_\infty(t) = 2A_\varphi + \overline{A_\psi}\frac{d\bar{t}}{dt} = 2A_\varphi + \overline{A_\psi}\exp(-2i\alpha_0), \tag{30.19}$$

where

$$A_\varphi = \frac{1}{4}(s_{yy} + s_{xx}), \quad A_\psi = \frac{1}{2}(s_{yy} - s_{xx} + 2is_{xy}). \tag{30.20}$$

Compare the hypersingular equation (30.17) with the equation (23.4), derived by the direct approach. We see that they coincide for an infinite plane without holes. Actually, we have obtained an extension of (23.4) enabling us to suppress the influence of eigenfunctions in problems for a finite system of blocks and/or for holes. This makes the hypersingular equation (30.17) very convenient for applications in micromechanics. An example illustrating its efficiency is given in § 32 after discussing stress intensity factors used in the example.

Holomorphicity representations for blocky systems. The holomorphicity representations take the form (Linkov [6]):

$$\frac{\chi_j + 1}{2\mu_j}\varphi(z) = \frac{1}{2\pi i}\int_L (\Delta u + a_1 f + \frac{1}{2}a_4\Delta f)\frac{d\tau}{\tau - z} + \frac{\chi_\infty + 1}{2\mu_\infty}A_\varphi z, \quad z \in D_j, \tag{30.21}$$

$$\frac{\chi_j + 1}{2\mu_j}\psi(z) = \frac{1}{2\pi i}\int_L [-\overline{\Delta u} + (a_3 - a_1)\bar{f} + \frac{1}{2}(a_2 - a_4)\Delta\bar{f}]\frac{d\tau}{\tau - z} +$$
$$+ \frac{1}{2\pi i}\int_L (\Delta u + a_1 f + \frac{1}{2}a_4\Delta f)\, d\frac{\bar{\tau}}{\tau - z} + \frac{\chi_\infty + 1}{2\mu_\infty}A_\psi z, \quad z \in D_j, \tag{30.22}$$

$$\frac{\chi_j + 1}{2\mu_j}\varphi'(z) = \frac{1}{2\pi i}\int_L (\Delta u' + a_1\sigma + \frac{1}{2}a_4\Delta\sigma)\frac{d\tau}{\tau - z} + \frac{\chi_\infty + 1}{2\mu_\infty}A_\varphi, \quad z \in D_j, \tag{30.23}$$

$$\frac{\chi_j + 1}{2\mu_j}\psi'(z) = \frac{1}{2\pi i}\int_L \left\{\left[-\overline{\Delta u'} + (a_3 - a_1)\bar{\sigma} + \frac{1}{2}(a_2 - a_4)\overline{\Delta\sigma}\right]\frac{d\bar{\tau}}{\tau - z} - \right.$$

$$-\bar{\tau}(\Delta u' + a_1\Delta\sigma + \frac{1}{2}a_4\Delta\sigma)\frac{d\tau}{(\tau - z)^2}\Bigg\} + \frac{\chi_\infty + 1}{2\mu_\infty}A_\psi, \quad z \in D_j, \qquad (30.24)$$

where, as above, in accordance with (30.20) the terms containing A_φ and A_ψ account for the stresses at infinity for blocks embedded into a matrix.

As usual, the integral representations (30.21)-(30.24) may give alternative forms of BIE not employing the direct values of singular and hypersingular integrals. In contrast with the equations containing principal value and finite-part integrals, they correspond to the numerical procedure of coming to the limit *after* integration. They will also serve us in chapter 9 to derive CV-BIE for bonded half-planes.

Calculation of displacements. The approach of this paragraph allows us to find the displacements within the blocks after any of the equations is solved. It is sufficient to substitute the representations (30.21), (30.22) of the complex functions into K-M formula for displacements (25.2). The result is

$$(\chi_j + 1)u(z) = \frac{1}{2\pi i}\int_L \Big\{(\chi_j - 1)\Delta u + \big[(\chi_j - 1)a_1 + a_3\big]f +$$

$$+\frac{1}{2}\big[(\chi_j - 1)a_4 + a_2\big]\Delta f\Big\}\frac{d\tau}{\tau - z} + \frac{1}{2\pi i}\int_L \Big\{\Big[\Delta u + (a_1 - a_3)f +$$

$$+\frac{1}{2}(a_4 - a_2)\Delta f\Big]dk_1 + \left(\overline{\Delta u} + a_1\bar{f} + \frac{1}{2}a_4\Delta\bar{f}\right)dk_2\Big\} +$$

$$+\frac{\chi_\infty + 1}{2\mu_\infty}\Big[(\chi_j - 1)A_\varphi z - \overline{A_\psi}\,\bar{z}\Big], \quad t = z \in D_j. \qquad (30.25)$$

When the traction is used instead of the resultant force, we take into account that $d\Delta f = \Delta\sigma d\tau$ and integrate the terms containing the resultant force in the integrals in (30.25), by parts.

§ 31. STRESS INTENSITY FACTORS

Stress intensity factors (SIF) play a significant role in micromechanics, fracture and rock mechanics. These values with the inconvenient dimension of 'stress multiplied by square root of length' characterize a local stress field near a crack tip. What is especially important is that their squares serve to evaluate the energy release induced by crack propagation in a body, or by advance of mining, or by shear rupture on a geological fault in the Earth's crust. Such evaluation is very useful for prediction of crack growth and for conclusions regarding the hazard of rockbursts and earthquakes.

These properties of the SIF make them popular in various fields related to fracture and geomechanics. For many particular problems they are calculated and compiled in handbooks (see, e. g. Murakami [1]). Nevertheless, it is impossible to

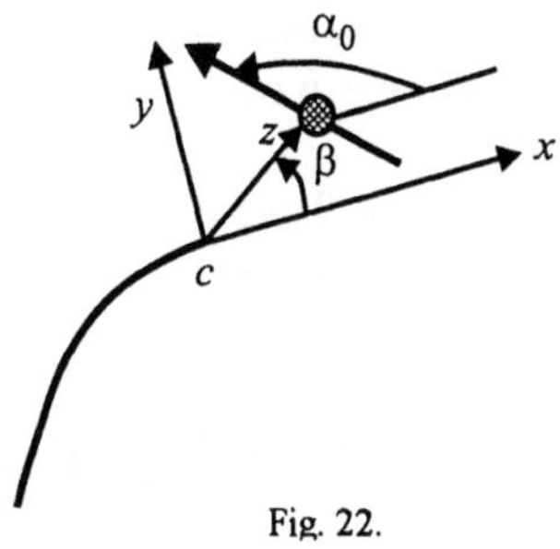

Fig. 22.

include all the configurations of bodies, cracks and their mutual positions in a handbook. Thus, employing a convenient computer program instead of a handbook suggests much greater facilities. The equations of the previous and the following paragraphs allow us to construct the needed programs (see part IV). The formulae derived below serve to calculate the SIF in such programs.

Consider a tip c of a crack (Figure 22). For definiteness, assume that the travel path leads to the tip, thus it is an end point rather than a starting point of the crack. (We shall comment below on minor changes needed when c is a starting point). To simplify discussion, locate the origin at the point c and take the x-axis in the direction of the tangent *out* of the crack. From the general theory (Muskhelishvili [4]) and from the complex equations derived above, for instance from (29.8), it follows that near the point c the functions $\Delta\varphi(t)$ and $\Delta\varphi'(t)$ have asymptotic representations

$$\Delta\varphi(t) = 2A\sqrt{t} + O(t), \quad \Delta\varphi'(t) = \frac{A}{\sqrt{t}} + O(1), \tag{31.1}$$

where A is a constant depending on the solution of a concrete problem; it differs from a linear combination of the normal and shear SIF only by a constant multiplier. The notations of the type $O(t)$, $O(1)$, as usual, characterize the behavior of a residual. In particular, $O(t)$ means that the residual term decreases at least as fast as t when t tends to zero.

From (25.8), (25.9) it follows:

$$(\chi+1)\Delta\varphi(t) = 2\mu\,\Delta u(t) + \Delta f(t), \quad (\chi+1)\Delta\varphi'(t) = 2\mu\,\Delta u'(t) + \Delta\sigma(t). \tag{31.2}$$

Thus, if the traction belongs to the H class in the vicinity of the tip $c = 0$, then the asymptotic behavior of the displacement discontinuity and its derivative is the same as that of $\Delta\varphi(t)$ and $\Delta\varphi'(t)$ respectively:

$$\Delta u(t) = 2A\frac{\chi+1}{2\mu}\sqrt{t} + O(t), \quad \Delta u'(t) = A\frac{\chi+1}{2\mu}\frac{1}{\sqrt{t}} + O(1). \tag{31.3}$$

In the case when the tractions on crack surfaces near the tip are equal in magnitude and opposite in direction, we have $\Delta\sigma = 0$. Then the residual in Δu decreases even more rapidly, as $O(t^{3/2})$.

Use the expression (25.6) for stresses. Take its complex conjugate, remembering that $\sigma = \sigma_{nn} + i\sigma_{nt}$, $\Phi(z) = \varphi'(z)$, $\Psi(z) = \psi'(z)$:

$$\sigma_{nn} - i\sigma_{nt} = \varphi'(z) + \overline{\varphi'(z)} + [\bar{z}\varphi''(z) + \psi'(z)]\frac{dz}{d\bar{z}}. \tag{31.4}$$

Then by using the holomorphicity representations (29.12) and the results of the general theory of complex singular integrals (Muskhelishvili [4, § 23-26]), after some algebra, we obtain from (31.1)-(31.4)[1]:

$$\sigma_{nn} - i\sigma_{nt} = \frac{1}{2\sqrt{r}}\left\{A[\gamma - \frac{1}{2}(\gamma^5 + \gamma)\delta] + \bar{A}(\gamma^{-1} + \gamma\delta)\right\}, \tag{31.5}$$

where $\gamma = \exp(-i\beta/2)$, $\delta = \exp(2i\alpha_0)$, α_0 is the angle between the element on which the stresses are found and the x-axis (Figure 22). If $\beta = \alpha_0 = 0$, we have $\sigma_{nn} = \sigma_{yy}$, $\sigma_{nt} = -\sigma_{xy}$, and (31.5) reads

$$\sigma_{yy} + i\sigma_{xy} = \frac{\bar{A}}{\sqrt{r}}.$$

From this

$$\sigma_{yy} = \frac{\operatorname{Re} A}{\sqrt{r}}, \qquad \sigma_{xy} = -\frac{\operatorname{Im} A}{\sqrt{r}}. \tag{31.6}$$

Usually (see, e. g. Rice [1]) these asymptotic formulae are written as

$$\sigma_{yy} = \frac{k_{\mathrm{I}}}{\sqrt{2\pi r}}, \qquad \sigma_{xy} = \frac{k_{\mathrm{II}}}{\sqrt{2\pi r}}, \tag{31.7}$$

and the coefficients k_{I} and k_{II} are termed *stress intensity factors* (SIF). k_{I} is the stress intensity factor of the *normal* stress, k_{II} of the *shear* stress. Comparison of (31.7) and (31.6) shows that $\operatorname{Re} A = k_{\mathrm{I}}/\sqrt{2\pi}$, $\operatorname{Im} A = -k_{\mathrm{II}}/\sqrt{2\pi}$, that is

$$A = \frac{k_{\mathrm{I}} - ik_{\mathrm{II}}}{\sqrt{2\pi}}. \tag{31.8}$$

Thus we may write the general formula (31.5) in terms of stress intensity factors:

---

[1] A detailed derivation of (31.5) is presented in the Russian edition of this book.

$$\sigma_{nn} - i\sigma_{nt} = \frac{1}{2\sqrt{2\pi r}}\left\{(k_{\mathrm{I}} - ik_{\mathrm{II}})[\gamma - \frac{1}{2}(\gamma^5 + \gamma)\delta] + (k_{\mathrm{I}} + ik_{\mathrm{II}})(\gamma^{-1} + \gamma\delta)\right\}. \quad (31.9)$$

In particular, for stresses in the system of coordinates connected with the crack tip (Figure 22), by taking in (31.9) successively $\alpha_0 = 0$ ($\sigma_{nn} = \sigma_{yy}$, $\sigma_{nt} = - \sigma_{xy}$) and $\alpha_0 = \pi/2$ ($\sigma_{nn} = \sigma_{xx}$, $\sigma_{nt} = \sigma_{xy}$), we arrive at the known formulae of fracture mechanics (see, e. g. Rice [1], Sec. III, A):

$$\begin{aligned}
\sigma_{xx} &= \frac{1}{\sqrt{2\pi r}}[k_{\mathrm{I}}\gamma_1(1-\gamma_2\gamma_4) - k_{\mathrm{II}}\gamma_2(2+\gamma_1\gamma_3)],\\
\sigma_{yy} &= \frac{1}{\sqrt{2\pi r}}[k_{\mathrm{I}}\gamma_1(1+\gamma_2\gamma_4) + k_{\mathrm{II}}\gamma_1\gamma_2\gamma_3)],\\
\sigma_{xy} &= \frac{1}{\sqrt{2\pi r}}[k_{\mathrm{I}}\gamma_1\gamma_2\gamma_3) + k_{\mathrm{II}}\gamma_1(1-\gamma_2\gamma_4)],
\end{aligned} \quad (31.10)$$

where $\gamma_1 = \cos\frac{\beta}{2}$, $\gamma_2 = \sin\frac{\beta}{2}$, $\gamma_3 = \cos\frac{3\beta}{2}$, $\gamma_4 = \sin\frac{3\beta}{2}$.

Analogously, by employing (25.7) we obtain the asymptotic formula for the derivative of the displacements in the direction dz prescribed by the angle $\alpha_0 = \beta$:

$$u'(z) = \frac{1}{2\mu\sqrt{z}}\left[\frac{1}{2}(\chi+1)A - \gamma_1(i\bar{A}\gamma_2 + A\gamma)\right]. \quad (31.11)$$

Integration (31.11) with respect to z from $z = c = 0$ yields

$$u(z) = u_c + \frac{1}{2\mu}\sqrt{\frac{r}{2\pi}}\left\{(k_{\mathrm{I}} - ik_{\mathrm{II}})[(\chi+1)\gamma^{-1} - 2\gamma_1] - 2i(k_{\mathrm{I}} + ik_{\mathrm{II}})\gamma_1\gamma_2\gamma^{-1}\right\} \quad (31.12)$$

where $u_c = u(c)$. By separating the real and imaginary parts in (31.12), we obtain asymptotic values of displacements in the form used in fracture mechanics (see Rice [1], Sec. III, A):

$$\begin{aligned}
u_x(z) - u_{cx} &= \frac{1}{2\mu}\sqrt{\frac{r}{2\pi}}[k_{\mathrm{I}}\gamma_1(\chi - 1 + 2\gamma_2^2) + k_{\mathrm{II}}\gamma_2(\chi + 1 + 2\gamma_1^2)],\\
u_y(z) - u_{cy} &= \frac{1}{2\mu}\sqrt{\frac{r}{2\pi}}[k_{\mathrm{I}}\gamma_2(\chi + 1 - 2\gamma_1^2) - k_{\mathrm{II}}\gamma_1(\chi - 1 - 2\gamma_2^2)].
\end{aligned} \quad (31.13)$$

Usually, equations (31.10), (31.13) are derived for the particular case of a straight crack. The discussion above presents a direct mathematical proof that they are true for arbitrary curvilinear cracks. Besides, we revealed the connection between the SIF and the solutions of the derived CV-BIE.

Formulae to calculate SIF. To simplify derivations, we used a special global coordinate system, shown in Figure 22. This system is very simply connected with the local system (**n**,**t**), that we agreed to use for vectors in § 11 (Figure 8). This local system differs from that used in derivations by a clockwise rotation with the angle $\pi/2$. Hence, to obtain a complex vector in this system from its expression in the system shown in Figure 22, it is sufficient to multiply it by $\exp(i\pi/2)$ that is i. Then for the vector of displacement discontinuity $\Delta v = \Delta v_n + i\Delta v_t$ in the local coordinates (**n**, **t**), from (31.3) by using (31.8) and accounting for $t = -\Delta s + O(\Delta s^2)$ we have

$$\Delta v(t) = -\sqrt{\frac{2}{\pi}}(k_{\mathrm{I}} - ik_{\mathrm{II}})\frac{\chi+1}{2\mu}\sqrt{\Delta s} + O(\Delta s), \tag{31.14}$$

$$\Delta v'(t) = \frac{1}{\sqrt{2\pi}}(k_{\mathrm{I}} - ik_{\mathrm{II}})\frac{\chi+1}{2\mu}\frac{1}{\sqrt{\Delta s}} + O(1). \tag{31.15}$$

Equations (31.14), (31.15) are basic for calculation of the SIF. They imply

$$k_{\mathrm{I}} - ik_{\mathrm{II}} = -\sqrt{\frac{\pi}{2}}\frac{2\mu}{\chi+1}\lim_{\Delta s\to 0}\frac{\Delta v(t)}{\sqrt{\Delta s}}, \tag{31.16}$$

$$k_{\mathrm{I}} - ik_{\mathrm{II}} = \sqrt{2\pi}\frac{2\mu}{\chi+1}\lim_{\Delta s\to 0}\left[\Delta v'(t)\sqrt{\Delta s}\right]. \tag{31.17}$$

Equation (31.16) is especially convenient to calculate the SIF. Its advantage is that it is true for any choice of the global system and, importantly, it does not depend on the direction of travel along a crack. If we change the direction, the components of a vector in the local system change their sign. Simultaneously the sign is changed in the discontinuity $\Delta v = v^+ - v^-$, because the left (right) vicinity becomes the right (left) one when we change the travel direction. As a result, the vector Δv stays the same. Thus equation (31.16) is equally applicable both to the starting and end point of a crack (in the direction of travel).

Equation (31.17), in contrast with (31.16), does not have this advantage. It is sensitive to the choice of the global system because the derivative is taken in the special global system shown in Figure 22. When using another global coordinates, the derivative is multiplied by $\exp(i\alpha_c)$ where α_c is the angle between the tangent and the x-axis (Figure 23). The derivative changes its sign when we change the travel direction. Hence, in a general case (31.17) becomes

$$k_{\mathrm{I}} - ik_{\mathrm{II}} = \pm\sqrt{2\pi}\frac{2\mu}{\chi+1}\exp(i\alpha_c)\lim_{\Delta s\to 0}\left[\Delta v'(t)\sqrt{\Delta s}\right]. \tag{31.18}$$

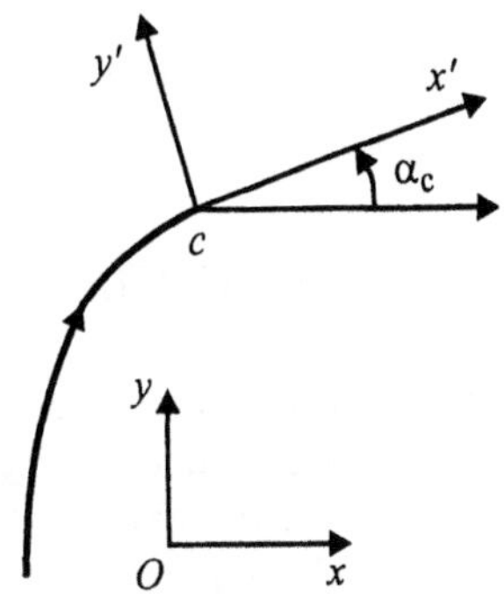

Fig. 23.

where the upper (lower) sign is taken for the end (start) tip of a crack.

Employing (31.16) or (31.18) allows us to avoid mistakes in the signs of the SIF, which may arise when taking the root from complex values. We take arithmetic (positive) value of the root from the positive value Δs. We may also write these formulae by using the displacement discontinuity $\Delta u = \Delta u_x + i\Delta u_y$ in the global system (Figure 23). Accounting for (21.9) we have:

$$k_{\mathrm{I}} - ik_{\mathrm{II}} = -i\sqrt{\frac{\pi}{2}}\,\frac{2\mu}{\chi+1}\exp(-i\alpha_c)\lim_{\Delta s\to 0}\frac{\Delta u(t)}{\sqrt{\Delta s}}, \tag{31.19}$$

$$k_{\mathrm{I}} - ik_{\mathrm{II}} = \pm i\sqrt{2\pi}\,\frac{2\mu}{\chi+1}\lim_{\Delta s\to 0}\left[\Delta u'(t)\sqrt{\Delta s}\,\right]. \tag{31.20}$$

Remember that Δs is the distance from the crack tip; in (31.20) the upper (lower) sign is taken for the end (starting) tip of a crack.

Practical application of any of equations (31.16)-(31.20) for the SIF calculation is as follows. We choose some CV-BIE and solve it with respect to the unknown complex displacement discontinuity or its derivative. After this, the solution is used in one of the formulae for the SIF. Accuracy grows significantly if we approximate unknown discontinuities by accounting for their asymptotic behavior in accordance with (31.3). For instance, when using the complex variable boundary element method (CV-BEM), we may approximate the displacement discontinuity Δu on the tip element as

$$\Delta u = \sqrt{t-c}\,P(t), \tag{31.21}$$

where $P(t)$ is an approximating polynomial; its coefficients are found from the solution of an CV-BIE. Then substitution of (31.21) into (31.19) and tending to the limit yield

$$k_{\mathrm{I}} - ik_{\mathrm{II}} = \sqrt{\frac{\pi}{2}}\,\frac{2\mu}{\chi+1}\exp\left(-\frac{i\alpha_c}{2}\right)P(c),$$

where $P(c)$ is the value of the polynomial at the tip c. We will discuss evaluation of the SIF in more detail in § 54, 55.

§ 32. APPLICATIONS TO MICROMECHANICS

It is appropriate to interrupt the string of formulae and to comment on applications of the tools. Note first that there is an area where using the boundary integral equations, in particular CV-BIE, has advantages over such powerful and popular methods as the finite element method (FEM) and the finite difference method. The latter methods, due to their ability to account for plastic deformations in a material, dominate many engineering fields. But they lose their superiority when irreversible deformations are localized on multiple surfaces of grains and cracks whereas the material itself is elastic. This is the case in many problems of micromechanics.

We have consistently emphasized that *hypersingular* boundary integral equations are a natural means for solving problems in this area because (i) they are formulated on surfaces of discontinuities and (ii) they involve the values which characterize contact interaction: the displacement discontinuities and the traction. As shown in the previous section, they also facilitate calculation of stress intensity factors needed to trace crack propagation.

In 2D, the hypersingular equation (30.17) is especially convenient for these applications. Its numerical implementation is easy on a conventional computer by employing the complex variable boundary element method (CV-BEM). (The computational details of this highly efficient method will be presented in Chapters 12, 13). As a result, one obtains a program which traces crack propagation, accounts for irreversible processes on surfaces of cracks and joints, and handles up to thousands of structural elements of arbitrary forms and locations (see, e. g. Dobroskok et al. [1], Wang et al. [1]). To illustrate the advantage of the approach in these problems, consider an example. It shows that the hypersingular CV-BEM is more efficient than the FEM even for a single growing crack. Naturally, the advantage increases for greater number of cracks and interacting structural elements.

Sliding crack with growing wings. Consider an initial sliding crack of length $2a$ (Figure 24, *a*). The angle between the crack and the x-axis is α. At infinity, the compressive stresses are $s_{yy} = -p$, $s_{xx} = -q$, $s_{xy} = 0$ and the axial pressure p is greater than the lateral one q ($p > q$). Assume that the surfaces of the initial crack are sufficiently smooth so that the shear traction is negligible: $\sigma_{n\tau} = 0$. The initial crack being closed by the external pressure, the normal displacement discontinuity on its surfaces is zero: $\Delta u_n = 0$. Wing cracks, starting at its tips, propagate due to tensile stresses arising ahead of the current position of a crack tip. The criterion

$$k_1 = k_{1C} \tag{32.1}$$

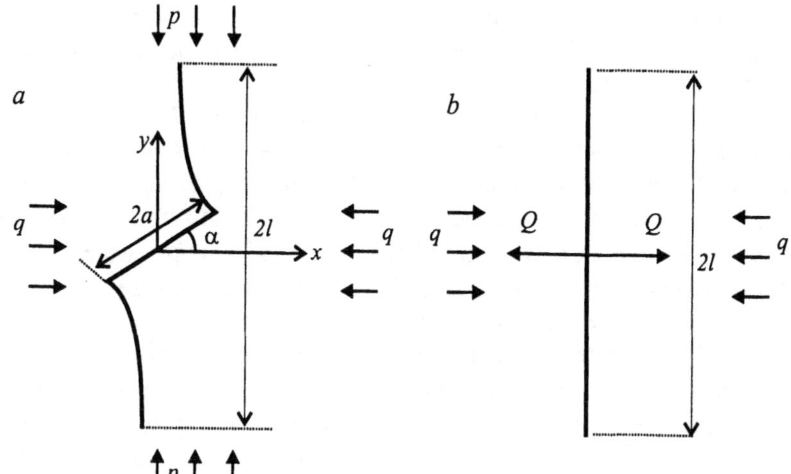

Fig. 24.

defines external loads providing the propagation; here, k_{IC} is the critical value of the *normal* stress intensity factor. The direction of the propagation is defined by the condition that the tensile stress is maximum on the area along this direction. The condition is equivalent to zero *shear* stress intensity factor on smooth parts of the trajectory (see, e. g. Cotterell and Rice [1]):

$$k_{11} = 0. \tag{32.2}$$

The wing cracks are open, hence, the tractions on their surfaces are zero: $\sigma_{nn} = 0$, $\sigma_{n\tau} = 0$. We need to find the trajectory satisfying the described conditions on the surfaces of the initial and wing cracks and the conditions (32.1), (32.2) at each current position of the crack tip. This requires repeated solution of the problem for small increments of each of the wings, by probing various directions to satisfy (32.2). Thus, besides the incremental steps of the wing propagation, we must use iterations to satisfy (32.2) within each step.

To simplify the problem, many authors (see, e.g. Fairhurst and Cook [1], Kemeny and Cook [1]) replaced the physically significant condition $\Delta u_n = 0$ on the initial crack by the plausible suggestion that the normal traction through it is equal to that in a medium without cracks: $\sigma_{nn} = -(p\cos^2\alpha + q\sin^2\alpha)$. Then the total normal force on the initial crack is $P_a = 2a(p\cos^2\alpha + q\sin^2\alpha)$. Its projection $Q = 2a(p\cos^2\alpha + q\sin^2\alpha)\sin\alpha$ to the x-axis is taken as a point force, which provides crack growth in accordance with the scheme shown in Figure 24, *b*. The SIF for this scheme (see, e.g. Murakami [1]) is:

$$k_{\rm I} = \frac{Q}{\sqrt{\pi l}} - q\sqrt{\pi l} = \frac{2a(p\cos^2\alpha + q\sin^2\alpha)\sin\alpha}{\sqrt{\pi l}} - q\sqrt{\pi l}, \tag{32.3}$$

where $2l$ is the vertical length of the crack including its wings. Then (32.1) provides the external load while the direction of the propagation defined by (32.2) is predetermined to be vertical.

The CV-BEM applied to the hypersingular equation (30.17) allows us to solve the problem easily without the simplifying assumptions. In this particular case, the l. h. s. of the equation contains only integrals from the displacement discontinuities because $a_1 = 0$, $a_3 = 0$, $\Delta\sigma = 0$ and there is no holes, so that $A_k = 0$. The coefficient a_2 on the r. h. s. of (30.17) according to (30.12) is $a_2 = (\chi^\infty + 1)/\mu^\infty$, where χ^∞ and μ^∞ are elastic constants of the infinite plane. To account for the stresses at infinity we add the term (30.18) to the r. h. s. of (30.17).

Figures 25-27 illustrate the numerical results obtained for uniaxial compression ($q = 0$) of a plane with a crack having the angle $\alpha = \pi/4$ with the x-axis.[2] The initial crack was represented by five boundary elements with quadratic polynomial approximation of the displacement discontinuities; for tip elements the approximation (31.21) was always used with a quadratic polynomial $P(t)$. The trajectories of the wing cracks were calculated in steps with a crack increment presented by a circular-arc element having the length $\Delta s = 0.1a$. After the wing had passed a tip element the latter was transformed into an ordinary element with the quadratic polynomial approximation. Within a step, a special iterative procedure served to find a new tip position which satisfied the condition (32.2) with prescribed tolerance (Dobroskok [2]). For each position tested in the iteration, the equation (30.17) was solved anew. Normally it took about seven iterations to satisfy (32.2) with prescribed tolerance. The accuracy of calculations provided not less than three correct significant digits for k_{I} and k_{II}.

Figure 25 shows the crack with the calculated wings. As could be expected, after the initial kink, the crack tends to grow in the direction of the applied compression p. This tendency agrees with the simplified scheme of Figure 24, *b*.

Meanwhile, both the calculated load P at the surface of the initial crack, and the external pressure needed for wings to propagate, differ from the values suggested by the simplified scheme. In particular, as indicated by Figure 26, the relative error $(P_a - P)/P$ of the assumption that the normal load is $P_a = 2a(p\cos^2\alpha + q\sin^2\alpha)$ becomes 63% when $\Delta l/a = 3$ (Δl is the increase of the wing length). Similar error occurs also in the corresponding external pressure p, calculated by using (32.3) in (32.1). This implies that the SIF is notably overestimated, while the external pressure is underestimated.

Figure 27 presents the values of the calculated SIF k_{I} for various positions of the wing tip. The SIF is normalized by the quantity $k_{\mathrm{I}0} = p\sqrt{\pi a}$ corresponding to uniaxial tension normal to the initial crack. The triangles present the results obtained by the complex variable *boundary* element method. For comparison, the solid line reproduces the results by Germanovich et al. [1] obtained by using the program FRANC2D of the *finite* element method.

---

[2] The calculations for the example were carried out by A. A. Dobroskok.

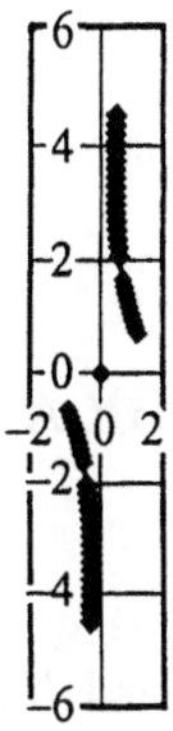

Fig. 25.

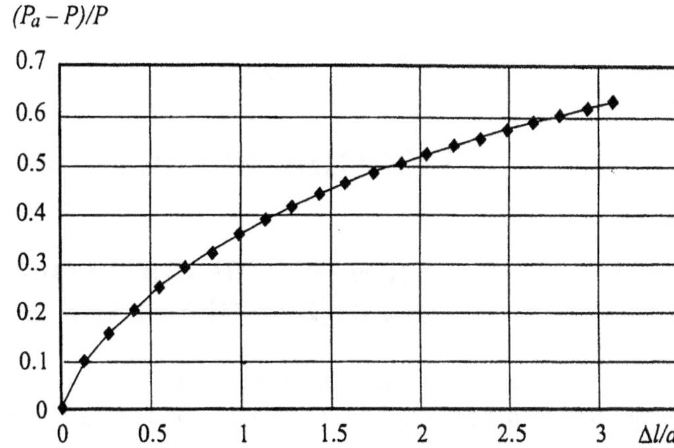

Fig. 26.

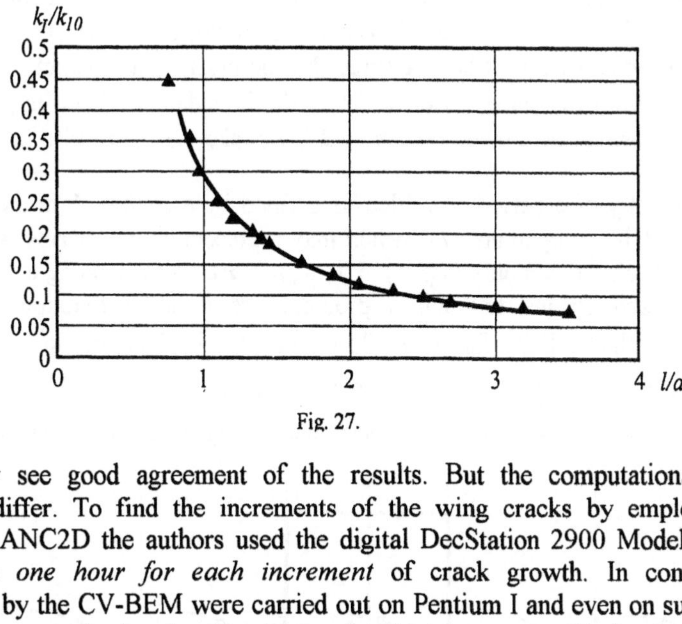

Fig. 27.

We may see good agreement of the results. But the computational efforts drastically differ. To find the increments of the wing cracks by employing the program FRANC2D the authors used the digital DecStation 2900 Model 600 and spent about *one hour for each increment* of crack growth. In contrast, the calculations by the CV-BEM were carried out on Pentium I and even on such a poor computer they required only *40 minutes for 44 increments*. Obviously, the high efficiency of the CV-BEM is due to the fact that the hypersingular BIE is a natural tool to handle discontinuities and calculate SIF.

The program used in the example, being universal, allows one to solve problems for multiple cracks, inclusions, voids and internal grains. Here we cannot dwell on numerous other examples. The reader may find them elsewhere (see, e. g. Mogilevskaya [1-4], Mogilevskaya, Rothenburg and Dusseault [1], Linkov and Koshelev [1,2], Dobroskok [2], Dobroskok et al. [1], Wang et al. [1]).

Chapter 7

PERIODIC PROBLEMS

§ 33. FORMULATION OF PERIODIC PROBLEMS FOR A HOMOGENEOUS PLANE

Consider an infinite plane with periodically repeated holes and/or cracks (Figure 8). Later we will comment on how to extend the results to inclusions and blocky systems.

For definiteness, we assume that the period is along the real axis and equal to π. As mentioned in § 27, in a case of arbitrary complex period 2ω, it is sufficient to multiply co-ordinates by $\pi/(2\omega)$. Assume that boundary and contact conditions are also periodic with the same period π. Then the boundary problem is periodic with period π. Following the conventions of § 27, we take the strip $-\pi/2 < x < \pi/2$ as the main strip.

The geometry of a periodic problem is defined by a contour L in the main strip. The contour L has p open arcs L_k, which may represent cracks or ribs ($k = 1,\ldots, p$), and m smooth closed contours L_j ($j = p + 1,\ldots, p + m$), corresponding to holes.

Stresses and strains in such a problem are periodic functions. Hence, in accordance with the K-M formulae (25.6), (25.7), the function $\varphi'(z)$ is periodic, that is $\varphi'(z + k\pi) = \varphi'(z)$ for any integer k. The function $\varphi(z)$, which is an integral from $\varphi'(z)$, is generally merely quasi-periodic. However, under the assumptions

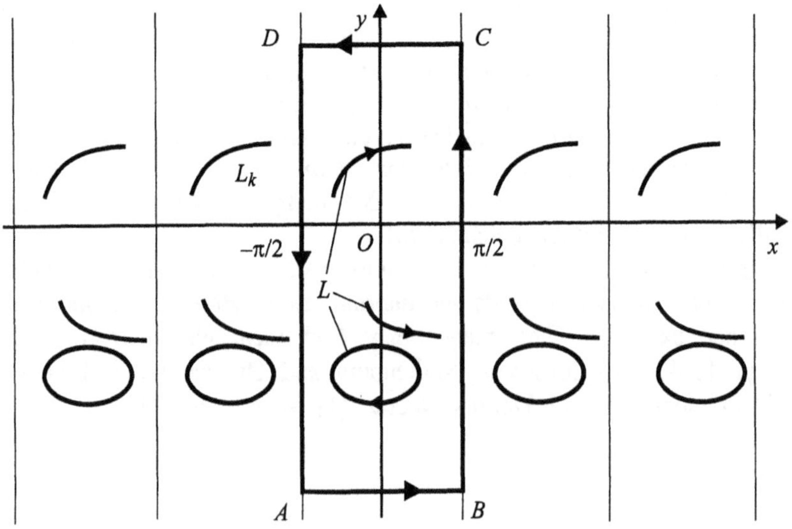

Fig. 28.

that (i) stresses and rotation at infinity are zero and (ii) the total resultant force applied to the entire contour L is zero, the function $\varphi(z)$ is periodic. (We will not dwell on a proof of this statement; it employs the equilibrium of a rectangle $ABCD$ with two sides along the boundaries of the main strip (Figure 28)). These assumptions, allowing us to shorten discussion and formulae, are not essential. The first of them is easily removed by employing a homogeneous field of stresses. As for the second, we will even strengthen it: we assume that the total resultant force applied to *each* arc or hole is zero:

$$\int_{L_j} \Delta\sigma \, d\tau = 0, \quad j = 1,\dots,p+m, \tag{33.1}$$

where $\Delta\sigma = \sigma^+ - \sigma^-$ is the traction discontinuity; as before, for holes we take $\sigma^+ = \sigma$, $\sigma^- = 0$. Recall that the conditions (33.1) do not influence equations containing the traction instead of the resultant force and that it is always possible to transform equations for the resultant force into equations for the traction by using integration by parts. Hence, finally we may always use the latter equations and neglect (33.1). Dealing with non-zero total resultant forces leads to more lengthy derivations (see Linkov [4]).

In a periodic problem, we do not impose restrictions on total moments acting on each contour and on their totality. We will show that the total moment is always equilibrated by the moment of forces acting on sides AB and CD of a rectangle $ABCD$ enclosing all arcs and holes.

Resultant force. In the problem, the resultant force $f(z)$ is defined within an arbitrary constant. We may fix the constant in various ways. We may assume that the resultant force is zero at $z = -i\infty$ or at $z = +i\infty$. We prefer to fix it by the symmetry condition: $f(\pm i\infty) = \pm f^\infty$. This will allow us to use holomorphicity theorem as it is formulated in § 27.

Displacements. We will also use the symmetry rule $u(\pm i\infty) = \pm u^\infty$.

Function $\varphi(z)$. From (25.8) it follows that K-M function $\varphi(z)$ is also fixed by similar symmetry condition, $\varphi(\pm i\infty) = \pm B$, where

$$B = \frac{2\mu\, u^\infty + f^\infty}{\chi + 1}. \tag{33.2}$$

The function $\varphi(z)$ is piece-wise holomorphic in the whole plane except for points on L and contours congruent to L. Besides, from (25.8) it follows that, as with the displacements and the resultant force, the function $\varphi(z)$ is periodic with period π. Hence, in accordance with the holomorphicity theorem of § 27, we may represent it by a Cauchy type integral:

$$\varphi(z) = \frac{1}{2\pi i}\int_L \Delta\varphi \cot(\tau - z)d\tau, \tag{33.3}$$

where $\Delta\varphi(t) = \varphi^+(t) - \varphi^+(t)$ is the discontinuity of $\varphi(z)$ on the contour L; as usual, for closed contours we assume $\varphi^+(t) = \varphi(t)$, $\varphi^-(t) = 0$. Besides, from expansion of $\cot z$ when $z \to \pm i\infty$ it follows

$$\frac{1}{2\pi}\int_L \Delta\varphi\, d\tau = B, \tag{33.4}$$

where B is the limiting value of $\varphi(z)$ when $z \to +i\infty$; it is connected to the physical values at infinity by equation (33.2).

***Function* $\psi(z)$.** In contrast to $\varphi(z)$, the function $\psi(z)$, although being piece-wise holomorphic by the theory of K-M, is not periodic. It is even not quasi-periodic. From (25.1) and the periodicity of $\varphi(z)$ and $f(z)$ we infer that

$$\psi(z + \pi) = \psi(z) - \pi\varphi'(z). \tag{33.5}$$

From the same equation (25.1) and the behavior of $\varphi(z)$ and $f(z)$ when $z \to \pm i\infty$ it follows that $\psi(z)$ should be symmetric at $z = \pm i\infty$ in the same sense as $\varphi(z)$ and $f(z)$. We will see that this condition is easy to satisfy. Meanwhile, to obtain an integral representation of $\psi(z)$ satisfying (33.5) we need a special function, which, being simple, satisfies an equation similar to (33.5).

***Function* $Q(z)$.** Consider the function $Q(z)$ defined by

$$Q(z) = -(\cot z + z\cot' z). \tag{33.6}$$

This function does not have poles, and is therefore holomorphic in the main strip. Elementary derivation shows that $Q(z)$ really satisfies an equation similar to (33.5)

$$Q(z + \pi) = Q(z) - \pi\cot'(z). \tag{33.7}$$

From the definition (33.6) and equation (33.7) it follows that we may easily construct a particular function $\psi_*(z)$, which meets (33.5) when $\varphi(z)$ is given by a Cauchy type integral (33.3). Indeed, define $\psi_*(z)$ by the formula

$$\psi_*(z) = \frac{1}{2\pi i}\int_L \Delta\varphi Q(\tau - z)d\tau. \tag{33.8}$$

Then, employing (33.7) and the result of differentiation (33.3) with respect to z, that is

$$\varphi'(z) = -\frac{1}{2\pi i}\int_L \Delta\varphi \cot'(\tau - z)d\tau,$$

we see that the function $\psi_*(z)$ really satisfies (33.5). It is holomorphic in the main strip and piece-wise holomorphic in other strips. Besides, as it is clear from the definitions (33.6), (33.8), the function $\psi_*(z)$ is symmetric at $z = \pm i\infty$. It is completely defined by the difference of the limiting values of the function $\varphi(z)$ on the contour L.

General representation of the function $\psi(z)$**.** The general representation of a piece-wise holomorphic function, satisfying (33.5), is given by a sum of a particular function $\psi_*(z)$ and an arbitrary *periodic* piece-wise holomorphic function. Denote the latter $\Omega(z)$. Thus, in a general form the K-M function $\psi(z)$ may be written as

$$\psi(z) = \Omega(z) + \psi_*(z) = \Omega(z) + \frac{1}{2\pi i}\int_L \Delta\varphi Q(\tau - z)d\tau. \tag{33.9}$$

We fix the value of $\Omega(z)$ at infinity by the same symmetry condition when $z \to \pm i\infty$: $\psi(\pm i\infty) = \pm\psi^\infty$. Then we may apply the holomorphicity theorem in its formulation of § 27 to the function $\Omega(z)$. To obtain a CV-BIE, it will suffice to substitute the representation (33.9) into boundary condition (25.14) or (25.15) and to apply the holomorphicity theorem to the functions $\Omega(z)$ and $\varphi(z)$ (Linkov [4]). We will proceed this way in the next paragraph, while, in concluding this one, we consider a condition to be satisfied by the total moment acting on all contours in the main strip.

Total moment. The moment applied to the contour L in the main strip is given by equation (see Muskhelishvili [5, § 102])

$$M_L = \mathrm{Re}\int_L \Delta f d\bar{\tau}, \tag{33.10}$$

where $\Delta f = f^+ - f^-$ is the discontinuity of the resultant force. As usual, we assume that for holes $f^+ = f$, $f^- = 0$. This moment is equilibrated by the moment M_{ext}, acting on the boundaries of any rectangle $ABCD$ enclosing the contour L (Figure 28):

$$M_{ext} = \mathrm{Re}\int_{ABCD} f d\bar{\tau}.$$

The integrals over the sides of the main strip vanish because, by periodicity, we have $f(t + \pi) = f(t)$. Two other sides AB and CD have the same length π. Then letting them tend to infinity and accounting for $f(\pm i\infty) = \pm f^\infty$, we obtain:

$$M_{ext} = -2\pi \operatorname{Re} f^{\infty}. \tag{33.11}$$

Inserting (33.10), (33.11) into the equilibrium equation $M_L + M_{ext} = 0$ yields

$$\operatorname{Re}\int_L \Delta f \, d\bar{\tau} - 2\pi \operatorname{Re} f^{\infty} = 0. \tag{33.12}$$

Later, we will show that a solution of complex variable BIE really satisfies this condition.

§ 34. COMPLEX VARIABLE BIE FOR PERIODIC PROBLEMS

Use the boundary condition in terms of the resultant force (25.14):

$$\varphi^{\pm}(t) + t\overline{\varphi'^{\pm}(t)} + \overline{\psi^{\pm}(t)} = f^{\pm}(t), \tag{34.1}$$

where

$$f^{\pm}(t) = f_c^{\pm}(t) + C^{\pm}(t),$$

$$f_c^{\pm}(t) = \int_{a_j}^{t} \Delta\sigma^{\pm} \, d\tau, \quad j = 1,\ldots,p+m,$$

a_j is a start point on a contour L_j; for an open arc it coincides with its beginning (in the direction of travel); for a hole it is taken arbitrarily; $C^{\pm}(t) = C^{\pm}_j$ are constants on any one contour; for an open arc L_j we assume $C^{+}_j = C^{-}_j = C_j$, for a hole $C^{+}_j = C_j$, $C^{-}_j = 0$; thus, on any particular contour L_j we have just one constant C_j. The constant C_j is an unknown value of the resultant force at the start point of the j-th contour ($j = 1,\ldots, p+m$). For holes, as before, we assume $f_c^{+} = f_c$, $f_c^{-} = 0$.

Substitution of (33.9) into (34.1) gives the limiting values of the piece-wise holomorphic function $\Omega(z)$:

$$\Omega^{\pm}(t) = \overline{f^{\pm}(t)} - \overline{\varphi^{\pm}(t)} - \bar{t}\varphi'^{\pm}(t) - \frac{1}{2\pi i}\int_L \Delta\varphi Q(\tau - t)\, d\tau.$$

Now we will use a derivation quite similar to that in § 29. Applying the holomorphicity theorem (27.2) to $\Omega(z)$ we have:

$$\overline{f^{+}} + \overline{f^{-}} - \left(\overline{\varphi^{+}} + \overline{\varphi^{-}}\right) - \bar{t}\left(\varphi'^{+} + \varphi'^{+}\right) - \frac{2}{2\pi i}\int_L \Delta\varphi Q(\tau - t)\, d\tau =$$

$$= \frac{1}{2\pi i}\int_L \left(\overline{\Delta f} - \overline{\Delta\varphi} - \bar{\tau}\Delta\varphi'\right)\cot(\tau - t)d\tau. \tag{34.2}$$

Here, we have used the fact that the function $\Omega(z)$ is holomorphic and therefore continuous in the entire main strip. To exclude sums $\varphi^{+}(t) + \varphi^{-}(t)$ and $\varphi^{+'}(t) + \varphi^{-'}(t)$, we may use equations (27.2) and (27.6) for $\varphi(t)$ and $\varphi'(z)$. From these equations we have:

$$\overline{\varphi^{+}} + \overline{\varphi^{-}} = \overline{\frac{1}{2\pi i}\int_L \Delta\varphi\cot(\tau - t)d\tau}, \quad \varphi'^{+} + \varphi'^{-} = \frac{1}{2\pi i}\int_L \Delta\varphi'\cot(\tau - t)d\tau. \tag{34.3}$$

Inserting (34.3) into (34.2), integrating by parts, and applying elementary transformations, we find:

$$\frac{1}{2\pi i}\int_L \left[2\Delta\varphi\cot(\tau - t)d\tau - \Delta\varphi\, dk_{\pi 1}(\tau,t) - \overline{\Delta\varphi}\, dk_{\pi 2}(\tau,t)\right] =$$

$$= \frac{1}{2}\left(f^{+} + f^{-}\right) + \frac{1}{2\pi i}\int_L \Delta f\cot(\tau - t) - \frac{1}{2\pi i}\int_L \Delta f\, dk_{\pi 1}(\tau,t), \tag{34.4}$$

where

$$dk_{\pi 1}(\tau,t) = \cot(\tau - t)d\tau - \overline{\cot(\tau - t)d\tau}, \quad dk_{\pi 2}(\tau,t) = d\left\{\left[(\tau - t) - (\bar{\tau} - \bar{t})\right]\overline{\cot(\tau - t)}\right\}.$$

Comparing (34.4) with the BIE (29.6) we see that, as expected, in a periodic problem the singular kernel $1/(\tau - t)$ is changed to $\cot(\tau - t)$. This yields changes in the kernels $k_{\pi 1}(\tau, t)$, $k_{\pi 2}(\tau, t)$. They are defined by

$$k_{\pi 1}(\tau,t) = \ln\frac{\sin(\tau - t)}{\overline{\sin(\tau - t)}}, \quad k_{\pi 2}(\tau,t) = \left[(\tau - t) - (\bar{\tau} - \bar{t})\right]\overline{\cot(\tau - t)}. \tag{34.5}$$

The common substitution (25.8), which expresses $\Delta\varphi(t)$ via physical values, yields equation analogous to (29.8):

$$\frac{1}{2\pi i}\int_L \Big[2\Delta u\, \cot(\tau - t)d\tau - \Delta u\, dk_{\pi 1} - \overline{\Delta u}\, dk_{\pi 2} -$$

$$-(\chi - 1)\frac{\Delta f}{2\mu}\cot(\tau - t)d\tau + \chi\frac{\Delta f}{2\mu}dk_{\pi 1} - \frac{\overline{\Delta f}}{2\mu}dk_{\pi 2}\Big] = \frac{\chi + 1}{2\mu}\,\frac{f^{+} + f^{-}}{2}. \tag{34.6}$$

The constants C_j ($j = 1,\dots, p + m$) entering $f^{\pm}(t)$ are defined by p conditions of zero displacement discontinuity at crack tips:

$$\Delta u(a_j) = \Delta u(b_j) = 0, \quad j = 1,\dots,p$$

and by m conditions of continuity of the resultant force when traveling form the origin to each hole.

Just as in § 29, we can obtain equations resulting from (34.6) by differentiating with respect to t. Write down two of such equations, singular and hypersingular, analogous respectfully to (29.9) and (29.10):

$$\frac{1}{2\pi i}\int_L \Big[2\Delta u'\cot(\tau - t)d\tau + \Delta u'\frac{\partial k_{1\pi}}{\partial t}d\tau + \overline{\Delta u}\frac{\partial k_{2\pi}}{\partial t}d\bar{\tau} -$$

$$-(\chi-1)\frac{\Delta\sigma}{2\mu}\cot(\tau - t)d\tau - \chi\frac{\Delta\sigma}{2\mu}\frac{\partial k_{1\pi}}{\partial t}d\tau + \frac{\overline{\Delta\sigma}}{2\mu}\frac{\partial k_{2\pi}}{\partial t}d\bar{\tau}\Big] = \frac{\chi+1}{2\mu}\frac{\sigma^+ + \sigma^-}{2}, \quad (34.7)$$

$$\frac{1}{2\pi i}\int_L \Big[2\Delta u\frac{d\tau}{\sin^2(\tau - t)} - \Delta u\frac{\partial}{\partial t}dk_{\pi 1} - \overline{\Delta u}\frac{\partial}{\partial t}dk_{\pi 2} -$$

$$-(\chi-1)\frac{\Delta\sigma}{2\mu}\cot(\tau - t)d\tau - \chi\frac{\Delta\sigma}{2\mu}\frac{\partial k_{\pi 1}}{\partial t}d\tau + \frac{\overline{\Delta\sigma}}{2\mu}\frac{\partial k_{\pi 2}}{\partial t}d\bar{\tau}\Big] = \frac{\chi+1}{2\mu}\frac{\sigma^+ + \sigma^-}{2}. \quad (34.8)$$

A solution of the first of these should satisfy additional conditions for displacements to be single-valued:

$$\int_{L_j}\Delta u'd\tau = 0, \quad j = 1,\dots,p.$$

The second, hypersingular, equation does not need additional conditions: from the beginning, we search a solution among functions which do not have discontinuities at the tips of open arcs. We may see that these equations may be obtained from equations (29.9), (29.10) for non-periodic systems of cracks and holes simply by changing the singular kernel $1/(\tau - t)$ to $\cot(\tau - t)$, the hypersingular kernel $1/(\tau - t)^2$ to $1/\sin^2(\tau - t)$, and the kernels $k_1(\tau, t)$, $k_2(\tau, t)$ to the kernels (34.5).

We have dwelt on the equations that follow from the boundary condition in terms of loads. We could also start from the boundary condition (25.15) in terms of displacements. Then derivation quite similar to that leading to (34.4) gives equations with displacements or their derivatives on the r. h. s. We will not write down these quite obvious equations.

Resultant force at infinity and equilibrium of moments. The K-M formula (25.1) and representation (33.9) of the function $\psi(z)$ give:

$$f(z)=\varphi(z)+z\overline{\varphi'(z)}+\overline{\Omega(z)}+\overline{\frac{1}{2\pi i}\int_L \Delta\varphi Q(\tau-z)d\tau}. \qquad (34.9)$$

Take into account that

$$\varphi(z)=\frac{1}{2\pi i}\int_L \Delta\varphi\cot(\tau-z)d\tau, \qquad (34.10)$$

$$\Omega(z)=\frac{1}{2\pi i}\int_L \Delta\Omega\cot(\tau-z)d\tau=\frac{1}{2\pi i}\int_L \left(\overline{\Delta f}-\overline{\Delta\varphi}-\overline{\tau}\Delta\varphi'\right)\cot(\tau-z)d\tau. \quad (34.11)$$

Turn to limits when $z\to\pm i\infty$. Note that $\cot(\tau-z)\to\pm\ i$; $z/\sin^2 z\to 0$; $Q(z)\to\mp i$;

$\varphi(z)\to\pm B$, where B is defined by (33.2); $\Omega(z)\to\pm\frac{1}{2\pi}\int_L [\Delta f d\tau-2i\mathrm{Im}(\Delta\varphi d\tau)]$.

By using these limits we obtain:

$$f(\pm i\infty)=\pm\left[2i\,\mathrm{Im}\left(B-\frac{1}{2\pi}\int_L \Delta\varphi d\overline{\tau}\right)+\frac{1}{2\pi}\int_L \Delta f\, d\overline{\tau}\right]. \qquad (34.12)$$

We see that, as expected, the values of the resultant force at $z=\pm i\infty$ differ only by their sign. Thus, equation (34.12) gives us the value $f^{\infty}=f(+i\infty)$. Accounting for (33.4) we may write this value as

$$f^{\infty}=\frac{i}{\pi}\mathrm{Im}\int_L \Delta\varphi(d\tau-d\overline{\tau})+\frac{1}{2\pi}\int_L \Delta f\, d\overline{\tau}.$$

The first term is purely imaginary; we have for the real part of f^{∞}:

$$\mathrm{Re}\, f^{\infty}=\frac{1}{2\pi}\mathrm{Re}\int_L \Delta f\, d\overline{\tau}$$

which proves that equation (33.12) of the moment equilibrium is satisfied.

For the imaginary part of f^{∞}, expressed in terms of physical values, we have:

$$\mathrm{Im}\, f^{\infty}=\frac{1}{\pi(\chi+1)}\mathrm{Im}\int_L \left(2\mu\Delta u+\Delta f\right)(d\tau-d\overline{\tau}).$$

Hence, the resultant force at $z=+i\infty$ is

$$f^{\infty} = \frac{1}{2\pi}\mathrm{Re}\int_L \Delta f\, d\bar{\tau} + \frac{i}{\pi(\chi+1)}\mathrm{Im}\int_L (2\mu\Delta u + \Delta f)(d\tau - d\bar{\tau}). \qquad (34.13)$$

When there are only open arcs (cracks) with continuous traction on their surfaces ($\Delta f = 0$), the real part of f^{∞} is zero. As mentioned, this means that the total moment acting on all cracks is equal to zero. In this case f^{∞} is purely imaginary and, as it follows from (34.13), it is

$$f^{\infty} = i\frac{2\mu}{\pi(\chi+1)}\mathrm{Im}\int_L \Delta u(d\tau - d\bar{\tau}). \qquad (34.14)$$

Displacements at infinity. One may obtain the displacements at infinity from the K-M formula (25.2) just as the resultant force was obtained from the K-M formula (25.1). However, it is simpler to derive them from (33.2) and using (33.4). Expressing, as usual, $\Delta\varphi$ via physical values, we obtain after simple transformations:

$$2\mu u^{\infty} = -f^{\infty} + \frac{1}{2\pi}\int_L (2\mu\Delta u + \Delta f)d\tau,$$

where f^{∞} is given by (34.13) in the general case, or by (34.14) for cracks with continuous traction.

In the last case, we have

$$u^{\infty} = \frac{1}{2\pi}\left(\int_L \Delta u d\tau - i\frac{4}{\chi+1}\mathrm{Re}\int_L \Delta u \sin\alpha\, ds\right),$$

where α is the angle between the tangent to a crack and the x-axis. Separation of the real and imaginary parts yields

$$2u_x^{\infty} = \frac{1}{\pi}\int_L (\Delta u_x \cos\alpha - \Delta u_y \sin\alpha)ds,$$

$$2u_y^{\infty} = \frac{1}{\pi}\int_L \left(\frac{\chi-3}{\chi+1}\Delta u_x \sin\alpha + \Delta u_y \cos\alpha\right)ds. \qquad (34.15)$$

The l. h. s. of first in (34.15) represents mutual shear of the upper side CD of the rectangle $ABCD$ with respect to its lower side AB when these sides go to infinity (Figure 28). The l. h. s. of the second expresses elongation of the strip between these sides. The right hand sides of (34.15) also have simple interpretations if all the cracks are collinear to the x-axis. When traveling the cracks in the positive x-direction, we have $\sin\alpha = 0$, $\cos\alpha = 1$ and the previous equations become

$$2u_x^{\infty} = \frac{1}{\pi}\int_L \Delta u_x ds, \quad 2u_y^{\infty} = \frac{1}{\pi}\int_L \Delta u_y ds.$$

The r. h. s. of the first of them expresses the average (over the period π) mutual shear of crack surfaces. The r. h. s. of the second expresses the change of the volume of the cracks, also related to the period.

Integral representations for periodic problems. Equations (34.10), (34.11) and (33.9) serve as integral representations for periodic problems analogous to the holomorphicity representations (29.11). Extensions for the potentials $\Phi(z) = \varphi'(z)$, $\Psi(z) = \psi'(z)$ are clear. We may obtain them by direct differentiation of the integral representations of φ (z), ψ (z) accompanied by integration by parts to have the density $\Delta\varphi'(t)$. Savruk et al. successfully used this way in problems for periodic *straight* cracks (Datsishin and Savruk [2], Panasiuk et al. [1]). Ioakimidis and Theocaris [1] used it for arbitrary curvilinear cracks.

Cracks along the real axis of periodicity. We saw that for non-periodic problems the BIE became especially simple in two cases: for a contour along a *single* straight line and for a contour along a *single* circumference. It is clear that the second case is impossible in a periodic problem: even if the contour L is located on a single circumference in the main strip, congruent contours in other strips are on different circumferences. For a contour along a straight line, there arises a restriction induced by the same reason: the line should be along the axis of periodicity. Otherwise, contours in different strips will belong to different straight lines. Thus, in periodic problems with a period along the x-axis, we may expect notable simplification only when the contour L in the main strip lays along the x-axis.

Consider a periodic problem for a system of cuts (cracks) along the x-axis (Figure 29). Denote as above L their contour in the main strip. It is comprised of p cuts L_j $(j = 1,\ldots,p)$.

On the axis x we have $\bar{\tau} = \tau$. Then kernels (34.5) are zero on L, therefore, the equations contain only one integral: singular in the singular BIE (34.4), (34.6), (34.7) or hypersingular in the hypersingular equation (34.8). The change of variables

$$w_z = \exp(2iz)$$

transforms them to the simplest form having an analytical solution. The resulting K-M functions $\Phi(z)$, $\Psi(z)$ are

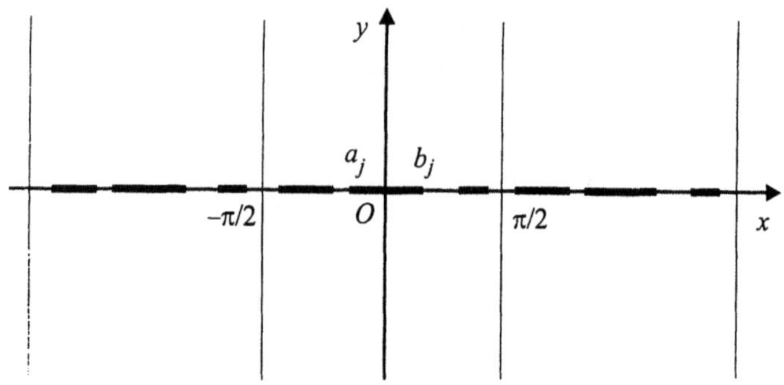

Fig. 29.

$$\Phi(z)=\varphi'(z)=\frac{1}{2\pi i Z(w_z)}\int_{L_w}\sigma\frac{Z(w_\tau)dw_\tau}{w_\tau-w_z}+\frac{1}{4\pi i}\int_{L_w}\Delta\sigma\frac{dw_\tau}{w_\tau-w_z}+\frac{P_{p\text{-}1}(w_z)}{Z(w_z)},$$

$$\Psi(z)=\bar{J}(z)-\Phi(z)-z\Phi'(z),$$

where $Z(w) = \Pi Z_j(w)$, $Z_j(w) = (w - w_{aj})^{1/2}(w - w_{bj})^{1/2}$; $w_{aj} = \exp(2ia_j)$ is the transformed co-ordinate of the start point a_j on the j-th crack; $w_{bj} = \exp(2ib_j)$ is the transformed co-ordinate of its end, L_w is the transformed contour L of the cracks; $P_{p-1}(w_t)$ is a polynomial of order $p - 1$; its coefficients are defined by the p conditions of zero displacement discontinuity at the crack tips:

$$\int_{a_j}^{b_j}\Phi(\tau)\,d\tau=0, \qquad j=1,\ldots,p;$$

$\sigma = ½(\sigma^+ + \sigma^-)$ is the mean traction, $\Delta\sigma = \sigma^+ + \sigma^-$ is the traction discontinuity;

$$J(z)=\Phi(z)-\frac{1}{2\pi i}\int_{L_w}\Delta\sigma\frac{dw_\tau}{w_\tau-w_z},$$

the notation $\bar{J}(z)$ means $\bar{J}(z)=\overline{J(\bar{z})}$.

In the particular case of a single crack in the main strip ($p = 1$, $a_1 = -a$, $b_1 = a$), when the traction is continuous through the crack surfaces ($\sigma^+ = \sigma^- = \sigma$) and symmetric with respect to the middle of the crack ($\sigma(t) = \sigma(-t)$), the formulae yield the classical result (Koiter [1]). In this case, if the traction is constant along the crack ($\sigma = \text{const}$), we have:

$$\Delta u'=-\frac{\chi+1}{2\mu}\sigma\frac{\sin x}{\sqrt{\sin^2 x-\sin^2 a}}, \qquad -a<x<a. \tag{34.16}$$

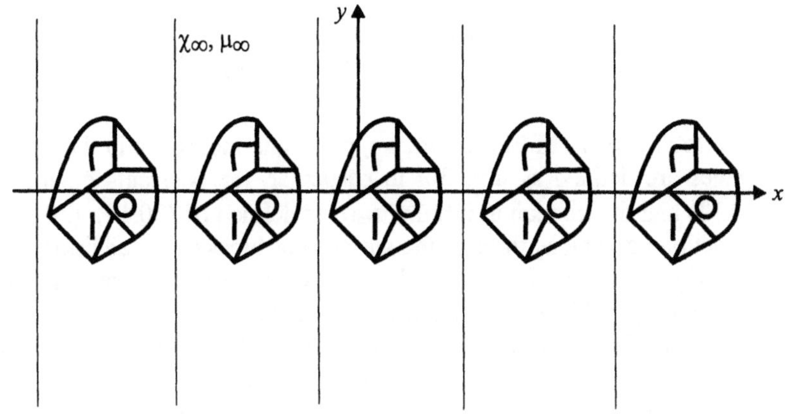

Fig. 30.

Then (31.20) gives a well known formula for the stress intensity factors (see, e.g., Murakami [1]):

$$k_1 - ik_{11} = -\sigma\sqrt{\pi \tan a}. \tag{34.17}$$

The simple analytical formulae (34.16), (34.17) allow us to evaluate mutual influence of collinear cracks and accuracy of calculations (see § 58). Additional results on cracks symmetrically located in the main strip may be found in Linkov [4].

§ 35. BIE FOR PERIODIC SYSTEMS OF BLOCKS

Consider a periodic system of m elastic blocks with elastic constants μ_j, χ_j (j = 1, ..., m) embedded into the infinite matrix with elastic constants μ_∞, χ_∞ (Figure 30). The blocks may contain internal cracks, holes, and/or inclusions.

Note that integrals from the displacement discontinuities in (34.6)-(34.8) do not contain elastic constants. Their counterparts given in § 29 had the analogous property what served us to obtain equations for blocky systems. Now we may use this property to derive analogous equations for the periodic systems of blocks. The equations for periodic systems of blocks immediately follow from the results of § 30 if we change the singular kernel $1/(\tau - t)$ to $\cot(\tau - t)$, the hypersingular kernel $1/(\tau - t)^2$ to $1/\sin^2(\tau - t)$, and the kernels $k_1(\tau, t)$, $k_2(\tau, t)$ to the kernels (34.5).

Hypersingular equation. Having in mind applications to micromechanics, we will write down explicitly only the hypersingular equation:

$$\frac{1}{2\pi i}\int_L \left[\frac{2\Delta u d\tau}{\sin^2(\tau - t)} - \Delta u \frac{\partial}{\partial t} dk_{\pi 1} - \overline{\Delta u} \frac{\partial}{\partial t} dk_{\pi 2} + (2a_1 - a_3)\sigma \cot(\tau - t) d\tau + \right.$$

$$+\left(a_1-a_3\right)\sigma\frac{\partial k_{\pi 1}}{\partial t}d\tau+a_1\bar{\sigma}\frac{\partial k_{\pi 2}}{\partial t}d\bar{\tau}\Bigg]=\frac{1}{2}a_2\sigma-\frac{\chi_\infty+1}{2\mu_\infty}\sigma_\infty,\quad t\in L, \qquad (35.1)$$

where L is the total contour of blocks, cracks, holes and inclusions in the main cell. All other values are defined as in § 30; in particular, the constants a_1, a_2, a_3 are given by (30.12) and the traction at infinity $\sigma_\infty(t)$ is defined by (30.19), (30.20). We assume the traction continuous through the surfaces of blocks. The additional terms, excluding the rigid motion of fictitious blocks when there are holes within the system, may be introduced in a way similar to that used in § 30.

Form of CVH-BIE convenient for calculations. Equation (35.1) resembles equation for a non-periodic system of blocks with $\cot(\tau - t)$ instead of $1/(\tau - t)$, $1/\sin^2(\tau - t)$ instead of $1/(\tau - t)^2$, and $k_{\pi 1}(\tau, t)$, $k_{\pi 2}(\tau, t)$ instead of $k_1(\tau, t)$, $k_2(\tau, t)$. This similarity has deep roots and allows us to obtain a form of the CVH-BIE very convenient for calculations. Indeed, from the definitions of the kernels (27.1), (34.5), we have following expressions:

$$\frac{1}{\sin^2(\tau-t)}=\sum_{j=-\infty}^{j=\infty}\frac{1}{[(\tau-j\pi)-t]^2},$$

$$dk_{\pi 1}(\tau,t)=\sum_{j=-\infty}^{j=\infty}dk_1(\tau-j\pi,t),\quad dk_{\pi 2}(\tau,t)=\sum_{j=-\infty}^{j=\infty}dk_2(\tau-j\pi,t),$$

$$\frac{\partial k_{\pi 1}(\tau,t)}{\partial t}=\sum_{j=-\infty}^{j=\infty}\frac{\partial k_1(\tau-j\pi,t)}{\partial t},\quad \frac{\partial k_{\pi 2}(\tau,t)}{\partial t}=\sum_{j=-\infty}^{j=\infty}\frac{\partial k_2(\tau-j\pi,t)}{\partial t}.$$

We see that each kernel in (35.1) is the sum of the corresponding kernels of a non-periodic equation taken in congruent points of all the strips. Then (35.1) may be written as

$$\lim_{N\to\infty}\sum_{j=-N}^{j=N}\frac{1}{2\pi i}\int_L\Bigg[2\Delta u\frac{d\tau}{[(\tau-j\pi)-t]^2}-\Delta u\frac{\partial}{\partial t}dk_1(\tau-j\pi,t)-\Delta\bar{u}\frac{\partial}{\partial t}dk_2(\tau-j\pi,t)+$$

$$+\left(2a_1-a_3\right)\sigma\frac{d\tau}{(\tau-j\pi)-t}+\left(a_1-a_3\right)\sigma\frac{\partial k_1(\tau-j\pi,t)}{\partial t}d\tau+a_1\bar{\sigma}\frac{\partial k_2(\tau-j\pi,t)}{\partial t}d\bar{\tau}\Bigg]=$$

$$=\frac{1}{2}a_2\sigma(t)-\frac{\chi_\infty+1}{2\mu_\infty}\sigma_\infty(t),\qquad t\in L, \qquad (35.2)$$

The form (35.2) implies that (35.1) may be considered as a particular case of (30.17) corresponding to the infinite number of congruent contours. The displacement discontinuities and tractions are considered the same at congruent points. This form allows us easily to adjust a computer program developed for a non-periodic system of blocks to analogous periodic system. To this end, we may truncate the series on the l. h. s. of (35.2)[1]. Then, since $(\tau - j\pi) - t = \tau - (t + j\pi)$, $k_1(\tau - j\pi, t) = k_1(\tau, t + j\pi)$, $k_2(\tau - j\pi, t) = k_2(\tau, t + j\pi)$, it is sufficient to sum standard integrals of a non-periodic problem at $2N + 1$ field points $t + j\pi$ ($j = -N,...,0,...,N$). Note that the quadrature rules of Chapter 12 provide a means for simple and accurate evaluation of the standard integrals at any field point. Hence, the summing is readily implemented into a computer code. This transforms a program for a system of non-periodic blocks into a program for periodic systems of blocks.

Holomorphicity representations. The integral representations corresponding to (35.1) are

$$\frac{\chi_j+1}{2\mu_j}\varphi(z)=\frac{1}{2\pi i}\int_L(\Delta u+a_1 f)\cot(\tau-z)\,d\tau+\frac{\chi_\infty+1}{2\mu_\infty}A_\varphi z,\quad z\in D_j,$$

$$\frac{\chi_j+1}{2\mu_j}\psi(z)=\frac{1}{2\pi i}\int_L\{-[\overline{\Delta u}+(a_1-a_3)\overline{f}\,]\cot(\tau-z)\,d\tau+ \tag{35.3}$$

$$+(\Delta u+a_1 f)d\,[\overline{\tau}\cot(\tau-z)-(\tau-z)\cot(\tau-z)]\}+\frac{\chi_\infty+1}{2\mu_\infty}A_\psi z,\quad z\in D_j,$$

where A_φ, A_ψ, are given by (30.20). Differentiation of (35.3) with respect to z gives the K-M functions $\Phi(z)$, $\Psi(z)$, which by (25.6) define stresses within blocks. The resulting formula for the stresses coincides with (35.1) taken for $t = z$, $dt = dz$, where dz is taken in the direction of an element at which we find the stress; $a_2 = (\chi_j + 1)/\mu_j$ for $z \in D_j$. Being used in the boundary condition (25.16), the representations (35.3) naturally lead to (35.1). As usual, they may serve for using numerical procedures of coming to the limit *after* integration (cf. § 7, 8, 15, 30).

§ 36. EXAMPLE: ECHELONS OF CRACKS WITH GROWING WINGS

We noted that the form (35.2) of the hypersingular equation (35.1) allows us easily to adjust a code for non-periodic problems to periodic ones. Thus, the CV-BEM code, which served in § 32 to present the results for sliding cracks with growing wings, may be employed to consider periodic systems of such cracks.

In practice, cracks often tend to form echelons along some direction. By using the code we may examine the influence of the number of cracks in an echelon on their propagation. Compare the cases of one (Figure 31, *a*), three (Figure 31, *b*) and infinite number (Figure 31, *c*) of cracks. As in § 32, the initial sliding cracks of the length $2a$ comprise the angle $\pi/4$ with the loading direction. The distance between the initial cracks in an echelon is $2d$. The shear traction and the normal displacement discontinuity on their surfaces is zero ($\sigma_{n\tau} = 0$, $\Delta u_n = 0$). The wing

---

[1] We will comment on the number N sufficient to guarantee results with at least three correct digits in Chapter 13.

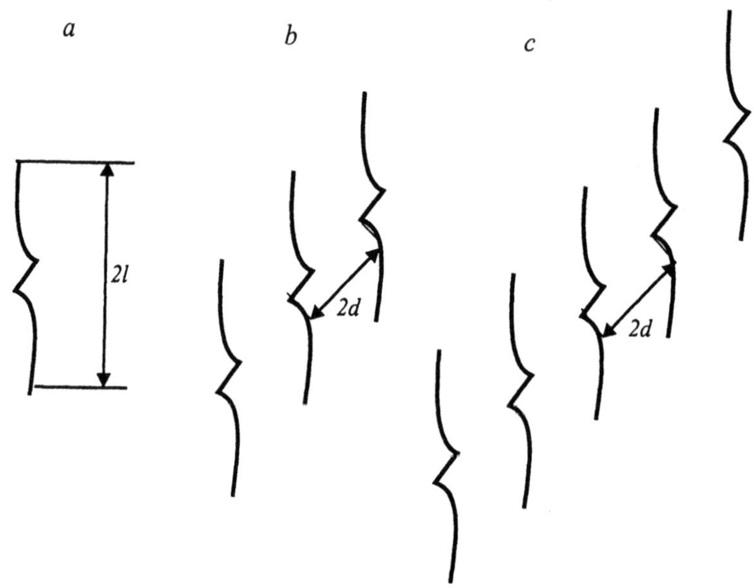

Fig. 31.

cracks, starting at their tips, propagate due to tensile stresses in accordance with the criterion $k_I = k_{IC}$ and the condition $k_{II} = 0$ (cf. (32.1), (32.2)).

Consider, for example, the case when the angle of the period is also $\pi/4$, the ratio $d/a = 2$ and the uniaxial pressure p acts along the vertical axis ($\sigma_{yy} = - p$). Calculations for the scheme Figure 31, *b* show that first the wings are initiated at the central crack.[1] Their growth decreases the SIFs k_I at their tips; this hampers the propagation, while the wings of the peripheral cracks start to grow. Thus, the process is as follows: the wings of all the cracks in an echelon grow with increasing compressive stress, the central crack is slightly ahead of the peripheral. (Similar picture was observed for an echelon of five cracks).

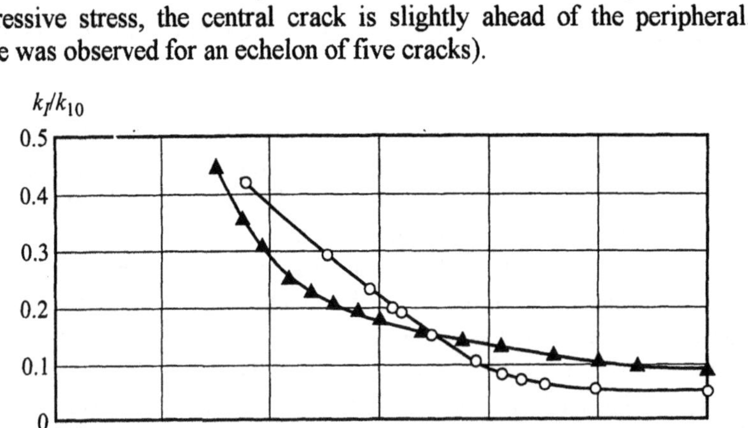

Fig. 32.

[1] Calculations were carried out by A. A. Dobroskok.

Figure 32 presents the dependence between the normalized vertical length l/a and the normalized SIF k_I/k_{I0}, where $k_{I0} = p\sqrt{\pi a}$, for the isolated crack (triangles) and for the infinite echelon of cracks (circles). We conclude that, although the results for two cases do not differ significantly, the wings in the infinite echelon start to propagate under lower external load, but when they become long enough, there appears shielding effect: the external load for their growth is greater than for an isolated crack. The calculations show also that the SIFs for the case Figure 31, *b* are intermediate between the two extreme cases *a* and *c* for any of the wings.

Chapter 8

DOUBLY PERIODIC PROBLEMS

We are interested in doubly periodic problems for two reasons. First, these problems arise in practice when dealing with perforated plates, regular composites and regular structures in solids and constructions. Secondly, as is especially important today, they arise from a general tendency of modern physics and applied science to account for details of the internal structure of a medium. A new branch of science, termed micromechanics, is swiftly growing up (see, e.g. Kemeny and Cook [1], Liu et al. [1], Napier and Pierce [1], Li and Wisnow [1], Tashkinov et al. [1], Dobroskok et al. [1]). Numerical simulation of doubly (in 2D) or triply (in 3D) periodic systems provides a unique opportunity to develop this science (Linkov and Koshelev [2]). It allows us to complement or substitute expensive and limited physical experiments by numerical ones. This approach has three advantages: (i) the cyclic constants of a displacement field provide *rigorous definition* of average (macroscopic) strains for prescribed average stresses, (ii) the ability to account for *at least two hierarchical levels* of structure, that of whole cells which may interact on their boundaries and that of internal elements of cells (grains, cracks, voids, etc.), (iii) calculations involve only a *restricted area* represented by the main cell. In 2D, the approach is strongly supported by the CV-BIE presented in this chapter.

§ 37. FORMULATION OF DOUBLY PERIODIC PROBLEMS

Consider an infinite plane with a doubly periodic system of cracks and/or holes (Figure 33). Afterwards, we will show how to extend the results to inclusions and blocky systems.

The complex vectors $2\omega_1$ and $2\omega_2$ prescribe the periods. Not being collinear they satisfy the inequality

$$\operatorname{Im}\left(\overline{\omega_1}\omega_2\right) \neq 0. \tag{37.1}$$

As in § 28, we assume the origin to be within the main parallelogram *ABCD*. The geometry of a doubly periodic problem is given by a contour L in this parallelogram, which we will also call the *main cell.* The contour L has p open arcs L_k, which may represent cracks or ribs ($k = 1,\ldots, p$), and m smooth closed contours L_j ($j = p + 1,\ldots, p + m$), corresponding to holes. Boundary conditions on congruent contours in other cells reproduce those on L. Hence, stresses and strains in such a problem are doubly periodic functions. Then K-M formulae (25.6), (25.7) imply that the function $\varphi'(z)$ is doubly periodic, that is $\varphi'(z + k\omega_j) = \varphi'(z)$ for any

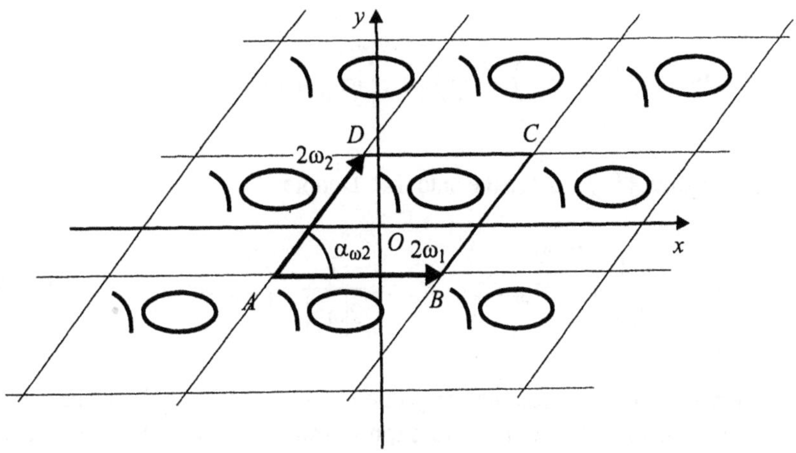

Fig. 33.

integer k (j = 1, 2). The function $\varphi(z)$ being an integral from $\varphi'(z)$ is only *quasi-periodic* in general:

$$\varphi(z + 2\omega_j) = \varphi(z) + 2\alpha_j, \tag{37.2}$$

where $2\alpha_j$ is a cyclic constant in the direction of the period $2\omega_j$ (j = 1, 2).

Not to digress into discussion of multi-valued functions, we assume that the total resultant force applied to each open and closed contour is zero:

$$\int_{L_k} \Delta\sigma \, d\tau = 0, \quad k = 1,\ldots, p + m. \tag{37.3}$$

Here, $\Delta\sigma = \sigma^+ - \sigma^-$ is the traction discontinuity; as usual, we assume that for holes $\sigma^+ = \sigma$, $\sigma^- = 0$. The assumption (37.3) does not influence equations which employ traction rather than the resultant force. However, in a doubly periodic problem, the sum of forces applied to the entire contour L should be equal to zero (self-equilibrated).

Resultant force. From the periodicity of stresses it follows that the resultant force is a quasi-periodic function

$$f(z + 2\omega_j) = f(z) + 2\gamma_j, \tag{37.4}$$

where $2\gamma_j$ is a cyclic constant in the direction of the period $2\omega_j$ (j = 1, 2).

We consider the physical meaning of the cyclic constants $2\gamma_j$. Consider the increment of the resultant force along the sides AB and AD of the main parallelogram $ABCD$. On these sides we have:

$$2\gamma_j = f(z+2\omega_j) - f(z) = i\int_z^{z+2\omega_j}(\sigma_{nx}+i\sigma_{ny})ds, \qquad (37.5)$$

where $j = 1$ for the side AB, $j = 2$ for the side AD. Denote

$$s_{nxj} = \frac{1}{2|\omega_j|}\int_z^{z+2\omega_j}\sigma_{nx}\,ds, \quad s_{nyj} = \frac{1}{2|\omega_j|}\int_z^{z+2\omega_j}\sigma_{ny}\,ds \qquad (37.6)$$

the average (over a period $2\omega_j$) traction projections on the axis x and y respectively. By using the notation (37.6) in (37.5) we express the cyclic constants by average tractions over periods:

$$2\gamma_j = i(s_{nxj} + i\,s_{nyj})2|\omega_j|. \qquad (37.7)$$

Consider now a homogeneous plane with the same average stresses but without cracks and holes. The stress field, being uniform, is periodic for arbitrary periods $2\omega_1$ and $2\omega_2$. From known Muskhelishvili's formulae ([5], § 8) we obtain that traction acting on an area having an angle $\alpha_{\omega j}$ with the x-axis is

$$s_{nxj} + is_{nyj} = -\frac{i}{2}\left[(s_{xx}+s_{yy}) + (s_{yy}-s_{xx}-2is_{xy})\exp(-2i\alpha_{\omega_j})\right]\exp(i\alpha_{\omega_j}). \qquad (37.8)$$

Substitution of (37.8) into (37.7) and accounting for $|\omega_j|\exp(i\alpha_{\omega j}) = \omega_j$ yields

$$2\gamma_j = [(s_{xx}+s_{yy}) + (s_{yy}-s_{xx}-2is_{xy})\exp(-2i\alpha_{\omega_j})]\omega_j. \qquad (37.9)$$

We see that the cyclic constants of the resultant force for a plane with cracks, holes, ribs and inclusions are expressed via average stresses in an entire (without internal structure) plane. These stresses are equal to those that arise in the initial doubly periodic problem.

Note that the r. h. s. of (37.9) contains only three real constants s_{xx}, s_{yy}, s_{xy}, while the complex quantities $2\gamma_1$, $2\gamma_2$ contain in total four real constants. We will show now that the odd (forth) constant is fixed by the equilibrium of the total moment acting on the main parallelogram.

Total moment. We do not impose restrictions on moments applied to particular contours. Meanwhile, for their totality L we assume that the total moment is zero. In accordance with a known formula for a moment (see, Muskhelishvili [5, § 102]), this equality reads

$$M_L = \mathrm{Re}\int_L \Delta f\, d\bar{\tau} = 0,$$

where $\Delta f = f^+ - f^-$ is the discontinuity of the resultant force (for holes $f^+ = f, f^- = 0$).

Therefore, the external moment acting on the boundary $ABCD$ of the main parallelogram (Figure 33) is also zero:

$$M_{ext} = \mathrm{Re}\int_{ABCD} f\, d\bar{\tau} = 0.$$

From this equation and (37.4) we have:

$$\mathrm{Re}\left(\gamma_1\overline{\omega_2} - \gamma_2\overline{\omega_1}\right) = 0. \tag{37.10}$$

It is easy to verify that equation (37.10) is satisfied by the expressions

$$2\gamma_j = a\omega_j + b\overline{\omega_j}, \tag{37.11}$$

where a is a real, b is a complex constant. Hence, the real condition (37.10) of the moment equilibrium leaves arbitrary only three of four real constants entering γ_1 and γ_2. By setting in (37.11) $a = s_{xx} + s_{yy}$, $b = s_{yy} - s_{xx} - is_{xy}$, we arrive at equation coinciding with (37.9). We see that equation (37.9) accounts for the equilibrium of the total moment and thus it contains only three real constants s_{xx}, s_{yy}, s_{xy}.

Displacements. Displacement $u(z)$ in a doubly periodic problem is a quasi-periodic function. Denote its cyclic constants $2\rho_j$ ($j = 1, 2$), so that

$$u(z + 2\omega_j) = u(z) + 2\rho_j. \tag{37.12}$$

Consider again an entire plane without internal structure. Suppose that there is a uniform field of strains ε_{xx}, ε_{yy}, ε_{xy} in it. This field is doubly periodic for arbitrary periods $2\omega_1$ and $2\omega_2$. A corresponding displacement field contains rigid rotation. Take this rotation to be zero. Then it is easy to verify that the corresponding displacement field is

$$u_0(z) = \frac{1}{2}[(\varepsilon_{xx} + \varepsilon_{yy})z + (\varepsilon_{xx} - \varepsilon_{yy} + 2i\varepsilon_{xy})\bar{z}]. \tag{37.13}$$

From (37.13) it follows that for the cyclic constants of this field $2\rho_{0j}$ along the periods $2\omega_j$ we have:

$$2\rho_{0j} = u_0(z + 2\omega_j) - u_0(z) =$$
$$= \frac{1}{2}[\varepsilon_{xx} + \varepsilon_{yy} + (\varepsilon_{xx} - \varepsilon_{yy} + 2i\varepsilon_{xy})\exp(-2i\alpha_{\omega_j})]2\omega_j.$$

These constants will be the same as those in the initial problem if we set

$$2\rho_j = 2\rho_{0j} = [\varepsilon_{xx} + \varepsilon_{yy} + (\varepsilon_{xx} - \varepsilon_{yy} + 2i\varepsilon_{xy})\exp(-2i\alpha_{\omega_j})]\omega_j. \quad (37.14)$$

We see that the cyclic constants of displacements are expressed via uniform (average) strains (in an entire homogeneous plane) that provide the same displacement increments along the periods as those in the initial doubly periodic problem.

Again note that the r. h. s. of (37.14) contains only three real constants ε_{xx}, ε_{yy}, ε_{xy}, while the complex cyclic constants $2\rho_1$, $2\rho_2$ include four real constants. The latter corresponds to rigid rotation without changes in a stressed state. Equation (37.14) fixes the rotation of the main cell: it is assumed to be zero. This implies zero rotation of the entire plane assumed in (37.13).

These physical meanings of the cyclic constants of the resultant force $2\gamma_1$, $2\gamma_2$ and of the displacements $2\rho_1$, $2\rho_2$ have far-reaching consequences for studying media with internal structure. They show that if, after solving a doubly periodic problem, we obtain a dependence between these constants, we actually find a dependence between average stresses and average strains in an equivalent entire (without structure) plane. For years, this has been used to give effective properties of perforated plates and regular composites (see, e.g. [Filshtinski [1], Grigoliuk and Filshtinski [1, 2]). Presently, with ever growing power of computers, this enables numerical simulation of macroscopic (effective) properties and evolutionary constitutive equations of a medium with a number of structural levels. The latter may include joints, cracks, voids, blocks, grains and inclusions interacting on their boundaries and changing their geometry in loading steps (Linkov and Koshelev [1, 2], Dobroskok [2], Dobroskok et al. [1]). Equations (37.9) and (37.14) are key formulae for obtaining the needed dependence.

Function **$\varphi(z)$.** The function $\varphi(z)$ being quasi-periodic and piece-wise holomorphic, in accordance with the holomorphicity theorem is represented by (28.10):

$$\varphi(z) = \frac{1}{2\pi i}\int_L \Delta\varphi(\tau)[\zeta(\tau - z) - \zeta(\tau)]d\tau - \frac{2}{\pi i}(\alpha_1\eta_2 - \alpha_2\eta_1)z, \quad (37.15)$$

where $\zeta(z)$ is Weierstrass' zeta-function for the periods $2\omega_1$, $2\omega_2$; $2\eta_j$ are its cyclic constants; $2\alpha_j$ are the cyclic constants of the function $\varphi(z)$ ($j = 1, 2$); $\Delta\varphi = \varphi^+ - \varphi^-$; as before, for holes we assume $\varphi^+ = \varphi$, $\varphi^- = 0$. Besides, in accordance with (28.9),

$$\frac{1}{2\pi i}\int_L \Delta\varphi d\tau = -\frac{2}{\pi i}\left(\alpha_1\omega_2 - \alpha_2\omega_1\right). \tag{37.16}$$

Take into account that $\varphi(z)$ is connected with the resultant force and displacements by (25.8). The latter implies that the cyclic constants $2\alpha_j$ of $\varphi(z)$ are connected with the cyclic constants of the resultant force $2\gamma_j$ and the cyclic constants of the displacements $2\rho_j$ by analogous equations:

$$2\alpha_j = \frac{2\mu 2\rho_j + 2\gamma_j}{\chi + 1}. \tag{37.17}$$

From (37.17) we see that for prescribed average stresses, when (37.9) defines the constants $2\gamma_j$, the cyclic constants $2\alpha_j$ define the cyclic constants $2\rho_j$ and by (37.14) average strains. Thus our aim is to find both the function $\varphi(z)$ and its cyclic constants $2\alpha_j$.

Function $\psi(z)$**.** Unlike to the function $\varphi(z)$, the function $\psi(z)$ is not quasi-periodic. Indeed, from (25.1) and the quasi-periodicity of $\varphi(z)$ and $f(z)$ it follows that

$$\psi\left(z + 2\omega_j\right) = \psi(z) - 2\overline{\omega_j}\varphi'(z) + \overline{2\gamma_j} - \overline{2\alpha_j}. \tag{37.18}$$

From (37.18) we see that the term $2\overline{\omega_j}\varphi'(z)$ makes this function not quasi-periodic. To overcome this difficulty, just as in § 33, we will use a special function $Q(z)$, which has a similar property.

Function $Q(z)$**.** Such a function is Natanzon's function defined by (28.5). The function $Q(z)$ satisfies (28.6). Note that it does not have poles in the main cell; therefore, it is holomorphic in the main cell.

From the definition (28.5) and equation (28.6) it follows that we may easily construct a particular function $\psi_*(z)$ having the property (37.18) if $\varphi(z)$ is defined by (37.15). Indeed, define $\psi_*(z)$ as

$$\psi_*(z) = \frac{1}{2\pi i}\int_L \Delta\varphi Q(\tau - z)d\tau. \tag{37.19}$$

Then by employing (28.5), (28.6), (37.15) and (37.16), after obvious transformations, we obtain:

$$+\frac{4}{\pi i}\left[\lambda_j\left(\alpha_1\omega_2 - \alpha_2\omega_1\right) - \overline{\omega_j}\left(\alpha_1\eta_2 - \alpha_2\eta_1\right)\right], \tag{37.20}$$

where $2\lambda_j$ ($j = 1, 2$) are the known constants of Natanzon's function for the periods $2\omega_1$, $2\omega_2$ (see § 28). The function $\psi_*(z)$ is holomorphic in the main parallelogram. Therefore, it is continuous through the contour L: $\Delta\psi_* = \psi_*^+ - \psi_*^- = 0$.

General representation of the function $\psi(z)$**.** A general form of a piece-wise holomorphic function having the property (37.18), is given by the sum of a particular representation $\psi_*(z)$ and an arbitrary *quasi-periodic* piece-wise holomorphic function. Denote the latter $\Omega(z)$ and its cyclic constants $2\beta_j$ ($j = 1, 2$). Thus, the K-M function $\psi(z)$ is expressed via a quasi-periodic piece-wise holomorphic function $\Omega(z)$ as

$$\psi(z) = \Omega(z) + \psi_*(z) = \Omega(z) + \frac{1}{2\pi i}\int_L \Delta\varphi Q(\tau - z)d\tau. \qquad (37.21)$$

Substitution of (37.21) into (25.1) and elementary transformation yield

$$\Omega(z) = \overline{f(z)} - \overline{\varphi(z)} - \bar{z}\,\varphi'(z) - \psi_*(z), \qquad (37.22)$$

where the function $\psi_*(z)$ is defined by (37.19). By employing (37.22) and (37.20), it is easy to express the cyclic constants $2\beta_j$ of the function $\Omega(z)$ via the constants of the functions $\varphi(z)$, $\zeta(z)$ and $Q(z)$. As a result we obtain:

$$2\beta_j = \frac{4}{\pi i}\left[\overline{\omega_j(\alpha_1\eta_2 - \alpha_2\eta_1) - \lambda_j(\alpha_1\omega_2 - \alpha_2\omega_1)}\right] + 2\overline{\gamma_j} - 2\overline{\alpha_j}. \qquad (37.23)$$

From (37.22) for the limiting values $\Omega^{\pm}(t)$ of $\Omega(z)$ on the contour L we have:

$$\Omega^{\pm}(t) = \overline{f^{\pm}(t)} - \overline{\varphi^{\pm}(t)} - \bar{t}\,\varphi'^{\pm}(t) - \psi_*(z). \qquad (37.24)$$

Note that in (37.24) we have written the function $\psi_*(z)$ without superscripts "plus" and "minus", labeling the limits. We took into account that, being holomorphic in the main cell, this function is continuous through the contour L. Now to obtain CV-BIE for the considered doubly periodic problems we need only to apply the holomorphicity theorem to the functions $\varphi(z)$ and $\Omega(z)$.

§ 38. COMPLEX VARIABLE BIE FOR DOUBLY PERIODIC PROBLEMS

Use (37.22) in terms of the resultant force to derive a CV-BIE. To this end, account for (37.19), and write (37.22) as

$$\Omega^{\pm}(t)=\overline{f^{\pm}(t)}-\overline{\varphi^{\pm}(t)}-\bar{t}\,\varphi'^{\pm}(t)-\frac{1}{2\pi i}\int_L \Delta\varphi Q(\tau-t)d\tau, \tag{38.1}$$

where

$$f^{\pm}(t)=f_c^{\pm}(t)+C^{\pm}(t),\quad f^{\pm}(t)=\int_{a_j}^{t}\sigma^{\pm}d\tau,$$

a_j is the start point on a contour L_j; for an open arc, it coincides with its beginning (in the direction of traveling); for a hole it is taken arbitrarily; $C^{\pm}(t) = C^{\pm}_j$ are constants on particular contours; for an open arc L_j we assume $C^+_j = C^-_j = C_j$, for a hole $C^+_j = C_j$, $C^-_j = 0$; thus, on any particular contour L_j we have only one constant C_j. The constant C_j is an unknown value of the resultant force at the start point of the j-th contour ($j = 1,\ldots, p + m$); for holes, as usual, we assume $f_c^+ = f_c$, $f_c^- = 0$.

Further derivation is quite similar to that in § 29, 34. Now we apply the holomorphicity theorem (28.8), (28.9) to $\Omega(z)$ and $\varphi(z)$. As a result we arrive at the CV-BIE and additional equations defining cyclic constants (Linkov [5]):

$$\frac{1}{2\pi i}\int_L \Big\{2\Delta\varphi[\zeta(\tau-t)-\zeta(\tau)]d\tau-\Delta\varphi\Big[(\zeta(\tau-t)-\zeta(\tau))d\tau-\big(\overline{\zeta(\tau-t)}-\overline{\zeta(\tau)}\big)d\bar{\tau}\Big]-$$
$$-\overline{\Delta\varphi}\Big[d\left((\tau-t)\overline{\zeta(\tau-t)}-\tau\overline{\zeta(\tau)}\right)+\overline{Q(\tau-t)}d\bar{\tau}\Big]\Big\}-\left(c_1 t+\bar{c}_2\bar{t}\right)=$$
$$=\frac{1}{2}\left(f^{+}+f^{-}\right)+\frac{1}{2\pi i}\int_L \Delta f\left[\overline{\zeta(\tau-t)}-\overline{\zeta(\tau)}\right]d\bar{\tau}, \tag{38.2}$$

$$\alpha_2\omega_1-\alpha_1\omega_2=\frac{1}{4}\int_L \Delta\varphi d\tau, \tag{38.3}$$

$$2i\,\mathrm{Im}\left(\alpha_2\overline{\omega_1}-\alpha_1\overline{\omega_2}\right)=\frac{1}{4}\int_L \overline{\Delta f}d\tau+\left(\overline{\gamma_1}\omega_2-\overline{\gamma_2}\omega_1\right)+\frac{1}{2}i\,\mathrm{Im}\int_L \Delta\varphi d\bar{\tau}. \tag{38.4}$$

In (38.2) we denoted

$$c_1=\frac{4}{\pi}\mathrm{Im}\left(\alpha_1\eta_2-\alpha_2\eta_1\right),\quad c_2=\frac{2}{\pi i}\left(\beta_1\eta_2-\beta_2\eta_1\right).$$

Recall that the constants $2\beta_1$, $2\beta_2$ are expressed via $2\alpha_1$, $2\alpha_2$ by (37.23).

We emphasize that since the l. h. s. of (38.4) is purely imaginary, this equation is solvable only when the real part of the first two terms on the r. h. s. is zero:

$$\operatorname{Re}\left[\int_L \Delta f\, d\bar{\tau} + 4\left(\gamma_1 \overline{\omega_2} - \gamma_2 \overline{\omega_1}\right)\right] = 0.$$

This condition, as it follows from the previous section, means that the total moment is zero. We have assumed that external loads satisfy this restriction. Then the three real equations (38.3), (38.4) for the four real constants entering α_1, α_2 are solvable. A solution of the corresponding homogeneous system is $\alpha_1 = ia\omega_1$, $\alpha_2 = ia\omega_2$, where a is an arbitrary real constant. The latter influences only rigid rotation and may be fixed arbitrarily. We choose it to exclude rigid rotation. In this case, we substitute (37.14) into (37.17), which gives

$$\alpha_j = \frac{\mu}{\chi+1}\left[\left(\varepsilon_{xx} + \varepsilon_{yy}\right)\omega_j + \left(\varepsilon_{xx} - \varepsilon_{yy} + 2i\varepsilon_{xy}\right)\overline{\omega}_j\right] + \frac{\gamma_i}{\chi+1}, \quad j = 1,2. \qquad (38.5)$$

Now insertion of (38.5) into the additional conditions (38.3), (38.4) leads to a solvable system with respect to the *three real* average values ε_{xx}, ε_{yy}, ε_{xy}. Taking into account (37.9) we may use real average stresses s_{xx}, s_{yy} and s_{xy}, instead of the complex cycling constants γ_1, γ_2.

We conclude that equations (38.3), (38.4) give average values ε_{xx}, ε_{yy}, ε_{xy} (or s_{xx}, s_{yy}, s_{xy}) under prescribed average values s_{xx}, s_{yy}, s_{xy} (or ε_{xx}, ε_{yy}, ε_{xy}). We will write down this important system explicitly in § 40, when discussing effective properties of a medium with doubly periodic structure. Naturally, equations (38.5) may be employed also to express constants α_1, α_2 in the CV-BIE (38.2) itself.

It is possible to avoid solving equation (38.2) together with equations (38.3), (38.4). To do this, exclude the cycling constants $2\alpha_1$, $2\alpha_2$, $2\beta_1$, $2\beta_2$ entering c_1, c_2 by using Legendre's identity (28.4), condition (37.16) including α_j and analogous condition including β_j, and (37.9). After lengthy, although simple transformations we obtain (Linkov and Koshelev [1, 2]):

$$c_1 = -\frac{1}{\pi}\operatorname{Im}\left(\frac{\eta_1}{\omega_1}\int_L \Delta\varphi\, d\tau\right) - a - 2A_\varphi,$$

$$c_2 = -\frac{1}{\pi}\frac{\eta_1}{\omega_1}\left(\operatorname{Im}\int_L \Delta\varphi\, d\bar{\tau} - \frac{i}{2}\int_L \overline{\Delta f}\, d\tau\right) + \frac{\overline{\omega}_1}{\omega_1}a + \frac{1}{2\pi i}\frac{\overline{\omega}_1}{\omega_1}\left(\frac{\eta_1}{\omega_1} - \frac{\lambda_1}{\overline{\omega}_1}\right)\int_L \Delta\varphi\, d\tau - A_\psi,$$

where A_φ, A_ψ are constants depending on the average stresses, which now play the role of the stresses at infinity; as usual, they are defined by (30.20); a and S are real quantities:

$$a = \frac{1}{S}\operatorname{Im}\left[\int_L \left(\Delta\varphi - \frac{1}{2}\Delta f\right)d\bar{\tau} - \left(\frac{\bar{\omega}_1}{|\omega_1|}\right)^2 \int_L \Delta\varphi d\tau\right],$$

$$S = 4\operatorname{Im}(\bar{\omega}_1\omega_2). \qquad (38.6)$$

Henceforth, we assume that the period $2\omega_2$ is turned *counter clockwise* with respect to the period $2\omega_1$ by an angle less than π. This gives a positive S: it represents the area of the parallelogram of periods. Recall that $2\eta_1$ is the cyclic constant of Weierstrass' function in the direction of the period $2\omega_1$; $2\lambda_1$ is Natanzon's constant for the same direction; these constants are given by the formulae in § 28. Without loss of generality, we may consider the period $2\omega_1$ to be real and positive; it is sufficient to choose the x-axis in the direction of $2\omega_1$. Then $\omega_1 = \bar{\omega}_1 = |\omega_1|$ and the formulae for c_1 and c_2 become slightly simpler.

In a case, when L is comprised only of closed contours, we have $\varphi^+ = \varphi$, $\varphi^- = 0$ on the whole contour and therefore $\Delta\varphi = \varphi$ on L. Then employing the holomorphicity formula (28.8) we may exclude the singular integral in (38.2). The equation becomes a Fredholm equation. Koiter [2] was the first to obtain this form for a doubly periodic system of *holes*. For a doubly periodic system of *arbitrary cracks*, it was derived in the paper to which we followed in this paragraph (Linkov [5]). The paper contains also analysis of the equation. It is shown that a unique solution exists when external loads belong to a physically significant Muskhelishvili's class H^*. The proof reproduces the classical scheme (Muskhelishvili [4]).

The usual substitution (25.8), which expresses $\Delta\varphi(t)$ via physical values, leads to the equation analogous to (29.8), (34.6) and to alternative equations resulting after differentiation with respect to t (Linkov and Koshelev [2]). We will write down only the hypersingular equation:

$$\frac{1}{2\pi i}\int_L \Big\{2\Delta u[-\varsigma'(\tau - t)]d\tau - \Delta u\frac{\partial}{\partial t}k_{d1}(\tau,t) - \overline{\Delta u}\,\frac{\partial}{\partial t}dk_{d2}(\tau,t) -$$

$$-(\chi-1)\frac{\Delta\sigma}{2\mu}\varsigma(\tau - t)d\tau - \chi\frac{\Delta\sigma}{2\mu}\frac{\partial}{\partial t}k_{d1}(\tau,t) + \frac{\overline{\Delta\sigma}}{2\mu}\frac{\partial}{\partial t}dk_{d2}(\tau,t)d\bar{\tau}\Big\} -$$

$$-\frac{\chi+1}{2\mu}\left(c_1 + \frac{\partial\bar{t}}{\partial t}\bar{c}_2\right) = \frac{\chi+1}{4\mu}\left(\sigma^+ + \sigma^-\right), \qquad (38.7)$$

where

$$\frac{\partial}{\partial t}dk_{d1}(\tau,t) = -\varsigma'(\tau - t) + \frac{\partial\bar{t}}{\partial t}\overline{\varsigma'(\tau - t)d\tau},$$

$$\frac{\partial}{\partial t}k_{d1}(\tau,t) = -\varsigma(\tau - t) + \frac{\partial\bar{t}}{\partial t}\overline{\varsigma(\tau - t)},$$

$$\frac{\partial}{\partial t}dk_{d2}(\tau,t)=\frac{\partial}{\partial t}d\left[(\tau-t)\overline{\varsigma(\tau-t)}\right]-\frac{\partial\bar{t}}{\partial t}d\overline{Q(\tau-t)},\qquad(38.8)$$
$$\frac{\partial}{\partial t}k_{d2}(\tau,t)=\frac{\partial}{\partial t}\left[(\tau-t)\overline{\varsigma(\tau-t)}\right]-\frac{\partial\bar{t}}{\partial t}\overline{Q(\tau-t)}.$$

The hypersingular equation does not need additional conditions. We solve it in a class of functions providing zero displacement discontinuity at the tips of open arcs.

Integral representations for doubly periodic problems. Combination of (37.15), (37.21) and the holomorphicity formula for $\Omega(z)$

$$\Omega(z)=\frac{1}{2\pi i}\int_L\left(\overline{\Delta f}-\overline{\Delta\varphi}-\bar{\tau}\Delta\varphi'\right)\left[\zeta(\tau-z)-\zeta(\tau)\right]d\tau-\frac{2}{\pi i}(\beta_1\eta_2-\beta_2\eta_1)z$$

provides holomorphicity representations for doubly periodic problems analogous to the representations of § 29. Extensions for the potentials $\Phi(z)=\varphi'(z)$, $\Psi(z)=\psi'(z)$ are clear: we may obtain them by direct differentiation of the integral representations of $\varphi(z)$, $\psi(z)$ accompanied by integration by parts to have the density $\Delta\varphi'(t)$.

§ 39. BIE FOR DOUBLY PERIODIC SYSTEMS OF BLOCKS

We see that the integrals from the displacement discontinuities in (38.7) do not contain elastic constants. This property was shared by their analogues derived in § 29 for non-periodic, and in § 34 for periodic systems of holes and cracks. This allows us to extend the equations to doubly periodic systems of *blocks*, which may contain cracks, holes and inclusions, as in § 30 for non-periodic systems of blocks. We will present only the most convenient hypersingular equation.

Hypersingular equation. Consider a doubly periodic system of n blocks. They, as well as the matrix embedding them, may contain cracks, holes and inclusions of a lower rank. Omitting lengthy derivation, we write down the final result (Linkov and Koshelev [1, 2]):

$$\frac{1}{2\pi i}\int_L\Big\{2\Delta u\left[-\zeta'(\tau-t)\right]d\tau-\Delta u\frac{\partial}{\partial t}dk_{d1}(\tau,t)-\overline{\Delta u}\frac{\partial}{\partial t}dk_{d2}(\tau,t)-$$
$$+(2a_1-a_3)\sigma\zeta(\tau-t)d\tau+(a_1-a_3)\sigma\frac{\partial}{\partial t}k_{d1}(\tau,t)d\tau+a_1\bar{\sigma}\frac{\partial}{\partial t}k_{d2}(\tau,t)d\bar{\tau}\Big\}-$$
$$-\left(2\operatorname{Re}A_\alpha+\frac{\partial\bar{t}}{\partial t}\overline{A_\beta}\right)+\frac{\chi_0+1}{2\mu_0}\sigma^\infty(t)=\frac{1}{2}a_2\sigma(t),\qquad(39.1)$$

where the contour L in the main cell is the boundary of blocks, inclusions, cracks and holes; the kernels are defined by (38.8); the coefficients a_1, a_2, a_3, characterizing properties of regions interacting along L, are given, as formerly, by equations (30.12); χ_0, μ_0 are elastic constants of the matrix enclosing blocks and inclusions; the contours of holes are traveled counter-clockwise; for them we may fix rigid rotations, as discussed in § 30;

$$A_\alpha = -\frac{1}{2\pi i}\frac{\eta_1}{\omega_1}\int_L (\Delta u - a_1\sigma\tau d\tau) -$$

$$-\frac{1}{2S}\{\mathrm{Im}\int_L \left[\Delta u d\bar{\tau} + \frac{1}{2}(a_3 - 2a_1)\sigma\bar{\tau}d\tau\right] + i\frac{\bar{\omega}_1}{\omega_1}\int_L (\Delta u - a_1\sigma\tau)d\tau\}, \qquad (39.2)$$

$$A_\beta = -\frac{1}{\pi}\frac{\eta_1}{\omega_1}[\mathrm{Im}\int_L (\Delta u d\bar{\tau} - a_1\sigma\bar{\tau}d\tau) + \frac{i}{2}\int_L (a_3\bar{\sigma}\tau d\bar{\tau})] + \frac{1}{2\pi i}\frac{\bar{\omega}_1}{\omega_1}\left(\frac{\eta_1}{\omega_1} - \frac{\lambda_1}{\bar{\omega}_1}\right)\int_L (\Delta u - a_1\sigma\tau)d\tau +$$

$$+\frac{\bar{\omega}_1}{\omega_1}\frac{1}{S}\mathrm{Im}\{\int_L \left[\Delta u d\bar{\tau} + \frac{1}{2}(a_3 - 2a_1)\sigma\bar{\tau}d\tau\right] - \left(\frac{\bar{\omega}_1}{\omega_1}\right)^2 \int_L (\Delta u - a_1\sigma\tau)d\tau\}, \qquad (39.3)$$

as usual, the traction $\sigma^\infty(t)$ is defined by (30.19), (30.20):

$$\sigma^\infty(t) = \frac{1}{2}[s_{xx} + s_{yy} + \frac{\partial\bar{t}}{\partial t}(s_{yy} - s_{xx} - 2is_{xy})];$$

it arises in an entire plane (without cracks, inclusions, etc.) under uniform stresses s_{xx}, s_{yy}, s_{xy}.

One can also easily derive singular equations, connected with (39.1) just as the singular equations (30.15), (30.16) are connected with the hypersingular equation (30.17). We will not write down these quite obvious but lengthy formulae given in Linkov and Koshelev [1].

Form of CVH-BIE convenient for calculations. We will represent equation (39.1) in a form practically coinciding with equation (30.17) for non-periodic problems. This form serves for using computer programs, developed for a *finite* system of blocks, inclusions, cracks and holes, to solve problems for *doubly periodic* systems.

On whole, equation (39.1) resembles the hypersingular equation (30.17) for a non-periodic system of blocks. Now the kernels contain Weierstrass' zeta-function $\zeta(z)$ and Natanzon's function $Q(z)$. These functions are defined by the series (28.2), (28.5) respectively. In practical calculations, it is sufficient to keep a finite number of terms because the series absolutely converge (their terms decrease as $1/w^3$). We take a reasonable number of terms to provide the needed accuracy[1]. For definiteness, assume that the integers j_1 and j_2 in $w=j_1 \cdot 2\omega_1 + j_2 \cdot 2\omega_2$ run through the

[1] We will comment on the number N sufficient to guarantee at least three correct digits in Chapter 13.

sequences $-N_1 \le j_1 \le N_1$, $-N_2 \le j_2 \le N_2$. In other words, account for $(2N_1+1)(2N_2+1)$ cells, including the main cell. Then we have

$$\varsigma(z) \approx \sum_{\substack{j_1=-N_1\\ j_2=-N_2}}^{N_1,N_2} \frac{1}{z-w} + \sum_{\substack{j_1=-N_1\\ j_2=-N_2}}^{N_1,N_2}{}' \frac{1}{w} + z \sum_{\substack{j_1=-N_1\\ j_2=-N_2}}^{N_1,N_2}{}' \frac{1}{w^2}$$

$$Q(z) \approx \sum_{\substack{j_1=-N_1\\ j_2=-N_2}}^{N_1,N_2}{}' \frac{\overline{w}}{(z-w)^2} - \sum_{\substack{j_1=-N_1\\ j_2=-N_2}}^{N_1,N_2}{}' \frac{\overline{w}}{w^2} - 2z \sum_{\substack{j_1=-N_1\\ j_2=-N_2}}^{N_1,N_2}{}' \frac{\overline{w}}{w^3}. \tag{39.4}$$

Each sum on the r. h. s. of (39.4), taken by itself, does not converge when N_1, $N_2 \to \infty$. Meanwhile, for the *finite* sums taken to represent $\zeta(z)$ and $Q(z)$ we still may use the decompositions (39.4). They allow us to obtain forms of CV-BIE, convenient in numerical implementation. Indeed, by differentiating the first of (39.4) and by using (39.4) in (38.8) we obtain after transformations:

$$-\varsigma'(\tau-t)d\tau \approx \sum_{\substack{j_1=-N_1\\ j_2=-N_2}}^{N_1,N_2} \frac{d\tau}{(\tau-w-t)^2} - b\,d\tau, \tag{39.5}$$

$$\frac{\partial}{\partial t} dk_{d1}(\tau,t) \approx \sum_{\substack{j_1=-N_1\\ j_2=-N_2}}^{N_1,N_2} \frac{\partial}{\partial t} dk_1(\tau-w,t) - b\,d\tau + \frac{\partial \bar{t}}{\partial t} \bar{b}\, d\bar{\tau}, \tag{39.6}$$

$$\frac{\partial}{\partial t} dk_{d2}(\tau,t) \approx \sum_{\substack{j_1=-N_1\\ j_2=-N_2}}^{N_1,N_2} \frac{\partial}{\partial t} dk_2(\tau-w,t) - \bar{b}\, d\bar{\tau} - \frac{\partial \bar{t}}{\partial t}(\bar{b}\, d\tau - 2\bar{b}_Q\, d\bar{\tau}), \tag{39.7}$$

where $k_1(\tau,t)$, $k_2(\tau,t)$ are kernels common in problems for non-periodic problems; they are defined by (12.18);

$$b = \sum_{\substack{j_1=-N_1\\ j_2=-N_2}}^{N_1,N_2}{}' \frac{1}{w^2}, \qquad b_Q = \sum_{\substack{j_1=-N_1\\ j_2=-N_2}}^{N_1,N_2}{}' \frac{\overline{w}}{w^3}. \tag{39.8}$$

In contrast to (39.8), the sums in (39.5) - (39.7) include w=0; for this reason, they are not labeled with a prime.

Equations (39.5) - (39.8) are of direct use for the terms in (39.1) containing the displacement discontinuities Δu. For the integrals containing tractions, their application is not so immediate. One may first use integration by parts to replace the traction to the resultant force; this transforms the kernels to forms similar to those for displacement discontinuities. After this, expressions (39.5) - (39.8) are inserted.

Finally, we again use integration by parts and return to the traction. As a result, equation (39.1) becomes

$$\frac{1}{2\pi i}\sum_{\substack{j_1=-N_1\\ j_2=-N_2}}^{N_1,N_2}\int_L\left\{2\Delta u\frac{d\tau}{(\tau-w-t)^2}-\Delta u\frac{\partial}{\partial t}dk_1(\tau-w,t)-\Delta\bar{u}\frac{\partial}{\partial t}dk_2(\tau-w,t)+\right.$$
$$\left.+(2a_1-a_3)\sigma\frac{d\tau}{\tau-w-t}+(a_1-a_3)\sigma\frac{\partial k_1}{\partial t}d\tau+a_1\bar{\sigma}\frac{\partial k_2}{\partial t}d\bar{\tau}\right\}- \tag{39.9}$$
$$-\left[\mathrm{Re}(A_\alpha+A)+\frac{\partial\bar{t}}{\partial t}(\bar{A}_\beta+\bar{B})\right]+\frac{\chi_0+1}{2\mu_0}\sigma^\infty(t)=\frac{1}{2}a_2\sigma(t),$$

where A and B are new constants:

$$A=\frac{1}{2\pi i}b(I_{u1}-I_{\sigma1}),\; B=\frac{b}{\pi}\mathrm{Im}(I_{u2}-I_{\sigma2})-\frac{b}{2\pi i}\bar{I}_{\sigma3}-\frac{b_Q}{\pi i}(I_{u1}-I_{\sigma1}),$$
$$I_{u1}=\int_L\Delta u d\tau,\; I_{u2}=\int_L\Delta u d\bar{\tau},\; I_{\sigma1}=\int_L a_1\,\sigma\tau\, d\tau,\; I_{\sigma2}=\int_L a_1\,\sigma\bar{\tau}\, d\tau,\; I_{\sigma3}=\int_L a_3\,\sigma\bar{\tau}\, d\tau,$$

the integrals I_{u1}, ..., $I_{\sigma3}$ are also present in A_α and A_β. We have used the equality sign in (39.9) assuming that the numbers N_1, N_2 are great enough to provide the necessary accuracy.

The CVH-BIE (39.9) resembles the common CVH-BIE (30.17) for a *non-periodic* problem involving $(2N_1+1)(2N_2+1)$ cells with two reservations:

(i) we consider Δu and σ to be repeated at congruent points;

(ii) (39.9) contains the additional term $-\left[2\mathrm{Re}(A_\alpha+A)+(\partial\bar{t}/\partial t)(\bar{A}_\beta+\bar{B})\right]$ which being multiplied by $2\mu_0/(\chi_0+1)$ may be interpreted as generated by some additional stresses at infinity.

The first feature allows us to solve equations only for the contour L in the main cell. The second presents an essential feature of doubly periodic problems. It is accounted for either directly in the influence matrix or by using successive steps starting from zero values of the constants A_α, A_β, A, B. Hence, one can easily adjust a CV-BEM program for a finite system of cracks, blocks, holes and/or inclusions in the main cell to solve problems for doubly periodic systems.

Holomorphicity representations. A solution of (39.1) or (39.9) makes known both the displacement discontinuity and the traction on L. Having these values, we immediately find stresses within blocks, inclusions and the embedding matrix by employing K-M formula (25.6). The functions $\Phi(z)$, $\Psi(z)$ entering (25.6) are:

$$\frac{\chi_j+1}{2\mu_j}\Phi(z)=\frac{1}{2\pi i}\int_L\{2\Delta u[-\zeta'(\tau-z)]+a_1\sigma\zeta(\tau-z)\}d\tau-A_\alpha+\frac{\chi_0+1}{2\mu_0}A_\varphi,$$

$$\frac{\chi_j+1}{2\mu_j}\Psi(z)=\frac{1}{2\pi i}\int_L\left\{-\overline{\Delta u}\left[-\zeta'(\tau-z)\right]d\tau+(a_3-a_1)\overline{\sigma}\zeta(\tau-z)d\overline{\tau}+\right.$$

$$\left.+\Delta u\left[-d\left(\overline{\tau}\zeta'(\tau-z)\right)+\frac{\partial}{\partial z}Q(\tau-z)d\tau\right]+a_1\sigma\overline{\tau}\zeta'(\tau-z)d\tau+a_1\sigma Q(\tau-z)d\tau\right\}-$$

$$-A_\beta+\frac{\chi_0+1}{2\mu_0}A_\psi, \tag{39.10}$$

where the point z is within the j-th block, inclusion (j = 1,..., n) or the matrix (j = 0); the constants A_α, A_β are defined by (39.2), (39.3); A_φ, A_ψ by (30.20).

The stresses corresponding to (39.10) may be also found directly from (39.1) or (39.9) if we apply $t = z$, $dt = dz$, $a_2 = (\chi_j + 1)/\mu_j$ when z is within the j-th block (j = 0,..., n). In this case, the direction dz is taken parallel to an element on which we calculate stresses; the normal to it points to the right of dz (cf. § 11). Being used in the boundary condition (25.16), the integral representations (39.10) lead to (39.1). As usual, this may serve for the numerical procedures of coming to the limit *after* integration (cf. § 7, 8, 15, 30).

§ 40. HOMOGENIZATION PROBLEM. CALCULATION OF EFFECTIVE COMPLIANCE

In applied problems we want to predict the macroscopic behavior of a medium by accounting for its microscopic features. Thus we need to solve a *homogenization* problem. This is a problem in which we replace an inhomogeneous medium having pores, cracks, joints, grains (blocks) and/or inclusions by a macroscopically equivalent homogeneous medium without internal structure. The equivalence means that average strains (stresses) are same in the real and homogeneous media under the same average stresses (strains). For irreversible processes on contacts and for growing cracks, the stresses and strains (both local and average) should be replaced by their *increments*. We will tacitly assume this throughout.

Equations connecting macroscopic values. From (37.9) we see that the average stresses s_{xx}, s_{yy}, s_{xy} define the cyclic constants of the resultant force, while by (37.14), the average strains ε_{xx}, ε_{yy}, ε_{xy} define the cyclic constants of displacements. Hence, the conditions (38.3), (38.4), connecting these values, may be used to find effective compliance. Use them first for a homogenous plane with cracks and holes.

Re-write (38.3), (38.4), taking into account that the real part of (38.4) is zero due to the moment equilibrium and that $\operatorname{Im} a = -\operatorname{Im}\overline{a}$:

$$\alpha_2\omega_1-\alpha_1\omega_2=\frac{1}{4}\frac{1}{\chi+1}\int_L(2\mu\Delta u+\Delta f)d\tau, \tag{40.1}$$

$$\operatorname{Im}\left(\alpha_2\overline{\omega_1}-\alpha_1\overline{\omega_2}\right)=\frac{1}{4}\frac{1}{\chi+1}\operatorname{Im}\int_L(2\mu\Delta u+\Delta f)d\overline{\tau}-\frac{1}{8}\operatorname{Im}\int_L\Delta f\,d\overline{\tau}-$$

$$-\frac{1}{2}\operatorname{Im}\left(\gamma_1\overline{\omega}_2-\gamma_2\overline{\omega}_1\right). \tag{40.2}$$

We have also used (25.8). Now use two identities:

$$\left(a\omega_2+b\overline{\omega}_2\right)\omega_1-\left(a\omega_1+b\overline{\omega}_1\right)\omega_2=b\left(\overline{\omega}_2\omega_1-\overline{\omega}_1\omega_2\right)=-\frac{-ibS}{2},$$
$$\left(a\omega_2+b\overline{\omega}_2\right)\overline{\omega}_1-\left(a\omega_1+b\overline{\omega}_1\right)\overline{\omega}_2=a\left(\overline{\omega}_1\omega_2-\overline{\omega}_2\omega_1\right)=\frac{iaS}{2}. \tag{40.3}$$

As before, S is the area of the main cell defined by (38.6). By (37.1), it is not zero.

Substitution of (38.5) into (40.1), (40.2) and employing (37.9), (40.3) yields the needed dependence between the average strains and stresses:

$$2\varepsilon_{xy}+i\left(\varepsilon_{yy}-\varepsilon_{xx}\right)=\frac{1}{S}\int_L\left(\Delta u+\frac{\Delta f}{2\mu}\right)d\tau+\frac{2s_{xy}+i\left(s_{yy}-s_{xx}\right)}{2\mu},$$
$$\varepsilon_{yy}+\varepsilon_{xx}=\frac{1}{S}\operatorname{Im}\int_L\left(\Delta u-\frac{\chi-1}{2}\frac{\Delta f}{2\mu}\right)d\overline{\tau}+\frac{\chi-1}{2}\frac{s_{yy}+s_{xx}}{2\mu}. \tag{40.4}$$

From (40.4) we have the strains:

$$\varepsilon_{xx}=\varepsilon_{xx}^0+\frac{1}{2S}\operatorname{Im}\int_L\left[\Delta u\left(d\overline{\tau}-d\tau\right)+\frac{\Delta\sigma}{2\mu}\left(\tau\,d\tau+\frac{\chi-1}{2}\overline{\tau}\,d\tau\right)\right],$$
$$\varepsilon_{yy}=\varepsilon_{yy}^0+\frac{1}{2S}\operatorname{Im}\int_L\left[\Delta u\left(d\tau+d\overline{\tau}\right)-\frac{\Delta\sigma}{2\mu}\left(\tau d\tau-\frac{\chi-1}{2}\overline{\tau}d\tau\right)\right], \tag{40.5}$$
$$\varepsilon_{xy}=\varepsilon_{xy}^0+\frac{1}{2S}\operatorname{Re}\int_L\left(\Delta u-\frac{\Delta\sigma}{2\mu}\tau\right)d\tau,$$

where ε_{xx}^0, ε_{yy}^0, ε_{xy}^0 are strains in a homogeneous (with the same elastic constants χ and μ) plane without cracks and holes under the average stresses s_{xx}, s_{yy}, s_{xy} (cf. (1.9)):

$$\varepsilon_{xx}^0=\frac{1}{8\mu}\left[(\chi+1)s_{xx}-(3-\chi)s_{yy}\right],\ \varepsilon_{yy}^0=\frac{1}{8\mu}\left[(\chi+1)s_{yy}-(3-\chi)s_{xx}\right],$$
$$\varepsilon_{xy}^0=\frac{1}{2\mu}s_{xy}. \tag{40.6}$$

Combination of (40.5) with any of the derived equations completely solves the homogenization problem.

The extension for a medium with a doubly periodic system of blocks (grains) and/or inclusions is given by formulae analogous to (40.5) (Koshelev and Linkov [1]):

$$\varepsilon_{xx} = \varepsilon_{xx}^0 + \frac{1}{2S}\operatorname{Im}\int_L \left\{ \Delta u \left(d\bar{\tau} - d\tau \right) + \sigma \left[a_1 \tau + \frac{1}{2}\left(a_3 - 2a_1 \right)\bar{\tau} \right] d\tau \right\},$$

$$\varepsilon_{yy} = \varepsilon_{yy}^0 + \frac{1}{2S}\operatorname{Im}\int_L \left\{ \Delta u \left(d\tau + d\bar{\tau} \right) - \sigma \left[a_1 \tau - \frac{1}{2}\left(a_3 - 2a_1 \right)\bar{\tau} \right] d\tau \right\}, \qquad (40.7)$$

$$\varepsilon_{xy} = \varepsilon_{xy}^0 + \frac{1}{2S}\operatorname{Re}\int_L \left(\Delta u - a_1 \tau \sigma \right) d\tau .$$

These equations together with the hypersingular equation (39.1) in the form (39.9) look the most attractive for solving problems for a medium with a number of structural levels.

Effective compliance. Often we are interested in the matrix of *effective compliance*. This is a matrix, which being multiplied by average stresses, provides average strains. For non-linear and/or irreversible processes, one uses the increments of the average values. Then the matrix connecting them is termed a *tangential* compliance matrix, while the matrix connecting the total values is termed a *secant* compliance matrix. The further discussion refers to both of these compliances but for certainty and to simplify notation we will use symbols of stresses and strains themselves.

We write the average stresses s_{xx}, s_{yy}, s_{xy} and average strains ε_{xx}, ε_{yy}, ε_{xy} and ε_{xx}^0, ε_{yy}^0, ε_{xy}^0 as column-vectors:

$$\mathbf{s} = (s_{xx}, s_{yy}, s_{xy})^{\mathrm{T}}, \quad \boldsymbol{\varepsilon} = (\varepsilon_{xx}, \varepsilon_{yy}, 2\varepsilon_{xy})^{\mathrm{T}}, \quad \boldsymbol{\varepsilon}^0 = (\varepsilon_{xx}^0, \varepsilon_{yy}^0, 2\varepsilon_{xy}^0)^{\mathrm{T}}.$$

The multiplier 2 in the last components of ε and ε^0 serves to account for symmetry of the stress tensor ($s_{xy} = s_{yx}$) when obtaining the compliance matrix **B** in accordance with its definition:

$$\boldsymbol{\varepsilon} = \mathbf{B}\,\mathbf{s}. \qquad (40.8)$$

From (40.8) it is clear that the j-th column of **B** corresponds to the strain vector induced by the unit j-th component of the vector **s**. Hence, it is sufficient to solve a BIE and to find unknown values of Δu or/and σ for the three sets of average stresses:

$$1)\ s_{xx} = 1,\ s_{yy} = 0,\ s_{xy} = 0;$$

$$2)\ s_{xx} = 0,\ s_{yy} = 1,\ s_{xy} = 0;$$

$$3)\ s_{xx}=0,\ s_{yy}=0,\ s_{xy}=1.$$

Substitution of the known Δu and σ for each set into (40.5) or (40.7) provides columns of the matrix **B**. By using the superscript 1 for the average strains for the first set, the superscript 2 for the second and 3 for the third, we can write the components b_{ij} $(i, j = 1, 2, 3)$ of the matrix **B** as

$$\begin{aligned} &b_{11}=\varepsilon_{xx}^{1}, \quad b_{12}=\varepsilon_{xx}^{2}, \quad b_{13}=\varepsilon_{xx}^{3},\\ &b_{21}=\varepsilon_{yy}^{1}, \quad b_{22}=\varepsilon_{yy}^{2}, \quad b_{23}=\varepsilon_{yy}^{3},\\ &b_{31}=2\varepsilon_{xy}^{1}, \quad b_{32}=2\varepsilon_{xy}^{2}, \quad b_{33}=2\varepsilon_{xy}^{3}. \end{aligned} \tag{40.9}$$

Naturally, in general, the equalities $b_{11} = b_{22}$, $b_{31} = b_{13} = b_{32} = b_{23} = 0$, characterizing isotropic medium, are not satisfied. Thus, except specific cases of symmetric cells and contours in them, the effective compliance corresponds to an anisotropic medium. For symmetric contact constitutive equations, describing processes on contacts, the matrix of effective compliance is symmetric (Dobroskok [2], Dobroskok et al. [1]). This allows us to check accuracy of calculations by comparing the components b_{kl} and b_{lk} for $k \neq l$.

As mentioned, the first terms on the r. h. s. of (40.5) and (40.7) correspond to Hooke's law (40.6) for a medium without cracks and holes. The corresponding compliance matrix $\mathbf{B}^0$, connecting the strains ε^0 and **s** as $\varepsilon^0 = \mathbf{B}^0\,\mathbf{s}$, has the components:

$$\begin{aligned} &b_{11}^0=\frac{\chi+1}{8\mu}, \quad b_{12}^0=-\frac{3-\chi}{8\mu}, \quad b_{13}^0=0,\\ &b_{21}^0=-\frac{3-\chi}{8\mu}, \quad b_{22}^0=\frac{\chi+1}{8\mu}, \quad b_{23}^0=0,\\ &b_{31}^0=0, \quad b_{32}^0=0, \quad b_{33}^0=\frac{1}{\mu}. \end{aligned} \tag{40.10}$$

We may always reformulate a problem in terms of additional effective strains $\varepsilon_{ij} - \varepsilon_{ij}^0$. It is sufficient to represent a solution as superposition of a homogeneous field s_{xx} s_{yy} s_{xy} and an additional field with zero average stresses. Boundary conditions are reformulated and a CV-BIE is solved with respect to *additional* values. Then equations (40.7) provide additional effective strains, while (40.9) gives the matrix of additional effective compliance $\mathbf{B}^{ad}$. Naturally, the sum

$$\mathbf{B}=\mathbf{B}^0+\mathbf{B}^{ad}$$

is the matrix of effective compliance of a medium with cracks and holes. This additivity makes the additional compliance a convenient measure of changes induced by internal structure.

In many applications to crack problems there is no traction discontinuity on crack surfaces. Then $\Delta f = 0$, and (40.5), written for additional strains, becomes especially simple:

$$\varepsilon_{xx}^{ad} = \frac{1}{2S}\operatorname{Im}\int_L \Delta u\left(d\bar{\tau} - d\tau\right),$$
$$\varepsilon_{yy}^{ad} = \frac{1}{2S}\operatorname{Im}\int_L \Delta u\left(d\tau + d\bar{\tau}\right), \qquad (40.11)$$
$$\varepsilon_{xy}^{ad} = \frac{1}{2S}\operatorname{Re}\int_L \Delta u\, d\tau.$$

In essence, equations (40.11) are relations well known in the theory of effective compliance of bodies with cracks. We have derived them in a purely formal way: they are a result of applying the holomorphicity theorem for quasi-periodic functions to the elasticity equations. In the theory of effective compliance they are written in real variables (see, e.g. Kachanov [1, Sec. VI, equation (6.3)]). Naturally, it is easy to re-write (40.11) in a real form noting that $d\tau - d\bar{\tau} = 2idy$, $d\tau + d\bar{\tau} = 2dx$. We will not do it because, as consistently noted, the complex form is more convenient for numerical calculations. Note that, from the first of (40.11), it is evident that for straight cracks collinear to the x-axis, we always have $\varepsilon_{xx}^{ad} = 0$. This means that effective compliance connected with the component s_{xx} is the same as in a medium without cracks: $b_{11} = b_{11}^{0}$, $b_{12} = b_{21} = b_{12}^{0}$, $b_{13} = b_{31} = b_{13}^{0}$. This conclusion is obvious from a physical point of view.

§ 41. EXAMPLES: DOUBLY PERIODIC CRACKS WITH GROWING WINGS

The form (39.9) of the hypersingular equation (39.1) provides easy implementation of the code, which served in § 32 to illustrate advantages of the CV-BEM, to doubly periodic systems of cracks. Now, due to the presence of the cyclic constants, we can examine not only local fields but also effective characteristics. In particular, we may trace changes in the effective compliance with the growth of crack wings.

In our discussion of the results for the examples, presented in Figure 34, we use the following normalized values. The geometrical sizes are normalized by the half-length a of the initial sliding cracks; the SIF by $k_{I0} = p\sqrt{\pi a}$, where p is the vertical compression; the effective additional compliance by the compliance b_{11}^{0}, given in (40.10), of a medium without internal structure. The figure presents trajectories for

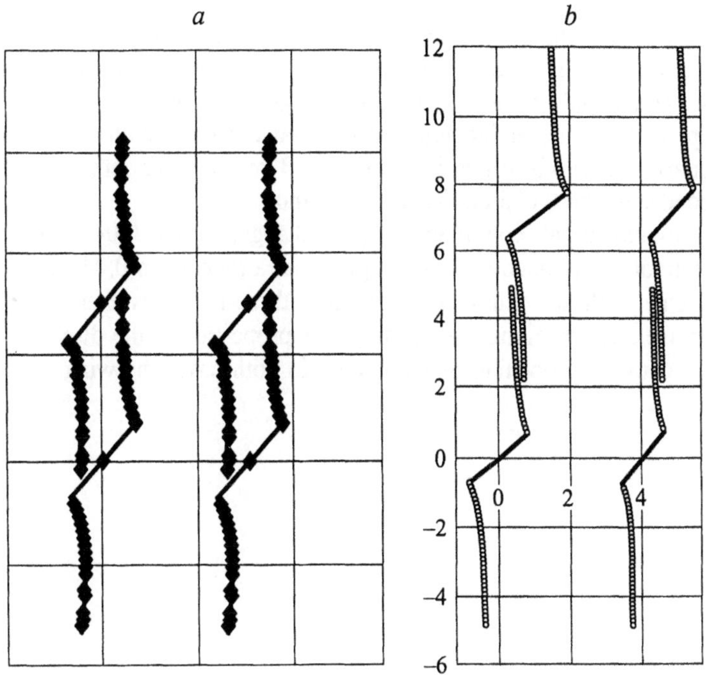

Fig. 34.

uniaxial compression of a plane with the initial cracks having the angle 45^0 with the loading direction.[1] The increments of the wing length are taken 0.1.

Figure 34, *a* shows trajectories for square cells with the length equal to three. In this case, the wings grow towards the centres of cracks in the neighbouring cells. The SIF changes as follows. For the fixed unit load $p = 1$, the SIF corresponding to growing wings, first decreases from 0.42 to 0.33 in the first four increments. On the fifth and the sixth step, when the distance between the wing tips is minimal, the SIF slightly grows and becomes 0.35 and 0.36 respectively. After this, it swiftly decreases from the value 0.33 on the eighth step to 0.16 on the fourteenth step, near the centre of the neighbouring crack. The external load being inversely proportional to the calculated k_I, the load needed to for wings propagation first grows, then slightly decreases and eventually quickly grows when the wing tip approaches the neighbouring crack.

The additional compliance b_{22}^{ad} grows from 0.15 for the initial cracks to 0.48 on the fourteenth step. The changes in the lateral compliance b_{11}^{ad} are even more significant: it grows from 0.15 to 1.42. This effect occurs due to opening of the wings.

Figure 34, *b* corresponds to the case when the centres of the initial cracks are shifted from the vertical, while their tips are located on the line approximately parallel to the applied load. The period along the x-axis is four; the second period has the angle 82^0 to the x-axis; its length is 7.2. In this case, the crack density is approximately three-fold less than in the previous; the wings develop closer to the

[1] Calculations were carried out by A. A. Dobroskok.

vertical direction and the growth of the SIF is more pronounced when the wing tips approach each other. Specifically, as before, k_I first decreases from 0.25 to 0.08 on the seventh and eighth steps. Then it starts growing and becomes 0.12 on the twelfth step, 0.18 on the eighteenth and 0.29 on the nineteenth. Afterwards, when the tips depart, it abruptly drops becoming 0.23 on the twentieth step, 0.08 on the twenty-second step and 0.035 on the twenty-eighth.

The additional axial compliance $b_{22}{}^{ad}$ changes from 0.049 to 0.077. The additional lateral compliance $b_{11}{}^{ad}$ again grows an order: from 0.048 to 0.43. Just as the crack density, these values approximately three-fold less then in the previous case. Again we can see significant changes of properties in the direction normal to compression caused by opening of the wings. In both cases the wings do not tend to coalesce.

Chapter 9

PROBLEMS FOR BONDED HALF-PANES AND CIRCULAR INCLUSION

Problems concerning cracks, holes, blocks and/or inclusions near the interface of media with different properties, in particular near the free or attached boundary of a body, are of great interest in material science, fracture mechanics and geomechanics. In this chapter we present general methods for reducing these problems to CV-BEM (Linkov [10]). We employ a device presented in the next paragraph. It works if the K-M functions are known for the whole plane: then standard transformations lead to K-M functions for bonded half planes.

We also consider problems for a plane with a circular inclusion. The derivations being essentially the same, we will not reproduce them. We will rather present brief comments and final formulae for circular inclusion problems at the ends of the following three sections.

§ 42. GENERAL FORMULAE FOR BONDED HALF-PLANES

Problem formulation. Consider two elastic half-planes S_1 and S_2 coupled along a straight boundary C (Figure 35).

In the general case, the shear moduli and Poisson's ratios are different: μ_1, ν_1 in the lower half-plane and μ_2, ν_2 in the upper half-pane. The position and configuration of cracks, holes, blocks and/or inclusions in the half planes may also differ. For inclusions, each may have different elastic characteristics. Let L_1 denote the total contour of the system of cracks, holes, blocks and/or inclusions in the lower half-plane, and let L_2 denote the total contour in the upper half-plane; $L = L_1 + L_2$ is the total contour. In the special case when $\mu_2 = 0$ we have only the lower half-plane with traction-free boundary. If $\mu_2 = \infty$, the problem corresponds to the lower half-plane being rigidly attached. The interpretations of the cases $\mu_1 = 0$ and $\mu_1 = \infty$ are similar. To simplify the discussion, we will assume that the individual contours comprising L_1 and L_2 are loaded so that the total resultant force acting on each of them is zero. A more general case is obtained by including special terms making allowance for the multi-valuedness of the functions. Further simplification, without significant loss of generality, will be achieved by assuming in non-periodic problems that the stresses vanish at infinity.

We introduce a global system of co-ordinates xOy with the x-axis pointing right along the common boundary of the half-planes and the y-axis pointing upwards (Figure 35).

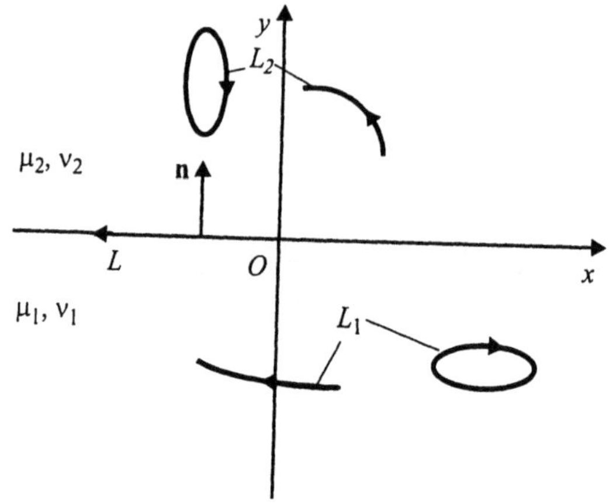

Fig. 35.

K-M functions for the whole plane. Let us assume that we have expressions for K-M functions in the case when the contour L_1 (or L_2) lies in the *whole* plane. Let $\varphi_1(z)$, $\psi_1(z)$, $\varphi_1'(z) = \Phi_1(z)$, $\psi_1'(z) = \Psi_1(z)$ denote K-M functions for the contour L_1 in the whole plane with the properties of the lower half-plane and with boundary conditions on L_1 of the same *type* as that stipulated on L_1 in the original problem. Let $\varphi_2(z)$, $\psi_2(z)$, $\varphi_2'(z) = \Phi_2(z)$, $\psi_2'(z) = \Psi_2(z)$ denote K-M functions for the problem on the contour L_2 in the whole plane with the properties of the upper half-plane.

Under the assumed conditions on the total resulting forces on each of the contours and at infinity, the functions with subscript 1 are holomorphic everywhere outside L_1, in particular in the whole upper half-plane. Functions with subscript 2 are holomorphic everywhere outside L_2, in particular, in the whole lower half-plane. All these functions vanish at infinity. The boundary values corresponding to these functions, of the traction $\sigma_{f1}(t)$ and displacements $u_{f1}(t)$ on L_1, and of the traction $\sigma_{f2}(t)$ and displacements $u_{f2}(t)$ on L_2, are assumed for a moment to be arbitrary; we will call them *fictitious* tractions and displacements, as indicated by the subscript f. In an important particular case, these functions may be prescribed by integral representations. Then the fictitious value plays the role of an unknown density.

The sums $\Phi_1(z) + \Phi_2(z)$, $\Psi_1(z) + \Psi_2(z)$ present a solution of an elasticity problem in both half-planes providing continuous traction $\sigma_{f1}(t)+\sigma_{f2}(t)$ but *discontinuous* displacements $u_{f1}(t) + u_{f2}(t)$ at the boundary C. To eliminate the discontinuity, we introduce additional functions: $\Phi_{a1}(z)$, $\Psi_{a1}(z)$ holomorphic in S_1 and $\Phi_{a2}(z)$, $\Psi_{a2}(z)$ holomorphic in S_2. We will seek a solution as the sums

$$\Phi(z) = \Phi_1(z) + \Phi_2(z) + \Phi_{aj}(z), \quad \Psi(z) = \Psi_1(z) + \Psi_2(z) + \Psi_{aj}(z), \quad (42.1)$$

where $j = 1$ in the lower half-plane and $j = 2$ for the upper half-plane. Our aim is for given functions $\Phi_j(z)$, $\Psi_j(z)$ to find such additional functions $\Phi_{aj}(z)$, $\Psi_{aj}(z)$, which make (42.1) to give both displacements and tractions continuous on the boundary C

of the half-planes. We will use the continuity of the derivative of the displacements instead of displacements themselves. Thus, we need to have

$$\sigma^+(t)=\sigma^-(t), \quad u'^+(t)=u'^-(t). \tag{42.2}$$

Problem solution. Use (42.1) in the K-M formulae (25.6), (25.7) and substitute the result into (42.2). By employing the equality $t = \bar{t}$ on C, we obtain two equations for $z = t \in C$:

$$\Phi_{a1}(z)+\overline{\Phi}_{a1}(z)+z\overline{\Phi}_{a1}{}'(z)+\overline{\Psi}_{a1}(z)=\Phi_{a1}(z)+\overline{\Phi}_{a2}(z)+z\overline{\Phi}_{a2}{}'(z)+\overline{\Psi}_{a2}(z),$$

$$\frac{1}{2\mu_1}\Big\{\chi_1[\Phi_1(z)+\Phi_2(z)+\Phi_{a1}(z)]-\Big[\overline{\Phi}_1(z)+\overline{\Phi_2}(z)+\overline{\Phi}_{a1}(z)\Big]-$$

$$-z\Big[\overline{\Phi}_1{}'(z)+\overline{\Phi}{}'_2(z)+\overline{\Phi}{}'_{a1}(z)\Big]-\Big[\overline{\Psi}_1(z)+\overline{\Psi_2}(z)+\overline{\Psi}_{a1}(z)\Big]\Big\}=$$

$$\frac{1}{2\mu_2}\Big\{\chi_2[\Phi_1(z)+\Phi_2(z)+\Phi_{a2}(z)]-\Big[\overline{\Phi}_1(z)+\overline{\Phi_2}(z)+\overline{\Phi}_{a2}(z)\Big]-$$

$$-z\Big[\overline{\Phi}_1{}'(z)+\overline{\Phi}{}'_2(z)+\overline{\Phi}{}'_{a2}(z)\Big]-\Big[\overline{\Psi}_1(z)+\overline{\Psi_2}(z)+\overline{\Psi}_{a2}(z)\Big]\Big\},$$

where, by definition, $\overline{g}(z)=\overline{g(\bar{z})}$ for any function $g(z)$.

Collect on the left hand sides of these equations the functions holomorphic in S_1: $\Phi_2(z)$, $\Psi_2(z)$, $\Phi_{a1}(z)$, $\Psi_{a1}(z)$, $\overline{\Phi}_1(z), \overline{\Psi}_1(z), \overline{\Phi}_{a2}(z), \overline{\Psi}_{a2}(z)$; on the right hand sides, the functions holomorphic in S_2: $\Phi_1(z)$, $\Psi_1(z)$, $\Phi_{a2}(z)$, $\Psi_{a2}(z)$, $\overline{\Phi}_2(z), \overline{\Psi}_2(z), \overline{\Phi}_{a1}(z), \overline{\Psi}_{a1}(z)$. Then the conditions (42.2) of continuity on C are written as

$$F_1(z)=F_2(z), \quad U_1(z)=U_2(z), \quad z=t\in C, \tag{42.3}$$

where

$$F_1(z)=-\Phi_{a1}(z)+\overline{\Phi}_{a2}(z)+z\overline{\Phi}_{a2}{}'(z)+\overline{\Psi}_{a2}(z),$$

$$F_2(z)=-\Phi_{a2}(z)+\overline{\Phi}_{a1}(z)+z\overline{\Phi}_{a1}{}'(z)+\overline{\Psi}_{a1}(z),$$

$$U_1(z)=\frac{1}{2\mu_2}\Big\{\chi_2\Phi_2(z)-\overline{\Phi}_1(z)-\overline{\Phi}_{a2}(z)-z\Big[\overline{\Phi}_1{}'(z)+\overline{\Phi}{}'_{a2}(z)\Big]-$$

$$-\overline{\Psi}_1(z)-\overline{\Psi}_{a2}(z)\Big\}-\frac{1}{2\mu_1}\Big[\chi_1\Phi_2(z)-\overline{\Phi}_1(z)+\chi_1\Phi_{a1}(z)-z\overline{\Phi}_1{}'(z)-\overline{\Psi}_1(z)\Big],$$

$$U_2(z)=\frac{1}{2\mu_1}\Big\{\chi_1\Phi_1(z)-\overline{\Phi}_2(z)-\overline{\Phi}_{a1}(z)-z\Big[\overline{\Phi}_2{}'(z)+\overline{\Phi}{}'_{a1}(z)\Big]-$$

$$-\overline{\Psi}_2(z)-\overline{\Psi}_{a1}(z)\Big\}-\frac{1}{2\mu_2}\Big[\chi_{21}\Phi_1(z)-\overline{\Phi}_2(z)+\chi_2\Phi_{a2}(z)-z\overline{\Phi}_2{}'(z)-\overline{\Psi}_2(z)\Big].$$

The function $F_j(z)$ is holomorphic in S_j and equal to zero at infinity ($j = 1,2$). The first of (42.3) implies that $F_2(z)$ continuously extends $F_1(z)$ to the entire plane. Similarly, $U_j(z)$ is holomorphic in S_j and equal to zero at infinity ($j = 1,2$), and $U_2(z)$ continuously extends $U_1(z)$ to the entire plane. Then by Liouville's theorem, we have:

$$F_1(z)=0,\quad F_2(z)=0,\quad U_1(z)=0,\quad U_2(z)=0. \tag{42.4}$$

These four equations define the four unknown functions $\Phi_{aj}(z)$, $\Psi_{aj}(z)$ ($j = 1, 2$). We immediately find the solution of (42.4):

$$\Phi_{aj}(z)=(-1)^{j+1}\Big\{K_{jj}\Big[\overline{\Phi}_j(z)+z\overline{\Phi}_j{}'(z)+\overline{\Psi}_j(z)\Big]+K_{js}\Phi_s(z)\Big\},\ (j=1,2),$$

$$\Psi_{aj}(z)=-z\Phi_{aj}{}'(z)-\Phi_{aj}(z)+\overline{\Phi}_{as}(z), \tag{42.5}$$

where $s=2$ if $j = 1$; $s=1$ if $j = 2$;

$$K_{11}=\frac{\mu_2-\mu_1}{\mu_1+\chi_1\mu_2},\ K_{21}=\frac{\chi_2\mu_1-\chi_1\mu_2}{\mu_2+\chi_2\mu_1},\ K_{12}=\frac{\chi_2\mu_1-\chi_1\mu_2}{\mu_1+\chi_1\mu_2},\ K_{22}=\frac{\mu_2-\mu_1}{\mu_2+\chi_2\mu_1}. \tag{42.6}$$

Integration of (42.5) with respect to z yields the additional functions φ_{aj}, ψ_{aj}:

$$\varphi_{aj}(z)=(-1)^{j+1}\Big\{K_{jj}\Big[z\overline{\varphi}_j{}'(z)+\overline{\psi}_j(z)\Big]+K_{js}\varphi_s(z)\Big\},$$

$$\psi_{aj}(z)=-z\varphi_{aj}{}'(z)+\overline{\varphi}_{as}(z). \tag{42.7}$$

By their construction, the additional functions are holomorphic and thus continuous in each of the half-planes. Hence, the additional stresses and the displacements, defined by (42.5), (42.7) are *continuous* on the contour L. This property is significant when deriving CV BIE for bonded half-planes.

When the properties of the half-planes are the same, we actually have a homogeneous plane: $\mu_1 = \mu_2 = \mu$, $\chi_1 = \chi_2 = \chi$. Then all the coefficients (42.6) and consequently the functions $\Phi_{aj}(z)$, $\Psi_{aj}(z)$ are zero. Equations (42.1) turn into formulae for a homogeneous plane with the contours L_1 and L_2. If cracks, inclusions and/or holes are present only in the lower half-plane we have $\Phi_2(z) = 0$, $\Psi_2(z) = 0$, and the first of (42.5), (42.7) give:

$$\Phi_{a1}(z) = K_{11}\left[\overline{\Phi_1}(z) + z\overline{\Phi_1}'(z) + \overline{\Psi_1}(z)\right], \quad \Phi_{a2}(z) = -K_{12}\Phi_1(z),$$

$$\varphi_{a1}(z) = K_{11}\left[z\overline{\varphi_1}'(z) + \overline{\psi_1}(z)\right], \quad \varphi_{a2}(z) = -K_{12}\varphi_1(z).$$

For this case, the formulae (42.7) were first derived by Y. Z. Chen [2,3] in another way. He used the K-M functions $\varphi_j(z)$, $\psi_j(z)$ instead of the functions $\Phi_j(z)$, $\Psi_j(z)$. We prefer to start from the latter functions because they provide easier extension to the problem of a circular inclusion.

When the boundary is traction-free, by setting $\mu_2 = 0$ in (42.6) we have $K_{11} = -1$, $K_{12} = 1$. If the boundary is rigidly attached, by setting $\mu_2 = \infty$ in (42.6), we have $K_{11} = 1/\chi_1$, $K_{21} = -\chi_1$.

The formulae (42.1), (42.5), (42.7) completely solve the problem if the functions $\Phi_j(z)$, $\Psi_j(z)$ for the whole plane with the contour L_j are known (j = 1, 2). The solution of the original problem for traction $\sigma(z)$ and displacements $u_h(z)$ is given by the sums

$$\begin{aligned} \sigma(z) &= \sigma_{f1}(z) + \sigma_{f2}(z) + \sigma_{aj}(z), \\ u(z) &= u_{f1}(z) + u_{f2}(z) + u_{aj}(z). \end{aligned} \tag{42.8}$$

Using (42.8) and equating $\sigma(z)$ and $u(z)$ to the prescribed values on L, we obtain equations for the fictitious quantities $\sigma_f(t)$ and displacements $u_f(t)$:

$$\begin{aligned} \sigma_{f1}(t) + \sigma_{f2}(t) + \sigma_{aj}(t) &= \sigma(t), \quad t \in L_\sigma, \\ u_{f1}(t) + u_{f2}(t) + u_{aj}(t) &= u(t), \quad t \in L_u. \end{aligned} \tag{42.9}$$

Note that $\sigma_{aj}(t)$ and $u_{aj}(t)$ depend linearly on $\sigma_{f1}(t)$, $u_{f1}(t)$, $\sigma_{f2}(t)$ and $u_{f2}(t)$. Therefore, equations (42.9) are linear in the fictitious quantities. If we are using integral representations of K-M potentials with unknown density, then equations (42.9) are equations for the density. Once $\sigma_{f1}(t)$, $u_{f1}(t)$ or the density have been found from (42.9) for the contour L_1 in the lower half-plane, as well as $\sigma_{f2}(t)$, $u_{f2}(t)$ or the density for the contour L_2 in the upper half-plane, the functions $\Phi(z)$, $\Psi(z)$ are fully defined: formulae (42.1), (42.5) and (42.7) determine K-M functions of the original problem.

Circular inclusion in a plane. Quite similarly, we may consider problems for a circular inclusion of unit radius in an infinite plane (Figure 36). (In the general case of the inclusion with the radius R, it is sufficient to change the co-ordinates in the following formulae to the co-ordinates normalized by R). We place the origin in the center of the inclusion.

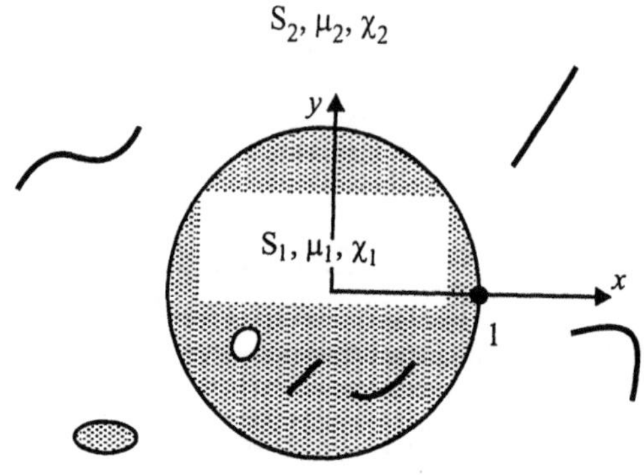

Fig. 36.

Values referring to the inclusion are labeled by thc subscript 1; to the plane by 2. Thus, S_1, μ_1, χ_1 is respectively the region, shear modulus and Muskhelishvili's parameter of the inclusion; S_2, μ_2, χ_2 are these values for the plane. In the special case when $\mu_2 = 0$ we have a circular disc with the boundary free of traction. If $\mu_2 = \infty$, the problem corresponds to the disc with rigidly attached circumference. The interpretations of the cases $\mu_1 = 0$ and $\mu_1 = \infty$ are similar: for $\mu_1 = 0$ we have a plane with a circular hole and zero traction on the boundary; for $\mu_1 = \infty$ the boundary is rigidly attached.

Both the inclusion and the embedding plane may have sources of perturbation. Let L_1 denote the total contour of the system of cracks, holes and blocks in the inclusion, and let L_2 denote the total contour in the embedding plane. To simplify formulae, we assume that there is either only the contour L_1, or only L_2. The general case is included by superposition.

Like for bonded half-planes, assume that we have expressions for the K-M functions $\Phi_1(z)$, $\Psi_1(z)$ (or $\Phi_2(z)$, $\Psi_2(z)$) in the case when the contour L_1 (or L_2) lies in the *whole* plane with the properties of the medium 1 (or 2). We again seek a solution as the sums (42.1), where unknown additional functions $\Phi_{aj}(z)$, $\Psi_{aj}(z)$ should be found to satisfy continuity conditions (42.2) on the circumference C.

The derivation similar to that for bonded half-planes employs the equality $t = \dfrac{1}{\bar{t}}$ on C. The only difference is that now we need to account for poles of functions analogous to the functions $F_j(z)$ and $U_j(z)$ used above. After lengthy derivations we obtain:

$$\Phi_{aj}(z) = (-1)^{j+1} K_{jj} \left\{ \overline{\Phi}_j\left(\frac{1}{z}\right) - \frac{1}{z^2}\left[z\overline{\Phi}_j{}'\left(\frac{1}{z}\right) + \overline{\Psi}_j\left(\frac{1}{z}\right)\right]\right\} + (-1)^{j+1} K_{js} \Phi_s(z) +$$

$$+ \delta_{1j} K_{11}\left(\overline{g} + \overline{\Phi}_{20} + \overline{a}_{\psi 1}\frac{1}{z}\right) + \delta_{2j} K_{22}\left(\overline{\Phi}_{20} + \gamma\frac{1}{z}\right), \qquad (42.10)$$

$$\Psi_{aj}(z) = -\frac{1}{z^2}\left[z\Phi_{aj}{}'(z) - \Phi_{aj}(z) + \overline{\Phi}_{as}\left(\frac{1}{z}\right) + g + \delta z\right],$$

where as above $s = 2$ if $j = 1$; $s = 1$ if $j = 2$; the constants K_{11}, K_{12}, K_{21}, K_{22} are defined by (42.6); the constants g, δ and γ serve to provide holomorphicity of $\Psi_{a1}(z)$ at the point $z = 0$ and zero value of $\Phi_{a2}(z)$ at infinity; besides, $\delta_{ij} = 1$ if $i = j$; $\delta_{ij} = 0$ if $i \neq j$; $\Phi_{20} = \Phi_2(0)$,

$$\operatorname{Re} g = \frac{1}{1 - K_{11}} \operatorname{Re}\left[-K_{11}\bar{b}_{\psi 1} + (1 + K_{12})\Phi_{20}\right] - \operatorname{Re}\Phi_{20},$$

$$\operatorname{Im} g = \frac{1}{1 + K_{11}} \operatorname{Im}\left[-K_{11}\bar{b}_{\psi 1} + (1 + K_{12})\Phi_{20}\right] - \operatorname{Im}\Phi_{20},$$

$$\delta = \frac{K_{21}\bar{a}_{\Phi 1} - K_{22}a_{\Psi 1}}{1 - K_{22}}, \gamma = \frac{\bar{a}_{\Psi 1} - K_{21}a_{\Phi 1}}{1 - K_{22}},$$

$a_{\Phi 1}$, $a_{\Psi 1}$, $b_{\Psi 1}$ are the coefficients of the expansions of $\Phi_1(z)$, $\Psi_1(z)$ at infinity:

$$\Phi_1(z) = \frac{a_{\Phi 1}}{z} + \ldots, \quad \Psi_1(z) = \frac{a_{\Psi 1}}{z} + \frac{b_{\Psi 1}}{z^2} + \ldots,$$

these coefficients are connected with the total resultant force $F_1 = X_1 + iY_1$ and the total moment M_1 acting within the inclusion:

$$a_{\Phi 1} = -\frac{1}{2\pi(\chi_1 + 1)}(X_1 + iY_1), \quad a_{\Psi 1} = \frac{\chi_1}{2\pi(\chi_1 + 1)}(X_1 - iY_1), \quad \operatorname{Im} b_{\Psi 1} = \frac{M_1}{2\pi}.$$

In the particular cases of a straight crack in a circular disc or in a plane with a circular hole, equations (42.10) give the results obtained by Y. Z. Chen [1-3].

By integration of (42.10) with respect to z we obtain additional functions $\varphi_{aj}(z)$, $\psi_{aj}(z)$:

$$\varphi_{aj}(z) = (-1)^{j+1}\left\{K_{jj}\left[z\bar{\varphi}_j{}'\left(\frac{1}{z}\right) + \bar{\psi}_j\left(\frac{1}{z}\right)\right] + K_{js}\varphi_s(z)\right\} +$$
$$+ \delta_{1j}\left[K_{11}\left(\bar{g} + \bar{\Phi}_{20}\right)z + \bar{a}_{\psi 1}\ln z\right] + \delta_{2j}K_{22}\left(\bar{\Phi}_{20}z + \gamma\ln z\right),$$
$$\psi_{aj}(z) = -\frac{1}{z}\varphi_{aj}{}'(z) + \bar{\varphi}_{as}\left(\frac{1}{z}\right) + g\frac{1}{z} - \delta\ln z, \qquad (42.11)$$

These expressions are more compact than (42.10);, they are easier in use when deriving CV-BIE for an inclusion and/or imbedding matrix containing cracks, holes and blocks.

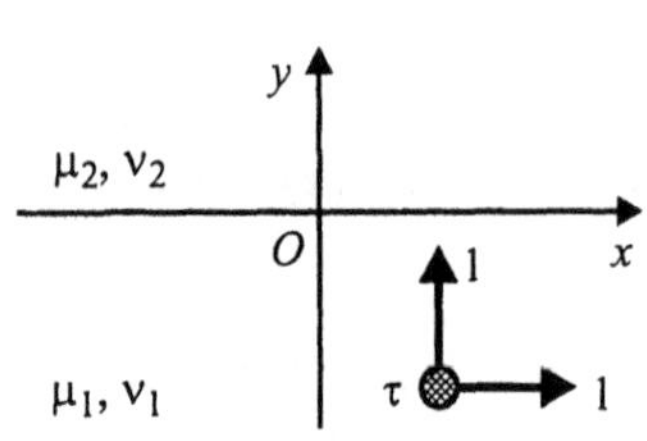

Fig. 37.

§ 43. SOLUTIONS FOR POINT FORCES

In part I we considered an alternative approach to deriving BIE, using the theory of real potentials. It uses complex forms of singular solutions that are solutions for point forces. This stimulates interest in obtaining such solutions for various regions. Presently, we have singular solutions for an entire plane (Muskhelishvili [5]), for a half-plane with traction-free boundary (Green and Zerna [1]), for a periodic system of point forces in a plane with traction-free or rigidly attached boundary (Grekov [3]) and for a plane with an elliptic hole (Denda and Kosaka [1]). Below we will use the formulae of § 42 to extend these results to bonded half-planes and for a circular inclusion in a plane.

Isolated force. Suppose a unit point force is applied at a point τ of the lower half-plane in the direction of the x (or y) axis (Figure 37). If such a force is applied in the whole homogeneous plane with the properties of the lower half-plane, K-M functions are (Muskhelishvili [5, § 57]):

$$\Phi_1(z,\tau) = -\frac{1}{2\pi(\chi_1+1)}\frac{F}{z-\tau_1},$$
$$\Psi_1(z,\tau) = \frac{1}{2\pi(\chi_1+1)}\left[\chi_1\overline{F}\frac{1}{z-\tau} - F\frac{\overline{\tau}}{(z-\tau)^2}\right], \tag{43.1}$$

where, $F = X + iY$; for a unit point force in the direction of the x-axis we set $X = 1$, $Y = 0$; for a unit point force in the direction of the y-axis we set $X = 0$, $Y = 1$. In the considered case of the force only in the lower half-plane, $\Phi_2(z) = 0$, $\Psi_2(z) = 0$. Hence, the solution for bonded half planes is given by (42.1) with $\Phi_1(z)$, $\Psi_1(z)$ defined by (43.1) and with $\Phi_2(z) = 0$, $\Psi_2(z) = 0$.

The additional functions entering (42.1) are defined by (42.5). By using (43.1) in (42.5) we obtain:

$$\Phi_{a1}(z,\tau) = -\frac{1}{2\pi(\chi_1+1)}K_{11}\left[-\chi_1 F\frac{1}{z-\overline{\tau}} + \overline{F}\frac{\tau-\overline{\tau}}{(z-\overline{\tau})^2}\right], \quad z = \in S_1,$$
$$\Psi_{a1}(z,\tau) = -\frac{1}{2\pi(\chi_1+1)}\left\{K_{11}F\left[\frac{\tau-\overline{\tau}-\chi_1\overline{\tau}}{(z-\overline{\tau})^2} + 2\overline{\tau}\frac{\tau-\overline{\tau}}{(z-\overline{\tau})^3}\right] - K_{21}\overline{F}\frac{1}{z-\overline{\tau}}\right\} \tag{43.2}$$

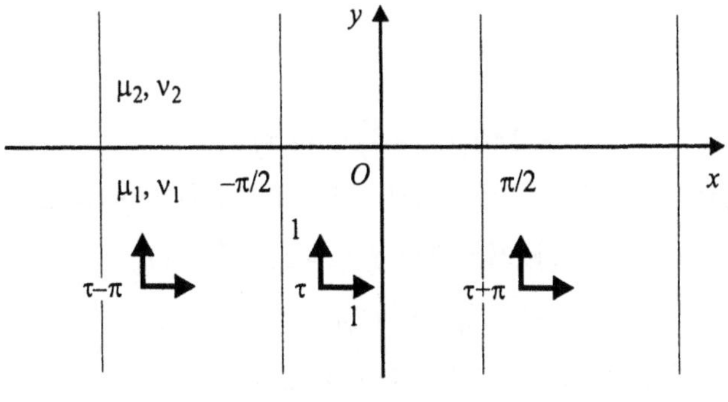

Fig. 38.

$$\Phi_{a2}(z,\tau)=\frac{1}{2\pi(\chi_1+1)}K_{21}F\frac{1}{z-\tau}, \quad z\in S_2,$$

$$\Psi_{a2}(z,\tau)=\frac{1}{2\pi(\chi_1+1)}\left\{K_{21}F\frac{\tau}{(z-\tau)^2}+K_{11}\overline{F}\left[\frac{\chi_1}{z-\tau}+\frac{\tau-\overline{\tau}}{(z-\tau)^2}\right]\right\}. \quad (43.3)$$

By integrating these functions or by using (42.7) we arrive at formulae for additional functions $\varphi_{aj}(z)$, $\psi_{aj}(z)$ (j = 1, 2). The additional displacements, resultant force and traction immediately follow from these expressions by their substitution into the K-M formulae. Since the resulting formulae are cumbersome, we will not present them explicitly[1].

Periodic system of forces (Figure 38). In exactly the same way, we obtain fundamental solutions for a π-periodic system of forces applied at the points $\tau = m\pi$ of the lower half-plane (m = ..., -2, -1, 0, 1, 2,...). (For an arbitrary real period a it will suffice to add the factor π/a before the variables z and τ - and hence also before $\overline{z}$ and $\overline{\tau}$ - in all subsequent formulae).

We begin with the formulae for a whole plane with the same disposition of periodic forces acting along the direction of the x- or y-axis (Linkov and Mogilevskaya [5]):

$$\Phi_1(z,\tau)=-\frac{F}{2\pi(\chi_1+1)}\cot(z-\tau),$$

$$\Psi_1(z,\tau)=\frac{1}{2\pi(\chi_1+1)}\left\{\chi_1\overline{F}\cot(z-\tau)-F[(\overline{\tau}-\tau+z)\operatorname{cosec}^2(z-\tau)-\cot(z-\tau)]\right\}.$$

Comparing these formulae with (43.1) we see that actually they contain cot(z-τ) instead of 1/(z-τ) and cosec(z-τ) instead of 1/(z-τ)^2. Hence, the needed formulae for additional functions follow from (43.2), (43.3) by formally replacing 1/ζ to cotζ and 1/ζ^2 to cosec^{2}ζ. There is no need to write them down explicitly. We rather present the final formulae for the additional displacements:

[1] The formulae for displacements are given in the Russian edition of the book.

$$-4\pi\mu_1(\chi_1+1)u_{a1}(z,\tau)=$$
$$=F\left\{K_{21}\overline{\ln\sin(z-\overline{\tau})}+K_{11}\left[-\chi_1^2\ln\sin(z-\overline{\tau})+(z-\overline{z})(\tau-\overline{\tau})\operatorname{cosec}^2(\overline{z}-\tau)\right]\right\}+$$
$$+\overline{F}K_{11}\left[-\chi_1(\tau-\overline{\tau})\cot(z-\overline{\tau})+\chi_1(z-\overline{z})\cot(\overline{z}-\tau)\right],\quad z\in S_1,$$
$$-4\pi\mu_2(\chi_1+1)u_{a2}(z,\tau)= \tag{43.4}$$
$$=F\left[-K_{21}\chi_2\ln\sin(z-\tau)+K_{11}\chi_1\overline{\ln\sin(z-\tau)}\right]+$$
$$+\overline{F}\left[K_{21}(z-\overline{z})\overline{\cot(z-\tau)}+K_{11}(\tau-\overline{\tau})\overline{\cot(z-\tau)}\right],\quad z=S_2.$$

Analysis shows that the complex displacements $u_{aj}(z,\tau)$, defined by thesc formulae ($j = 1, 2$), have a cyclic constant $A(i-1)$ in both half-planes, where

$$A=\frac{K_{21}+\chi_1{}^2K_{11}}{4\mu(\chi_1+1)}.$$

The constant may be removed by adding a linear field of displacements, but this generates non-zero uniform tangential stresses σ_{xx} in both half-panes.

In the cases of the lower half-plane with traction-free ($K_{11} = -1$, $K_{21} = 1$) or rigidly attached ($K_{11} = 1/\chi_1$, $K_{21} = -\chi_1$) boundary, the first of (43.4) gives fundamental solutions identical with the results by Grekov [3], apart from terms corresponding to the uniform stresses in a half-plane. (Grekov [3] satisfied additional condition of periodicity of $u_{aj}(z,\tau)$; as mentioned, this generates uniform tangential stresses).

Point forces in a circular inclusion or plane. Similarly we can find additional functions for point forces in a problem for a circular inclusion in a plane (Figure 36). It is sufficient to substitute the K-M functions for a pointed force in an infinite plane

$$\Phi_j(z,\tau)=-\frac{1}{2\pi(\chi_j+1)R}\frac{F}{z-\tau},\quad \Psi_j(z,\tau)=-\frac{1}{2\pi(\chi_j+1)R}\left[-\chi_j\frac{\overline{F}}{z-\tau}+\frac{F\overline{\tau}}{(z-\tau)^2}\right]$$

into (42.1), (42.10). For $j = 1$, the point force is in the inclusion; for $j = 2$ in the plane; the multiplier R in the denominators accounts for normalization of the co-ordinates z and τ by the radius of the inclusion. As above, τ is the point where the point force F acts; $F = X + iY$, where $X = 1$, $Y = 0$ for a unit force in the x-direction; $X = 0$, $Y = 1$ for a unit force in the y-direction. The substitution yields the following formulae.

For the point force in the inclusion ($|\tau| < 1$):

$$\Phi_{a1}(z,\tau)=\frac{zK_{11}}{2\pi(\chi_1+1)R}\left[\overline{F}\frac{\bar{\tau}\tau-1}{1-\bar{\tau}z}\left(1+\frac{1}{1-\bar{\tau}z}\right)-F\chi_1\frac{\bar{\tau}^2}{1-\bar{\tau}z}\right]+g,\quad |z|\le 1,$$

$$\Psi_{a1}(z,\tau)=-\frac{1}{2\pi(\chi_1+1)R}\left\{\overline{F}\frac{\bar{\tau}}{1-\bar{\tau}z}\left[K_{11}\frac{\bar{\tau}\tau-1}{1-\bar{\tau}z}\left(1+\frac{2}{1-\bar{\tau}z}\right)+K_{21}\right]-F\chi_1K_{11}\frac{\bar{\tau}^3}{(1-\bar{\tau}z)^2}\right\},$$

$$\Phi_{a2}(z,\tau)=\frac{FK_{21}}{2\pi(\chi_1+1)R}\frac{1}{z-\tau}+K_{22}\gamma\frac{1}{z},\qquad |z|\ge 1,$$

$$\Psi_{a2}(z,\tau)=-\frac{1}{z^2}\left\{-\frac{\overline{F}\chi_1K_{11}}{2\pi(\chi_1+1)R}\frac{\tau^2}{z-\tau}+\frac{F}{2\pi(\chi_1+1)R}\left[K_{11}\frac{\bar{\tau}\tau-1}{z-\tau}\left(1+\frac{z}{z-\tau}\right)-\right.\right.$$

$$\left.\left.-K_{21}\frac{2z-\tau}{(z-\tau)^2}\right]+g+\bar{g}+\delta z-2K_{22}\gamma\frac{1}{z}\right\},$$

where

$$\operatorname{Re}g=-\frac{K_{11}}{1-K_{11}}\operatorname{Re}b_{\psi1},\qquad \operatorname{Im}g=\frac{K_{11}}{1+K_{11}}\operatorname{Im}b_{\psi1},$$

$$\delta=\frac{K_{21}\bar{a}_{\Phi1}-K_{22}a_{\Psi1}}{1-K_{22}},\qquad \gamma=\frac{\bar{a}_{\Psi1}-K_{21}a_{\Phi1}}{1-K_{22}},$$

$$a_{\Phi1}=-\frac{F}{2\pi(\chi_1+1)R},\quad a_{\psi1}=\frac{\chi_1\overline{F}}{2\pi(\chi_1+1)R},\quad b_{\psi1}=\frac{\chi_1\overline{F}\tau-F\bar{\tau}}{2\pi(\chi_1+1)R}.$$

For the point force in the plane ($|\tau|>1$):

$$\Phi_{a1}(z,\tau)=-\frac{FK_{12}}{2\pi(\chi_2+1)R}\frac{z}{\tau(z-\tau)}+g,\quad |z\le 1|,$$

$$\Psi_{a1}(z,\tau)=-\frac{1}{2\pi(\chi_2+1)R}\left[\overline{F}\chi_2K_{22}\frac{1}{z-\tau}+F\frac{K_{22}(1-\bar{\tau}\tau)+K_{12}}{\tau(z-\tau)^2}\right],$$

$$\Phi_{a2}(z,\tau)=-\frac{K_{22}}{2\pi(\chi_2+1)R}\left[\overline{F}\frac{\bar{\tau}\tau-1}{\tau(1-\bar{\tau}z)^2}-F\chi_2\frac{1}{z(1-\bar{\tau}z)}\right],\quad |z\ge 1|,$$

$$\Psi_{a2}(z,\tau)=-\frac{1}{z^2}\left\{\frac{\overline{F}}{2\pi(\chi_2+1)R}\left[-K_{22}\frac{z^2\bar{\tau}(\bar{\tau}\tau-1)}{(1-\bar{\tau}z)^2}\left(1+\frac{2}{1-\bar{\tau}z}\right)+K_{22}\frac{\bar{\tau}\tau-1}{\bar{\tau}}-\right.\right.$$

$$\left.\left.-K_{12}\frac{1}{\bar{\tau}(1-\bar{\tau}z)}\right]+\frac{F\chi_2K_{22}}{2\pi(\chi_2+1)R}\left[\frac{z^2\bar{\tau}^3}{(1-\bar{\tau}z)^2}-\bar{\tau}-\frac{2}{z}\right]+g+\bar{g}\right\},$$

where now

$$\operatorname{Re} g = \frac{1}{2\pi(\chi_2+1)R}\frac{K_{12}+K_{11}}{1-K_{11}}\operatorname{Re}\frac{F}{\tau}, \quad \operatorname{Im} g = \frac{1}{2\pi(\chi_2+1)R}\frac{K_{12}-K_{11}}{1+K_{11}}\operatorname{Im}\frac{F}{\tau}.$$

In the particular case of a hole with traction-free circumference, the formulae for $|\tau| > 1$ turn into known results (Denda and Kosaka [1]). In the general case they are presented for the first time. Substitution of the derived fundamental solutions into the equations of chapters 2, 3, 4 immediately results in CV BIE (Fredholm, singular and hypersingular).

§ 44. CV-BIE FOR BLOCKY SYSTEMS

Bonded half-planes. Consider a blocky system in bonded half-planes. Assume for simplicity that cracks, holes and blocks are only in the lower half-plane. Then $\Phi_2(z) = 0$, $\Psi_2(z) = 0$, while for $\Phi_1(z)$, $\Psi_1(z)$ we can use integral representations such as (30.23), (30.24). Their substitution into (42.5) provides the additional functions $\Phi_{a1}(z)$, $\Psi_{a1}(z)$. Employing them in (42.1) and the latter in the boundary condition for traction (25.16), we arrive at CV-BIE with an additional term completely defined by the additional functions. Thus, the equations of § 21, 23, 29, 30 are valid with the additional term on their l. h. s. For instance, the singular equation (30.16) and the hypersingular equations (23.7), (30.17) acquire the additional term on their l. h. s.[1]:

$$\frac{\chi_1+1}{2\mu_1}\sigma_{a1}(t) = \frac{1}{2\pi i}\int_L \left\{ -\Delta u\frac{\partial}{\partial t}dk_3 - \overline{\Delta u}\frac{\partial}{\partial t}dk_4 + \right.$$

$$\left. +\sigma\left[a_1\frac{\partial k_3}{\partial t} + a_3K_{11}\frac{\partial}{\partial t}\ln(\bar\tau - t)\right]d\tau + \bar\sigma\left(a_1\frac{\partial}{\partial t}k_4 + a_3K_{11}\frac{\partial}{\partial t}\frac{t-\bar t}{\tau-\bar t}\right)d\bar\tau\right\}, \quad (44.1)$$

where

$$k_3(\tau,t) = K_{11}\left[-\ln(\bar\tau - t) + \frac{(t-\bar t)(\tau-\bar\tau)}{(\tau-\bar t)^2}\right] + K_{21}\ln(\tau-\bar t),$$

$$k_4(\tau,t) = K_{11}\left(\frac{\tau-t}{\bar\tau-t} - \frac{t-\bar t}{\tau-\bar t}\right).$$

The constants a_1, a_2, a_3 are defined, as usual, by (30.12); the constants K_{11}, K_{21} by (42.6). The stresses at infinity s_{xx}, s_{yy}, s_{xy} are accounted for by the term (30.18)

[1] We assume traction continuous through the contacts of blocks and surfaces of cracks.

with one reservation: for bonded half-planes the shear stress s_{xy} at infinity may be non-zero only if $\mu_1 = \mu_2$.

For the particular case of homogeneous half-planes with cracks, the corresponding singular equation, following from (21.6), was derived by Ioakimidis and Theocaris [5]. In the special case of a single half-plane with a traction-free boundary, the results are the same as those derived by Mogilevskaya by other means – using complex singular solutions (Mogilevskaya [3], see also Mogilevskaya and Linkov [2]).

Circular inclusion. Similarly, we obtain BIE for blocky systems in a circular inclusion or in the embedding plane. Now we employ the additional functions $\Phi_{a1}(z)$, $\Psi_{a1}(z)$ given by (42.10) or (42.11). Then the additional term on the l. h. s. of the singular equation (30.16) and of the hypersingular equations (23.7), (30.17) is:

$$\frac{\chi_j+1}{2\mu_j}\sigma_{aj}(t)=(-1)^{j+1}\frac{1}{2\pi i}\int_L\left\{-\Delta u\frac{\partial}{\partial t}dk_3-\overline{\Delta u}\frac{\partial}{\partial t}dk_4+\right.$$
$$+\sigma\left[a_1\frac{\partial k_3}{\partial t}+a_3K_{jj}\frac{\partial}{\partial t}\ln\left(\bar\tau-\frac{1}{t}\right)\right]d\tau+ \tag{44.2}$$
$$\left.+\bar\sigma\left(a_1\frac{\partial k_4}{\partial t}-a_3K_{jj}\frac{\partial}{\partial t}\frac{t-1/\bar t}{\bar t^2(\tau-1/\bar t)}\right)d\bar\tau\right\}+\frac{\partial G_j}{\partial t}+\frac{\chi_j+1}{2\mu_j}\sigma_{aj}^{\infty}(t),$$

where now

$$k_3(\tau,t)=K_{jj}\left[-\ln\left(\bar\tau-\frac{1}{t}\right)+\frac{(t-1/\bar t)(\tau\bar t^{-2}+\bar\tau-2\bar t)}{\bar t^{-2}(\tau-1/\bar t)^2}\right]+K_{sj}\ln\left(\tau-\frac{1}{\bar t}\right),$$
$$k_4(\tau,t)=K_{jj}\left(\frac{\tau-t}{\bar\tau-1/t}+\frac{1}{\bar t^2}\frac{t-1/\bar t}{\tau-1/\bar t}\right);$$

for blocks in the inclusion: $j=1, s=2$,

$$G_1(t)=\left[2tK_{11}-\frac{1}{\bar t}(K_{11}-1)\right]C_1,$$
$$C_1=\frac{K_{11}}{K_{11}-1}\frac{1}{2\pi}\operatorname{Im}\int_L[2\Delta u\,d\bar\tau-(2a_1-a_3)\sigma\bar\tau\,d\tau],$$
$$\sigma_{a1}^{\infty}(t)=-2\frac{K_{11}+K_{12}}{K_{11}-1}A_\varphi-K_{22}\bar A_\psi\frac{d\bar t}{dt};$$

for blocks in the matrix: $j=2, s=1$,

$$G_2(t) = tK_{22}(C_2 + \overline{C}_2) + \frac{1}{\overline{t}}\left[(K_{11}\overline{C}_2 - K_{22}C_2 + (K_{11}+1)D_2\right],$$

$$C_2 = \frac{1}{2\pi i}\int_L \left(\frac{\Delta u}{\tau^2} + a_1\frac{\sigma}{\tau}\right)d\tau, \quad D_2 = -\frac{K_{11}+K_{12}}{K_{11}-1}\operatorname{Re}C_2 + i\frac{K_{11}-K_{12}}{K_{11}+1}\operatorname{Im}C_2,$$

$$\sigma_{a2}^{\infty}(t) = 2K_{22}\operatorname{Re}\left(\overline{A}_\psi\frac{1}{t^2}\right) + \frac{1}{\overline{t}^2}\left[2\frac{K_{11}+K_{12}}{K_{11}-1}A_\varphi + K_{22}A_\psi\frac{1}{\overline{t}}\left(-2t+\frac{3}{\overline{t}}\right)\right]\frac{d\overline{t}}{dt},$$

where the constants A_φ and A_ψ account for the stresses s_{xx}, s_{yy} and s_{xy} at infinity; as usual they are given by (30.20).

K-M functions $\varphi(z)$, $\psi(z)$ for points within the blocks and embedding medium are:

$$\varphi(z) = \varphi_j(z) + \frac{2\mu^k}{\chi^k+1}\frac{\chi_j+1}{2\mu_j}\left[\varphi_{aj}(z) + \varphi^{\infty}(z) + \varphi_{aj}^{\infty}(z)\right], \quad z \in D^k,$$

$$\psi(z) = \psi_j(z) + \frac{2\mu^k}{\chi^k+1}\frac{\chi_j+1}{2\mu_j}\left[\psi_{aj}(z) + \psi^{\infty}(z) + \psi_{aj}^{\infty}(z)\right], \qquad (44.3)$$

where $\varphi_j(z)$, $\psi_j(z)$ are K-M functions corresponding to a similar system of blocks but embedded into an infinite plane with the properties of the medium j (cf. (30.21), (30.22)):

$$\varphi_j(z) = \frac{2\mu^k}{\chi^k+1}\frac{1}{2\pi i}\int_L \frac{\Delta u + a_1 f}{\tau - z}d\tau,$$

$$\psi_j(z) = \frac{2\mu^k}{\chi^k+1}\frac{1}{2\pi i}\int_L\left[-\frac{\overline{\Delta u} + (a_1 - a_3)\overline{f}}{\tau - z}d\tau + (\Delta u + a_1 f)d\frac{\overline{\tau}}{\tau - z}\right],$$

we use the superscript k to label a block while as above the subscript j serves to label values for the medium embedding the blocks; for the embedding medium we assume $k = 0$, $\mu^0 = \mu_j$, $\chi^0 = \chi_j$; the additional functions $\varphi_{aj}(z)$, $\psi_{aj}(z)$ are given by formulae:

$$\varphi_{aj}(z) = (-1)^{j+1}\frac{2\mu_j}{\chi_j+1}K_{jj}\frac{1}{2\pi i}\int_L\left[-\frac{\Delta u + (a_1 - a_3)f}{\overline{\tau} - 1/z}d\overline{\tau} - (\overline{\Delta u} + a_1\overline{f})d\frac{\tau - z}{\overline{\tau} - 1/z}\right] +$$
$$+ (-1)^{j+1}K_{jj}c_j z,$$

$$\psi_{aj}(z) = -\frac{1}{z}\varphi_{aj}'(z) + (-1)^{j+1}K_{sj}\frac{2\mu_j}{\chi_j+1}\frac{1}{2\pi i}\int_L\frac{\overline{\Delta u} + a_1\overline{f}}{\overline{\tau} - 1/z}d\overline{\tau} + (g_j + d_j)\frac{1}{z};$$

the last terms in the brackets on the r. h. s. of (44.3) account for the stresses at infinity:

$$\varphi^{\infty}(z) = A_{\varphi}z, \ \psi^{\infty}(z) = A_{\psi}z;$$

for blocks in the inclusion: $j = 1, s = 2, K_{sj} = K_{21}, \ c_1 = g_1 = -\dfrac{2\mu_1}{\chi_1 + 1}C_1, \ d_1 = 0,$

$$\varphi_{a1}^{\infty}(z) = -\frac{K_{11} + K_{12}}{K_{11} - 1}A_{\varphi}z, \quad \psi_{a1}^{\infty}(z) = -K_{22}A_{\psi}z;$$

for blocks in the matrix: $j = 2, s = 1, K_{sj} = K_{12}, c_2 = -\overline{\Phi}_{20}, \ d_2 = K_{11}(g_2 + \Phi_{20}),$

$$\Phi_{20} = \frac{2\mu_2}{\chi_2 + 1}C_2, \ g_2 = \frac{2\mu_2}{\chi_2 + 1}\overline{D}_2,$$

$$\varphi_{a2}^{\infty}(z) = -K_{22}\overline{A}_{\psi}\frac{1}{z}, \quad \psi_{a2}^{\infty}(z) = -2\frac{K_{11} + K_{12}}{K_{11} - 1}A_{\varphi}\frac{1}{z} - K_{22}\overline{A}_{\psi}\frac{1}{z^3};$$

the constants C_1, C_2, D_2 are defined above.

In the particular case of a plane with a circular hole ($\mu_1 = 0, j = 2, s = 1$), the formulae reduce to the equations obtained in Mogilevskaya et al. [1].

§ 45. CRACKS ALONG STRAIGHT LINE OR CIRCUMFERENCE

In two previous paragraphs we have seen that results for bonded half-panes and for a circular inclusion are obtained by a standard procedure from the results for a whole plane if we know the corresponding K-M functions $\Phi_j(z)$, $\Psi_j(z)$. As noted in § 29, this holds for problems concerning cracks along a single straight line or along a single circumference. In these cases, the functions $\Phi_j(z)$, $\Psi_j(z)$ are given by well known integrals (Muskhelishvili [5, § 120, 124]).

Naturally, we may immediately use the additional terms (44.1) or (44.2) in corresponding singular or hypersingular equations for these problems. Meanwhile, now there is also another option provided by a specific choice of the functions $\Phi_1(z)$ and $\Psi_1(z)$ for a straight or circular-arc crack in an infinite plane. Y. Z. Chen [1-3] and M. A. Grekov [1, 2] systematically used this option. It leads to non-singular equations of Fredholm type. Grekov [2] stated by direct numerical tests, that the equation for a half-plane with traction-free boundary is very easily solved, even when a crack is located close to the boundary. Specifically, the approximation of the fictitious traction by a polynomial of order less than 10, provided an error less than 0.1 %, even when the distance between the crack and the boundary was 0.01 of the crack length. Taking into account the small computational effort, this accuracy looks good.

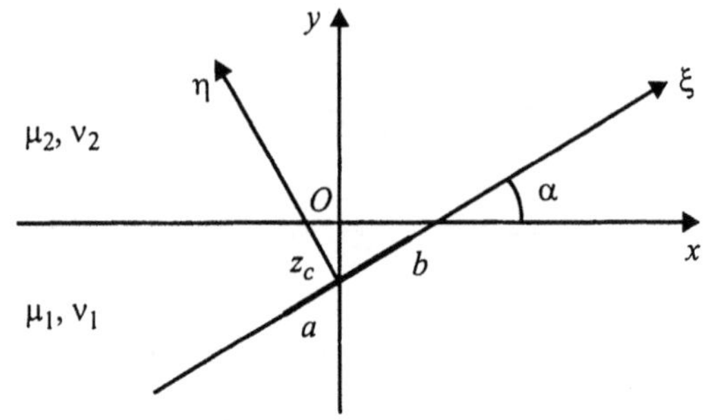

Fig. 39.

Consider for illustration the simplest case of a single straight crack in the lower half-plane (Figure 39). To shorten notation, assume that the tractions prescribed on the crack surfaces are continuous: $\sigma^+ = \sigma^- = \sigma$. Then the functions $\Phi_1(z)$, $\Psi_1(z)$ for a crack in an infinite plane are obtained from the Muskhelishvili's solution [5, § 120]:

$$\Phi_1(z) = \frac{1}{2\pi i X(z)} \int_a^b \frac{X(\tau)\sigma_{f1}(\tau)}{\tau - z} d\tau,$$

$$\Psi_1(z) = e^{-2i\alpha}\left[\overline{\Phi_1\left((\bar{z} - \bar{z}_c)e^{2i\alpha} + z_c\right)} - \Phi_1(z) - (z - z_c)\Phi_1'(z)\right] - \bar{z}_c\Phi_1'(z), \quad (45.1)$$

where $X(z) = (z - a)^{1/2}(z - b)^{1/2}$ and $X(z) = + z + O(1)$ at infinity, $X(t)$ presents $X(z)$ on the left edge of the crack; α is the angle between the direction of traveling along the crack and the x-axis: z_c is the co-ordinate of an arbitrary fixed point on a straight line along which the crack is located; σ_{f1} is the fictitious traction to be found from the Fredholm equation derived below.

It is convenient to choose z_c as the point of intersection of the y-axis with the straight line and to use the local system $\xi O_1 \eta$ shown in Figure 39. In this case, z_c is purely imaginary ($z_c = i\, y_c$) and the connection between the new complex variable $\zeta = \xi + i\eta$ and the old variable $z = x + iy$ is given by the formula $\zeta = (z - z_c)\exp(2i\alpha)$. Then after using (45.1) in (42.5) and the result in the K-M formula (25.16) for traction, we obtain the term $\sigma_{a1}(z)$ in (42.8):

$$\sigma_{a1}(z) = K_{11}\left\{e^{2i\alpha}\Phi_l(\zeta_0) + \left[-\frac{K_{21}}{K_{11}}\frac{d\bar{z}}{dz} + 2i\left(1 - \frac{d\bar{z}}{dz}\right)e^{-i\alpha}\sin\alpha\right]\Phi_l(\overline{\zeta_0}) + \right.$$

$$+\left(1 - \frac{d\bar{z}}{dz}\right)e^{-2i\alpha}\overline{\Phi_l(\zeta_0)} - 2ie^{i\alpha}\sin\alpha\,\overline{\Phi_l(\overline{\zeta_0})} - 2ie^{2i\alpha}\,\overline{z_*\Phi_l'(\overline{\zeta_0})} + 2ie^{-2i\alpha}\,\bar{z}_*\Phi_l'(\overline{\zeta_0}) +$$

$$+\frac{d\bar{z}}{dz}(z - \bar{z})\left[e^{-3i\alpha}\overline{\Phi_l'(\zeta_0)} + 4ie^{-2i\alpha}\sin\alpha\,\Phi_l'(\overline{\zeta_0}) + \right.$$

$$\left. + \frac{d\bar{z}}{dz}(z - \bar{z})\left[e^{-3i\alpha}\overline{\Phi_l'(\zeta_0)} + 4ie^{-2i\alpha}\sin\alpha\,\Phi_l'(\overline{\zeta_0}) + 2ie^{-3i\alpha}\,\bar{z}_*\Phi_l''(\overline{\zeta_0})\right]\right\}. \quad (45.2)$$

where $\Phi_l(\zeta)=\dfrac{1}{2\pi i X_l(\zeta)}\displaystyle\int_{a_l}^{b_l}\frac{X_l(\tau)\sigma_{f1}(\tau)}{\tau-\zeta}d\tau$; $X_l(\zeta)=(\zeta-a_l)^{1/2}(\zeta-b_l)^{1/2}$;

$\zeta_0 = (z - \overline{z_c})\exp(i\alpha)$; $z_* = z\sin\alpha + y_c\cos\alpha$; a_l and b_l are the co-ordinates of start and end point in the local system: $a_l = (a - z_c)\exp(-i\alpha)$, $b_l = (b - z_c)\exp(-i\alpha)$. The derivative Φ_l' of the function Φ_l is taken with respect to the entire argument in the brackets.

For a traction-free boundary, we set $\mu_2 = 0$. Then, by (42.6), $K_{11} = -1$, $K_{21} = 1$ and equation (45.2) coincides with the result by Grekov [2] obtained for this particular case. For a rigidly attached boundary, $\mu_2 = \infty$. Then $K_{11} = 1/\chi_1$, $K_{21} = -\chi_1$ and formula (45.2) corrects the result given in the paper by Grekov [2] with a misprint.

All the functions on the r. h. s. of (45.2) are non-singular in the lower half-plane. Hence, the first of equations (42.9), which in the considered case become

$$\sigma_{f1}(t)+\sigma_{a1}(t)=\sigma(t),$$

is non-singular; it is a Fredholm equation for the fictitious traction.

Comment on problems for circular inclusion. In the same way, one may derive equations for a straight or circular-arc crack in a circular inclusion or in an embedding plane. We will not write down these lengthy equations. For the particular cases of a circular disc ($\mu_2 = 0$) and a circular hole ($\mu_1 = 0$), they are given by Y. Z. Chen [1,3].

Fredholm equations obtained as described in this section are formulated for *fictitious* traction. This suggests their immediate extension to systems of straight and/or circular-arc cracks: it is sufficient to sum additional traction from each of the cracks. Y. Z. Chen [1-3] systematically employed this option.

PART III

THEORY OF COMPLEX INTEGRAL EQUATIONS

In the two previous parts we derived complex variable BIE of elasticity in two different ways. We saw that except for the equations of Muskhelishvili, the equations were singular or hypersingular. We also noted that the hypersingular equations are especially useful for the numerical solution of modern applied problems. Derivations were performed formally: we assumed that operations applicable for usual improper integrals were valid also for their extensions – singular and hypersingular integrals. We did not justify these operations rigorously, so as not to digress from our main theme, the presentation of methods serving to derive CV BIE, and interconnections between various BIE.

In this part we present the justification. We do not discuss *singular* equations because Muskhelishvili's monograph [4] contains the comprehensive theory, but *hypersingular* equations are discussed in some detail. Unlike singular ones, they have appeared quite recently; their theory is less established and less well known.

We recommend that readers read chapter 10, which contains a simple exposition of the theory of complex hypersingular integrals. At least, this will deliver them from hyper-fear of hyper-singularity! Chapter 11 contains the theory of hypersingular equations. Mathematical techniques used in this chapter stem from the basics of complex variable theory, but involve results not normally presented in standard courses on complex variables. A reader interested more in applications than in their rigorous justification, can skip over this chapter to the next part where we discuss numerical implementation of equations: it is impossible to have proofs in advance for all the problems arising in practice; as a rule, common sense and the skill of formulating a problem properly can replace sophisticated mathematical proofs. We present theorems and proofs rather for completeness of exposition: it would be unreasonable to ignore these results obtained by methods that are quite simple although not discussed in standard courses.

Chapter 10

COMPLEX HYPERSINGULAR AND FINITE-PART INTEGRALS

We want to justify the operations used when deriving equations, first of all hypersingular equations. What do we need to prove? In our discussion and derivations of hypersingular integrals, we employed the following formal elements:

a) we assumed that there is a definition of divergent integrals that endows them with the usual properties of integrals, specifically with linearity and additivity;

b) we used differentiation under the integral sign;

c) we used integration by parts;

d) we used formulae of Sokhotski-Plemelj type that connect direct values of divergent integrals with their limiting values.

For the direct values of *singular* integrals, i.e. the principal value (Cauchy) integrals, these elements have been comprehensively justified (see, e. g. Muskhelishvili [4, 5], Gahov [1]). Now we will extend them to *hypersingular* integrals of arbitrary order. This extension will show that there is no significant difference between hypersingular and singular integrals: the complex variable principal value integrals appear as a special case of complex hypersingular finite-part integrals.

§ 46. DEFINITION OF DIRECT VALUES OF DIVERGENT COMPLEX INTEGRALS

In the literature and in our previous discussion, the term "hypersingular integral" is used in three different ways:

1) as a *proper* integral when a field point is outside the contour of integration;

2) as a *limit* value in a normal limit process when a field point tends to a point on the integration contour;

3) as a *direct* value, i.e. in the sense of a finite-part (Hadamard) integral; in this case a field point is located on the contour of integration and we need to define how to interpret the integral.

The first two applications involve few complications. Consider the third application, following Linkov [8].

Let (a, b) be a sufficiently smooth arc in the complex plane (Figure 40). Consider an integral over it:

$$I_{k+1}^{a} = \int_{a}^{b} \frac{g(\tau)d\tau}{(\tau - a)^{k+1}}, \tag{46.1}$$

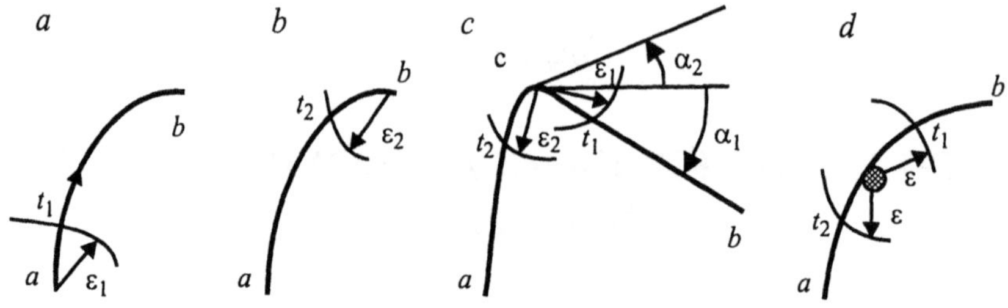

Fig. 40.

where k is a real number. For $-1 < k < 0$ we have a usual improper integral. For $k = 0$ the integral can be termed a (Cauchy) principal value integral. If $k > 0$ the integral will be called a finite-part (Hadamard) integral. We will call $1/(\tau - a)^{k+1}$ a *kernel*, $g(\tau)$ a *density*. The density is assumed to be sufficiently smooth; specifically, it has a $(k + 1)$-th Holder continuous derivative.

We need to define how to interpret the integral (46.1) for $k \geq 0$ when the r. h. s. does not exist as a usual improper integral. Denote $k_1 = [k]$, the integer part of k, that is the greatest integer which is less than or equal to k. The difference $\beta = k - k_1$ is the fractional part of k. Then we may write the kernel as a $(k_1 + 1)$-th derivative with respect to τ.

Specifically, when $\beta = 0$, $k = k_1$ is an integer and

$$\frac{1}{(\tau - a)^{k+1}} = \frac{(-1)^k}{k!}\frac{d^{k+1}}{d\tau^{k+1}}\ln(\tau - a). \tag{46.2}$$

When $\beta \neq 0$ $(0 < \beta < 1)$, the analogous representation reads

$$\frac{1}{(\tau - a)^{k+1}} = \frac{(-1)^{k_1+1}}{\beta(\beta+1)\ldots(\beta+k_1)}\frac{d^{k_1+1}}{d\tau^{k_1+1}}\frac{1}{(\tau - a)^{\beta}}. \tag{46.3}$$

We will term the integer $k_1 + 1$ the *order* of hypersingularity, and also call it the *order* of the corresponding hypersingular integral. Both for $\beta = 0$ and $\beta \neq 0$, the expression to the right of the differentiation symbol on the r. h. s. of (46.2) and (46.3) may be integrated as a usual improper integral. Hence, transferring differentiation from this expression onto density by integration by parts, we arrive at an integral which exists in the usual sense. We illustrate this procedure for the case $\beta = 0$, the most important for our applications.

Assume $\beta = 0$, i. e. $k = k_1$ is an integer. Then using (46.2), write (46.1) as

$$I_{k+1}^{a} = \int_a^b \frac{g(\tau)d\tau}{(\tau - a)^{k+1}} = \frac{(-1)^k}{k!}\int_a^b g(\tau)\frac{d^{k+1}}{d\tau^{k+1}}\ln(\tau - a)d\tau.$$

The meaning of the integral is still undefined. To define it, exclude a small arc (a, t_1) at the distance ε_1 from the point a (Figure 40, a). Then the integral

$$\int_{t_1}^{b} \frac{g(\tau)d\tau}{(\tau-a)^{k+1}} = \frac{(-1)^k}{k!} \int_{t_1}^{b} g(\tau) \frac{d^{k+1}}{d\tau^{k+1}} \ln(\tau-a) d\tau$$

becomes a usual proper integral. Under the assumptions concerning the contour and density, we may use integration by parts. By applying this operation $k+1$ times we obtain:

$$\int_{t_1}^{b} \frac{g(\tau)d\tau}{(\tau-a)^{k+1}} = \frac{(-1)^k}{k!} \left[g(\tau) \frac{d^k}{d\tau^k} \ln(\tau-a) - g'(\tau) \frac{d^{k-1}}{d\tau^{k-1}} \ln(\tau-a) + \right.$$

$$\left. + g''(\tau) \frac{d^{k-2}}{d\tau^{k-2}} \ln(\tau-a) - \ldots + (-1)^k g^{(k)}(\tau) \ln(\tau-a) \right] \Bigg|_{t_1}^{b} -$$

$$- \frac{1}{k!} \int_{t_1}^{b} g^{(k+1)}(\tau) \ln(\tau-a) d\tau, \tag{46.4}$$

where $\Big|_{t_1}^{b}$ denotes the usual double substitution of the limits of integration. Employing (46.2), we may write (46.4) as

$$\int_{t_1}^{b} \frac{g(\tau)d\tau}{(\tau-a)^{k+1}} = \left[\frac{1}{k!} g^{(k)}(\tau) \ln(\tau-a) - \frac{1}{k!} g^{(k-1)}(\tau) \frac{1}{\tau-a} - \ldots - \frac{1}{k} g(\tau) \frac{1}{(\tau-a)^k} \right] \Bigg|_{t_1}^{b} -$$

$$- \frac{1}{k!} \int_{t_1}^{b} g^{(k+1)}(\tau) \ln(\tau-a) d\tau. \tag{46.5}$$

When $t_1 \to a$, the r. h. s. of (46.5) is a sum of terms of two different types. Some of these, those corresponding to the upper integration limit, and the integral, are "good" terms. "Bad" terms are those corresponding to the lower limit of integration; they tend to infinity when $\tau = a$.

Collect all the "bad" terms on the l. h. s. of (46.5). When doing this, we will consider only the real part of the logarithm to be "bad". Actually, we could consider the entire logarithm to be "bad", but we take only its real part to obtain a definition coinciding with the classical definition of the (Cauchy) principal value integral in the special case of a singular integral. We obtain:

$$\int_{t_1}^{b}\frac{g(\tau)d\tau}{(\tau-a)^{k+1}}+\frac{1}{k!}g^{(k)}(t_1)\ln|t_1-a|-\frac{1}{k!}\frac{g^{(k-1)}(t_1)}{t_1-a}-\ldots-\frac{1}{k}\frac{g(t_1)}{(t_1-a)^k}=$$
$$=-\frac{1}{k!}g^{(k)}(t_1)i\arg(t_1-a)+\left[\frac{1}{k!}g^{(k)}(b)\ln(b-a)-\frac{1}{k!}\frac{g^{(k-1)}(b)}{b-a}-\ldots-\frac{1}{k}\frac{g(b)}{(b-a)^k}\right]-$$
$$-\frac{1}{k!}\int_{t_1}^{b}g^{(k+1)}(\tau)\ln(\tau-a)d\tau. \qquad (46.6)$$

The r. h. s. of (46.6) being "good", has a limit when $t_1 \to a$, that is when $\varepsilon_1 \to 0$. Hence, the l. h. s., comprised of "bad" terms, also has the same limit. We define this common limit to be the complex finite-part integral. Thus, by definition

$$I_{k+1}^{a}=\int_{a}^{b}\frac{g(\tau)}{(\tau-a)^{k+1}}d\tau\equiv\lim_{\varepsilon_1\to 0}\left[\int_{t_1}^{b}\frac{g(\tau)}{(\tau-a)^{k+1}}d\tau+\frac{1}{k!}g^{(k)}(t_1)\ln|t_1-a|-\right.$$
$$\left.-\frac{1}{k!}\frac{g^{(k-1)}(t_1)}{t_1-a}-\ldots-\frac{1}{k(k-1)}\frac{g'(t_1)}{(t_1-a)^{k-1}}-\frac{1}{k}\frac{g(t_1)}{(t_1-a)^k}\right]. \qquad (46.7)$$

According to (46.6), the r. h. s. of (46.7) can be also represented by a sum of "good" terms. Thus we obtain an equivalent definition

$$I_{k+1}^{a}=\int_{a}^{b}\frac{g(\tau)}{(\tau-a)^{k+1}}d\tau\equiv-\frac{1}{k!}g^{(k)}(a)i\alpha_a+\frac{1}{k!}g^{(k)}(b)\ln(b-a)-\frac{1}{k!}\frac{g^{(k-1)}(b)}{b-a}-\ldots-$$
$$-\frac{1}{k(k-1)}\frac{g'(b)}{(b-a)^{k-1}}-\frac{1}{k}\frac{g(b)}{(b-a)^k}-\frac{1}{k!}\int_{a}^{b}g^{(k+1)}(\tau)\ln(\tau-a)d\tau, \qquad (46.8)$$

where α_a is the limiting value of the argument of $t_1 - a$, when $t_1 \to a$, i. e. it is the angle between the tangent at the point a and the x-axis. We see that the equivalent definition (46.8) contains only a *finite part* of the expression, which is obtained by formal integration; infinite terms of this expression have been dropped. This explains why the *direct* value of a divergent integral is also termed a *finite-part* integral. We will use both these names.

When the singular point is at the upper limit of integration (Figure 40, b), we exclude a small ε_2- vicinity of the point b. As a result, we obtain the definition

$$I_{k+1}^{b}=\int_{a}^{b}\frac{g(\tau)d\tau}{(\tau-b)^{k+1}}\equiv\lim_{\varepsilon_2\to 0}\left[\int_{a}^{t_2}\frac{g(\tau)d\tau}{(\tau-b)^{k+1}}-\frac{1}{k!}g^{(k)}(t_2)\ln|t_2-b|+\right.$$

$$+\frac{1}{k!}\frac{g^{(k-1)}(t_2)}{t_2-b}+\ldots+\frac{1}{k(k-1)}\frac{g'(t_2)}{(t_2-b)^{k-1}}+\frac{1}{k}\frac{g(t_2)}{(t_2-a)^k}\Bigg]. \qquad (46.9)$$

The formula representing this definition through the finite part of a divergent integral is analogous to (46.8):

$$I_{k+1}^b=\int_a^b\frac{g(\tau)d\tau}{(\tau-b)^{k+1}}\equiv\frac{1}{k!}g^{(k)}(b)\,i\alpha_b-\frac{1}{k!}g^{(k)}(a)\ln(b-a)+\frac{1}{k!}\frac{g^{(k-1)}(a)}{a-b}+\ldots+$$

$$+\frac{1}{k(k-1)}\frac{g'(a)}{(a-b)^{k-1}}+\frac{1}{k}\frac{g(a)}{(a-b)^k}-\frac{1}{k!}\int_a^b g^{(k+1)}(\tau)\ln(\tau-b)d\tau, \qquad (46.10)$$

where α_b is the limiting value of the argument of $b-t_2$, when $t_2\to b$, i. e. it is the angle between the tangent (in the direction of traveling) at the point b and the x-axis. Note that in the first term on the r. h. s. of (46.10) we have used the argument $\alpha_b=\arg(b-t_2)$ instead of $\arg(t_2-b)$, while in the second term we have taken $b-a$ instead of $a-b$ under the logarithm symbol; this has served to eliminate the terms $\pm i\pi$.

It is clear that these definitions provide integrals having the usual property of being linear with respect to integrands. Besides, it follows from (46.7), (46.9) that, just as for ordinary integrals, a change in the integration direction changes the sign of the finite-part integral.

For the integral

$$I_{k+1}(t)=\int_a^b\frac{g(\tau)d\tau}{(\tau-t)^{k+1}}, \qquad (46.11)$$

where $t\in(a,b)$, we assume that the limits of the density and its derivatives are the same from the both sides of the point t, i. e. $\lim g^{(j)}(t\pm)=g^{(j)}(t)$ $(j=0,\ldots,k)$. (Extension to the case, when these equalities are not fulfilled, is immediate if the corresponding limits exist).

For simplicity, take $\varepsilon_1=\varepsilon_2=\varepsilon$. Then $|t_2-t|=|t_1-t|=\varepsilon$ and we obtain

$$I_{k+1}=\int_a^b\frac{g(\tau)d\tau}{(\tau-t)^{k+1}}\equiv\lim_{\varepsilon\to 0}\Bigg[\int_a^{t_2}\frac{g(\tau)d\tau}{(\tau-t)^{k+1}}+\int_{t_1}^b\frac{g(\tau)d\tau}{(\tau-t)^{k+1}}+$$

$$+\frac{1}{k!}g^{(k-1)}(t)\frac{-\exp(-i\alpha_2)-\exp(-i\alpha_1)}{\varepsilon}+\ldots+$$

$$+\frac{1}{k(k-1)}g'(t)\frac{(-1)^{k-1}\exp(-(k-1)i\alpha_2)-\exp(-(k-1)i\alpha_1)}{\varepsilon^{k-1}}+$$

$$+\frac{1}{k}g(t)\frac{(-1)^k \exp(-ik\alpha_2)-\exp(-ik\alpha_1)}{\varepsilon^k}\Bigg], \tag{46.12}$$

where α_1 and α_2 are angles which the two, in general different, tangents (in the direction of traveling from a to b) at the point t make with the x-axis (Figure 40, c). At an internal point with a continuous tangent (Figure 40, d), we have $\alpha_1 = \alpha_2 = \alpha$, and (46.12) yields:

$$I_{k+1} = \int_a^b \frac{g(\tau)d\tau}{(\tau-t)^{k+1}} \equiv \lim_{\varepsilon\to 0}\Bigg\{\int_a^{t_2}\frac{g(\tau)d\tau}{(\tau-t)^{k+1}} + \int_{t_1}^b \frac{g(\tau)d\tau}{(\tau-t)^{k+1}} - 2\frac{1}{k!}\frac{g^{(k-1)}(t)\exp(-i\alpha)}{\varepsilon} - \ldots -$$
$$-\left[1-(-1)^{k-1}\right]\frac{1}{k(k-1)}\frac{g'(t)\exp[-(k-1)i\alpha]}{\varepsilon^{k-1}} - \left[1-(-1)^k\right]\frac{1}{k}\frac{g(t)\exp(-ik\alpha)}{\varepsilon^k}\Bigg\}, \tag{46.13}$$

where α is the angle of the tangent at the point t in the direction of travel from a to b.

An equivalent formula, presenting the definitions (46.12), (46.13) through the finite parts, is

$$I_{k+1} = \int_a^b \frac{g(\tau)d\tau}{(\tau-t)^{k+1}} = \frac{1}{k!}g^{(k)}(t)i(\alpha_2-\alpha_1) + \frac{1}{k!}\left[g^{(k)}(b)\ln(b-t) - g^{(k)}(a)\ln(t-a)\right] -$$
$$-\frac{1}{k!}\left[\frac{g^{(k-1)}(b)}{b-t} - \frac{g^{(k-1)}(a)}{a-t}\right] - \ldots - \frac{1}{k(k-1)}\left[\frac{g'(b)}{(b-t)^{k-1}} - \frac{g'(a)}{(a-t)^{k-1}}\right] -$$
$$-\frac{1}{k}\left[\frac{g(b)}{(b-t)^k} - \frac{g(a)}{(a-t)^k}\right] - \frac{1}{k!}\int_a^b g^{(k+1)}(\tau)\ln(\tau-t)d\tau. \tag{46.14}$$

For $k = 0$, equation (46.13) gives the well-known (Cauchy) principal value integral:

$$I_1 = \int_a^b \frac{g(\tau)d\tau}{\tau-t} \equiv \lim_{\varepsilon\to 0}\left[\int_a^{t_2}\frac{g(\tau)d\tau}{\tau-t} + \int_{t_1}^b \frac{g(\tau)d\tau}{\tau-t}\right].$$

Hence, there is no essential difference between complex singular Cauchy integrals and complex hypersingular Hadamard (finite-part) integrals. Thus when discussing hypersingular integrals we include (Cauchy) singular integrals. All the usual properties of integrals, such as linearity with respect to a density, and additivity with respect to integration intervals, remain. There is a change in the

Newton-Leibnitz formula for *complex* finite-part integrals; this will be discussed below.

For $k = 1$, the complex hypersingular integral (46.13) was introduced by Linkov and Mogilevskaya [2] (see also [3]). At a point with continuous tangent, their result turns into the formula:

$$I_2 = \int_a^b \frac{g(\tau)d\tau}{(\tau - t)^2} \equiv \lim_{\varepsilon \to 0} \left[\int_a^{t_2} \frac{g(\tau)d\tau}{(\tau - t)^2} + \int_{t_1}^b \frac{g(\tau)d\tau}{(\tau - t)^2} - \frac{2g(t)\exp(-i\alpha)}{\varepsilon} \right].$$

One-sided complex integrals for $k = 0$ and $k = 1$ were suggested and employed by Mogilevskaya [2]. In a general case $k > 1$, the definitions (46.7), (46.9), (46.11)-(46.13) were given in the review by Linkov and Mogilevskaya [5]. The finite-forms of these definitions (46.8), (46.10), (46.14) were not given explicitly in those papers. It is important that they clearly reveal the connection with integration formulae of Newton-Leibnitz type. Indeed, assume that the integral (46.11) has (formally) an antiderivative $G(\tau)$. Then from (46.14) it follows:

$$I_{k+1} = \int_a^b \frac{g(\tau)d\tau}{(\tau - t)^{k+1}} = G(b) - G(a) + \frac{1}{k!} g^k(t) i (\pi + \alpha_2 - \alpha_1). \qquad (46.15)$$

From (46.15) we see that the Newton-Leibnitz formula must be modified for finite-part integrals: its r. h. s. includes the additional term $(1/k!)g^{(k)}(t)i(\pi + \alpha_2 - \alpha_1)$. Its appearance is a direct consequence of the fact that we considered only the real part of the logarithm to be "bad". Certainly, one could consider the entire logarithm to be "bad". Then the Newton-Leibnitz formula would not change for finite-part integrals. However, the payment for this advantage would be deviation from the traditional definition of the (Cauchy) principal value integral. Besides, as it will be clear from § 48, this definition would influence formulae of Sokhotski-Plemelj type, which connect the limiting values of a complex hypersingular integral with its direct (finite-part) values. This would also lead away from the tradition established in scientific literature.

In conclusion, note that Hue and Mukherjee [1] introduced an alternative definition of two-sided complex finite-part integrals by using derivatives of Cauchy integrals, but only in the special case of a *real straight line*. In the next paragraph we will see that this definition, although different from that presented above, is equivalent to our definition.

§ 47. REGULARIZATION FORMULAE

In this paragraph we will prove three formulae serving to decrease the order of a finite-part integral up to a common improper integral. We will term such formulae regularization formulae. They, as well as the modified Newton-Leibnitz formula

(46.15), facilitate evaluation of direct values, avoiding a procedure of coming to a limit. Consider two-sided integrals. The regularization formulae for one-sided integrals are obtained analogously.[1]

The first regularization formula. Consider the two-sided integral (46.11):

$$I_{k+1} = \int_a^b \frac{g(\tau)d\tau}{(\tau - t)^{k+1}}. \tag{47.1}$$

Represent the density near the point t by a Taylor expansion:

$$\begin{gathered} g(\tau) = g_k(\tau) + R(\tau), \\ g_k(\tau) = g(t) + g'(t)(\tau - t) + \ldots + \frac{1}{k!} g^{(k)}(t)(\tau - t)^k, \end{gathered} \tag{47.2}$$

where $R(\tau)$ is the remainder term, of order $O((\tau - t)^{k+1})$. Integration of the density $g_k(\tau)$ in accordance with the modified Newton-Leibnitz formula (46.15) yields:

$$\int_a^b \frac{g_k(\tau)d\tau}{(\tau - t)^{k+1}} = \frac{1}{k} g(t)\left[\frac{1}{(a - t)^k} - \frac{1}{(b - t)^k}\right] +$$

$$+ \frac{g'(t)}{k - 1}\left[\frac{1}{(a - t)^{k-1}} - \frac{1}{(b - t)^{k-1}}\right] + \ldots + \frac{g^{(k)}(t)}{k!}\left[\ln\frac{b - t}{a - t} + i(\pi + \alpha_2 - \alpha_1)\right].$$

Then subtracting and adding $g_k(\tau)$ to the density in (47.1), we arrive at *the first regularization formula*:

$$I_{k+1} = \int_a^b \frac{g(\tau) - g_k(\tau)}{(\tau - t)^{k+1}} d\tau + \frac{1}{k} g(t)\left[\frac{1}{(a - t)^k} - \frac{1}{(b - t)^k}\right] +$$

$$+ \frac{g'(t)}{k - 1}\left[\frac{1}{(a - t)^{k-1}} - \frac{1}{(b - t)^{k-1}}\right] + \ldots + \frac{g^{(k)}(t)}{k!}\left[\ln\frac{b - t}{a - t} + i(\pi + \alpha_2 - \alpha_1)\right]. \tag{47.3}$$

Now the r. h. s. contains only a usual proper integral because, as mentioned, the remainder $g(\tau)$ - $g_k(\tau)$ = $R(\tau)$ has the order $O((\tau - t)^{k+1})$. In the special case of integration over an interval (a, b) of the real axis, we have $\alpha_1 = \alpha_2 = 0$, $i\pi + \ln[(b - t)/(a - t)] = \ln[(b - t)/(t - a)]$. This result shows that for a *real* interval, the accepted definition of the complex finite-part integral leads to the formula for real variables and to the usual Newton-Leibnitz formula.

---

[1] The explicit formulae for one-sided finite-part complex variable integrals are given in the Russian edition of the book.

Formula (47.3) decreases the order of singularity up to zero. In the same way we may decrease the order of singularity only by unity: it is sufficient to keep only one term in the Taylor expansion. Then for $k \geq 1$ we have:

$$\int_a^b \frac{g(\tau)d\tau}{(\tau - t)^{k+1}} = \int_a^b \frac{g(\tau) - g(t)}{(\tau - t)^{k+1}} d\tau + \frac{1}{k} g(t) \left[\frac{1}{(a - t)^k} - \frac{1}{(b - t)^k} \right]. \tag{47.4}$$

We will also call formula (47.4) *the first regularization formula*. Both formulae (47.3) and (47.4) are obtained by the standard method "subtract and add".

The second regularization formula. Write down (46.14) by taking the derivative $g'(\tau)$ as the density and setting $k = s - 1$ ($s \geq 1$). We obtain:

$$\int_a^b \frac{g'(\tau)d\tau}{(\tau - t)^s} = \frac{1}{(s-1)!} g^{(s)}(t) i (\alpha_2 - \alpha_1) + \frac{1}{(s-1)!} \left[g^{(s)}(b) \ln(b - t) - g^{(s)}(a) \ln(t - a) \right] -$$

$$- \frac{1}{(s-1)!} \left[\frac{g^{(s-1)}(b)}{b - t} - \frac{g^{(s-1)}(a)}{a - t} \right] - \ldots - \frac{1}{(s-1)(s-2)} \left[\frac{g''(b)}{(b - t)^{s-2}} - \frac{g''(a)}{(a - t)^{s-2}} \right] -$$

$$- \frac{1}{s-1} \left[\frac{g'(b)}{(b - t)^{s-1}} - \frac{g'(a)}{(a - t)^{s-1}} \right] - \frac{1}{(s-1)!} \int_a^b \left[g^{(s+1)}(\tau) \ln(\tau - t) \right] d\tau. \tag{47.5}$$

When $s = k$, the r. h. s. of (47.5) differs from that of (46.14) only by the additional multiplier $1/k$ and the additional term $(1/k)[g(b)/(b - t)^k - g(a)/(a - t)^k]$ in (46.14). Hence, the finite-part integral (47.1) can be represented by *the second regularization formula*:

$$\int_a^b \frac{g(\tau)d\tau}{(\tau - t)^{k+1}} = \frac{1}{k} \int_a^b \frac{g'(\tau)}{(\tau - t)^k} d\tau - \frac{1}{k} \left[\frac{g(b)}{(b - t)^k} - \frac{g(a)}{(a - t)^k} \right]. \tag{47.6}$$

Formula (47.6) decreases the singularity order by unity. Applying it $k + 1$ times, we arrive at a regular integral. By writing (47.6) as

$$\int_a^b \frac{g'(\tau)d\tau}{(\tau - t)^k} = \frac{g(\tau)}{(\tau - t)^k} \Bigg|_{\tau=a}^{\tau=b} + k \int_a^b \frac{g(\tau)d\tau}{(\tau - t)^{k+1}}$$

we see that, in essence, the second regularization formula expresses the usual rule of integration by parts. Hence, like usual integrals, *finite-part integrals may be integrated by parts*.

The third regularization formula. Use again (46.14). By setting $k = s - 1$ ($s \geq 1$) we have:

$$\int_a^b \frac{g(\tau)d\tau}{(\tau-t)^s} = \frac{1}{(s-1)!} g^{(s-1)}(t) i(\pi+\alpha_2-\alpha_1) + \frac{1}{(s-1)!}\left[g^{(s-1)}(b)\ln(b-t) - \right.$$
$$\left. - g^{(s-1)}(a)\ln(a-t)\right] - \frac{1}{(s-1)!}\left[\frac{g^{(s-2)}(b)}{b-t} - \frac{g^{(s-2)}(a)}{a-t}\right] - \ldots - \frac{1}{(s-1)(s-2)}\left[\frac{g'(b)}{(b-t)^{s-2}} - \frac{g'(a)}{(a-t)^{s-2}}\right] -$$
$$- \frac{1}{s-1}\left[\frac{g(b)}{(b-t)^{s-1}} - \frac{g(a)}{(a-t)^{s-1}}\right] - \frac{1}{(s-1)!}\int_a^b g^{(s)}(\tau)\ln(\tau-t)d\tau. \quad (47.7)$$

Differentiate both parts of (47.7) with respect to t. We obtain:

$$\frac{d}{dt}\int_a^b \frac{g(\tau)d\tau}{(\tau-t)^s} = \frac{1}{(s-1)!} g^{(s)}(t)\, i(\alpha_2-\alpha_1) - \frac{1}{(s-1)!}\left[\frac{g^{(s-1)}(b)}{b-t} - \frac{g^{(s-1)}(a)}{a-t}\right] - \ldots -$$
$$- \frac{1}{s-1}\left[\frac{g'(b)}{(b-t)^{s-1}} - \frac{g'(a)}{(a-t)^{s-1}}\right] - \left[\frac{g(b)}{(b-t)^s} - \frac{g(a)}{(a-t)^s}\right] -$$
$$- \frac{1}{(s-1)!}\left[g^{(s)}(b)\ln(b-t) - g^{(s)}(a)\ln(t-a)\right] - \frac{1}{(s-1)!}\int_a^b g^{(s)}(\tau)\ln(\tau-t)d\tau. \quad (47.8)$$

When differentiating the integral on the r. h. s. of (47.7), we have used integration by parts by representing $\ln(\tau - t)d\tau$ as the differential of the integral from logarithm $d\int\ln(\zeta - t)d\zeta$. Besides, we again have used the equality $\ln(a-t) = i\pi + \ln(t-a)$. Comparing (47.8) taken for $s = k$ with (46.14), we see that the r. h. s. of (46.14) differs only by the multiplier $1/k$. Thus, for $k \geq 1$ we have the *third regularization formula*:

$$\int_a^b \frac{g(\tau)d\tau}{(\tau-t)^{k+1}} = \frac{1}{k}\frac{d}{dt}\int_a^b \frac{g(\tau)d\tau}{(\tau-t)^k}. \quad (47.9)$$

It decreases the order of singularity by unity. Re-writing it as

$$\frac{d}{dt}\int_a^b \frac{g(\tau)d\tau}{(\tau-t)^k} = k\int_a^b \frac{g(\tau)d\tau}{(\tau-t)^{k+1}},$$

we see that actually equation (47.9) implies that *finite-part integrals may be differentiated under the integral symbol.*

Applying (47.9) k times, we obtain an expression for a finite-part integral of an arbitrary order $k + 1$ through the k-th derivative of a principal value (Cauchy) integral:

$$\int_a^b \frac{g(\tau)d\tau}{(\tau - t)^{k+1}} = \frac{1}{k!}\frac{d^k}{dt^k}\int_a^b \frac{g(\tau)d\tau}{\tau - t}. \tag{47.10}$$

For $k = 0$ we have

$$\int_a^b \frac{g(\tau)d\tau}{\tau - t} = -\frac{d}{dt}\int_a^b g(\tau)\ln(\tau - t)d\tau.$$

In combination with (47.8) this equation allows us to replace a finite-part integral of the $(k + 1)$-th order by the $(k + 1)$-th derivative of a usual improper integral.

Note that Hui and Mukherjee [1] used formula (47.10) as a definition of a finite-part integral, although only for the real axis (the definition (27) and formula (26) of their cited paper). Now we see that this formula may serve as a general alternative definition. Certainly, with conventions accepted regarding the density and the integration contour, one can also use equivalent definitions (47.4), (47.6), (47.9). Moreover, by using well-known results for Cauchy integrals (e. g. Muskhelishvili [4], Gahov [1]) and the regularization formulae, it is easy to see that it is sufficient to assume that the density in the previous equations has only a k-th Holder continuous derivative.

§ 48. FORMULAE CONNECTING LIMIT AND DIRECT VALUES OF COMPLEX HYPERSINGULAR INTEGRALS

The *direct* values of complex hypersingular integrals, introduced in the two previous paragraphs, are still no more than an elegant formal construction, a pure mental exercise. In applications, as emphasized in chapters I and II, we need the *limiting* values of complex potentials to satisfy prescribed boundary conditions. Hence, in order to use direct values of divergent integrals we must define their connection with limiting values. This connection is rather simple.

Consider the complex hypersingular potential

$$\Phi_{k+1}(z) = \frac{1}{2\pi i}\int_a^b \frac{g(\tau)d\tau}{(\tau - z)^{k+1}} \tag{48.1}$$

when the point z does not belong to the integration contour (a, b). When $k + 1 = 1$, the integral on the r. h. s. is a Cauchy type integral. Its properties are comprehensively studied (see, e. g. Muskhelishvili [4]). For $k + 1 > 1$, it is a Hadamard type integral. The function $\Phi_{k+1}(z)$ defined by it, is holomorphic outside

(a, b). It has derivatives of any order and we may apply the usual rules of integration by parts and differentiation under the integral sign.

We are interested in the limiting values $\Phi_{k+1}^{+}(z)$, $\Phi_{k+1}^{-}(z)$ of the integral (48.1) when z tends to a point t on the contour (a, b) from the left (+) or right (-) of the traveling path. Assume that the point t does not coincide with the ends of an integration arc, and suppose for simplicity that it is not a corner point. Under the conventions of § 46 regarding the density and the contour, the limits $\Phi_{k+1}^{+}(z)$, $\Phi_{k+1}^{-}(z)$ exist. We may easily see this by writing the kernel of the integral as the k-th derivative of $1/(\tau - z)$:

$$\Phi_{k+1}(z)=\frac{(-1)^k}{k!}\frac{1}{2\pi i}\int_a^b g(\tau)\frac{\partial^k}{\partial\tau^k}\frac{1}{\tau - z}\,d\tau.$$

Then, applying integration by parts k times, we obtain:

$$\Phi_{k+1}(z)=-\frac{1}{2\pi i}\left[\frac{1}{k!}\frac{g^{(k-1)}(\tau)}{\tau - z}+\ldots+\frac{1}{k(k-1)}\frac{g'(\tau)}{(\tau - z)^{k-1}}+\right.$$
$$\left.+\frac{1}{k}\frac{g(\tau)}{(\tau - z)^k}\right]\Bigg|_{\tau=a}^{\tau=b}+\frac{1}{k!}\frac{1}{2\pi i}\int_a^b\frac{g^{(k)}(\tau)d\tau}{\tau - z}. \tag{48.2}$$

The result of double substitution on the r. h. s. of (48.2) is continuous across the integration arc at its internal points. The limits of the integral on the r. h. s. of (48.2) are given by the Sokhotski-Plemelj formulae (11.26). Finally, we obtain expressions for the limits $\Phi_{k+1}^{+}(z)$, $\Phi_{k+1}^{-}(z)$:

$$\Phi_{k+1}^{\pm}(t)=-\frac{1}{2\pi i}\left[\frac{1}{k!}\frac{g^{(k-1)}(\tau)}{\tau - t}+\ldots+\frac{1}{k(k-1)}\frac{g'(\tau)}{(\tau - t)^{k-1}}+\right.$$
$$\left.+\frac{1}{k}\frac{g(\tau)}{(\tau - t)^k}\right]\Bigg|_{\tau=a}^{\tau=b}+\frac{1}{k!}\left[\pm\frac{1}{2}g^k(t)+\frac{1}{2\pi i}\int_a^b\frac{g^{(k)}(\tau)d\tau}{\tau - t}\right], \tag{48.3}$$

where the integral on the r. h. s. is understood as a (Cauchy) principal value integral.

From (48.3) by (47.6) applied successively k times, we have:

$$\Phi_{k+1}^{\pm}(t)=\pm\frac{1}{2}\frac{g^{(k)}(t)}{k!}+\frac{1}{2\pi i}\int_a^b\frac{g(\tau)d\tau}{(\tau - t)^{k+1}}, \tag{48.4}$$

where the integral on the r. h. s. is a finite-part integral defined in § 46.

Formula (48.4) is the required result. It expresses the limiting values of a complex hypersingular integral through its direct value. Thus, the direct values of hypersingular integrals become involved in boundary value problems. When $k = 0$, this formula naturally turns into the classical Sokhotski-Plemelj formula (11.26), which served us for the derivation. For $k = 1$, it was first published by Linkov and Mogilevskaya [2] (see also the paper by these authors [3]). In the general case $k > 1$, it was derived in Linkov and Mogilevskaya [5]. An analogous formula that follows from (48.4) after replacing the integral on the r. h. s. by the r. h. s. of (47.10) was obtained by Hui and Mukherjee [1]. The paper by Ladopulos [1] contains analogous formula for the special case of the real axis and odd $k + 1$.

We now establish the rule for interchanging the order of differentiation and taking a limit. To derive this rule, write the hypersingular integral of the first (Cauchy) order

$$\Phi_1(z)=\frac{1}{2\pi i}\int_a^b\frac{g(\tau)d\tau}{\tau-z} \tag{48.5}$$

and the Sokhotski-Plemelj formula for it:

$$\Phi_1^{\pm}(t)=\pm\frac{1}{2}g(t)+\frac{1}{2\pi i}\int_a^b\frac{g(\tau)d\tau}{\tau-t}. \tag{48.6}$$

Differentiate (48.5) k times with respect to z. Then, by definition (48.1), we obtain:

$$\frac{d^k\Phi_1(z)}{dz^k}=\frac{k!}{2\pi i}\int_a^b\frac{g(\tau)d\tau}{(\tau-z)^{k+1}}=k!\Phi_{k+1}(z).$$

By taking the limit and accounting for (48.4) we have:

$$\left[\frac{d^k\Phi_1(z)}{dz^k}\right]^{\pm}=k!\left[\pm\frac{1}{2}\frac{g^k(t)}{k!}+\frac{1}{2\pi i}\int_a^b\frac{g(\tau)d\tau}{(\tau-t)^{k+1}}\right]. \tag{48.7}$$

Now differentiate (48.6) k times and take (47.9) into account. We obtain:

$$\frac{d^k\Phi_1^{\pm}(t)}{dt^k}=\pm\frac{1}{2}g^{(k)}(t)+k!\frac{1}{2\pi i}\int_a^b\frac{g(\tau)d\tau}{(\tau-t)^{k+1}}. \tag{48.8}$$

Comparison (48.7) with (48.8) yields

$$\left[\frac{d^k\Phi_1(z)}{dz^k}\right]^{\pm} = \frac{d^k\Phi_1^{\pm}(t)}{dt^k}. \tag{48.9}$$

From (48.9) it immediately follows that

$$\left[\frac{d\Phi_k(z)}{dz}\right]^{\pm} = \frac{d\Phi_k^{\pm}(t)}{dt}. \tag{48.10}$$

Equations (48.9) and (48.10) imply that one may interchange the order of differentiation and taking a limit in hypersingular integrals. We use this result when deriving complex BIE.

Chapter 11

COMPLEX VARIABLE HYPERSINGULAR INTEGRAL EQUATIONS (CVH-BIE)

As noted in chapter 10, hypersingular integrals can be obtained by differentiation of singular integrals and/or integration by parts. This gives a connection between complex variable hypersingular and singular equations. Hence, we can construct a theory of complex variable hypersingular equations (CVH-BIE) by employing the well-established theory of singular integral equations (SIE). We use such an approach in the theoretical discussion in this chapter. Paragraphs 49-51 reproduce the results briefly presented in papers by Linkov and Mogilevskaya (Linkov and Mogilevskaya [4], Linkov, Zoubkov and Mogilevskaya [1]). The general case of the plane elasticity problem for a set of open and closed contours is presented in § 52 for the first time.

§ 49. PROBLEM FORMULATION

Consider a complex variable hypersingular integral equation (CVH-BIE)

$$K_H\varphi \equiv A(t)\varphi(t)+\frac{1}{\pi i}\int_L\frac{K_H(\tau,t)\varphi(\tau)}{(\tau-t)^2}d\tau=f(t), \tag{49.1}$$

where the integral is understood as a finite-part integral; the contour L is a set of non-intersecting closed and open sufficiently smooth parts; $A(t)$, $f(t)$ are prescribed functions of the H_0 class. We assume that the kernel $K_H(\tau, t)$ may be represented as

$$K_H(\tau,t)=B_H(t)+B(t)(\tau-t)+k(\tau,t)(\tau-t)^2,$$

where $B_H(t) = dK_H(\tau, t)/d\tau\big|_{\tau=t}$, $k(\tau, t)$ is a function of the H_0 class in both variables, $k(\tau, t) = [K_H(\tau, t) - B_H(t) - B(t)(\tau - t)]/ (\tau - t)^2$. The equation (49.1) takes the form

$$K_H\varphi \equiv \frac{B_H(t)}{\pi i}\int_L\frac{\varphi(\tau)d\tau}{(\tau-t)^2}+K\varphi=f(t), \tag{49.2}$$

where K is the singular operator:

$$K\varphi = A(t)\varphi(t) + \frac{B(t)}{\pi i}\int_L \frac{\varphi(\tau)d\tau}{\tau - t} + k\varphi, \tag{49.3}$$

the non-singular operator k is defined as:

$$k\varphi = \frac{1}{\pi i}\int_L k(\tau,t)\varphi(\tau)d\tau.$$

Recall some definitions of the general theory of singular equations (Muskhelishvili [4]). The equation $K\varphi = 0$ should be satisfied at all *ordinary* points of L, which are points other than a finite number n of *nodes* on the contour L. The nodes, denoted henceforth c_k ($k = 1,\ldots, n$) or simply c, are the *start* and *end points* of open arcs and *discontinuity points* of $G(t)$ and $f_0(t)$. Besides, for each of closed contours we take one of its points to be a node; it serves both as the start and end point of this contour. We assume that each contour is a smooth line in the sense discussed in § 11. The characteristic part of a singular operator K

$$K^0\varphi = A(t)\varphi(t) + \frac{B(t)}{\pi i}\int_L \frac{\varphi(\tau)}{\tau - t}d\tau$$

defines *non-special* nodes and classes of functions. These are nodes where a solution of the homogeneous equation $K^0\varphi = 0$ may have singularity of the type $(t - c)^\alpha$ with $0 < \alpha < 1$. Let m be the total number of non-special nodes. Consider a function $\varphi(t)$, which turns to zero at some of them. For definiteness, we assume that these are the first q of the non-special nodes. Then $\varphi(t)$ is called the function of the *class* $h_q = h(c_1,\ldots, c_q)$. The class of functions which turn to zero at the remaining $q' = m - q$ non-special nodes $c_{q+1},\ldots, c_m$ is called the *adjoint class* and denoted $h_{q'} = h(c_{q+1},\ldots, c_m)$. The widest class, corresponding to $q = 0$, is denoted h_0; the narrowest class, corresponding to $q = m$, is denoted h_m. For hypersingular equations we introduce the *class* h_m^* of functions belonging to h_m and having a derivative of the class h_0.

The equation adjoint to (49.1) is obtained as usual by interchanging τ and t under the integral sign:

$$K_H{}'\psi \equiv A(t)\psi(t) + \frac{1}{\pi i}\int_L \frac{K_H(\tau,t)\psi(\tau)}{(\tau - t)^2}d\tau = g(t), \tag{49.4}$$

where $g(t)$ is a function of the H_0 class and a solution is in the class h_m^*. Note that here we differ from the theory of singular equations: for the latter, a solution of the adjoint equation is looked for in the *adjoint* class, whereas for a hypersingular adjoint equation we look for the function $\psi(t)$ belonging to the *same* class h_m^* as the solution of the initial CVH-BIE (49.2).

We can re-write equation (49.4) as

$$K_H{}'\psi \equiv \frac{1}{\pi i}\int_L \frac{B_H(\tau)\psi(\tau)}{(\tau-t)^2}\,d\tau + K'\psi = g(t), \tag{49.5}$$

where K' is the singular operator adjoint of the singular operator K:

$$K'\psi = A(t)\psi(t) - \frac{1}{\pi i}\int_L \frac{B(\tau)\psi(\tau)}{\tau - t}\,d\tau + k'\psi,$$

k' is a non-singular operator adjoint to k, it is defined as:

$$k'\psi = \frac{1}{\pi i}\int_L k(t,\tau)\psi(\tau)\,d\tau.$$

We term an *index* of the CVH-BIE the difference between the numbers of linearly independent solutions of the homogeneous CVH-BIE (49.2) and its adjoint CVH-BIE (49.5).

§ 50. CASE OF INTERMITTENT LINE

Reduction of CVH-BIE to equivalent singular equations. Let L be a set of p isolated open arcs (cuts, cracks) $L_k = a_k b_k$ $(k = 1,\ldots,p)$. Assume for simplicity that the nodes are only the start a_k and end b_k points of arcs, and all these nodes are non-special. Thus the total number of nodes is $m = 2p$.

We will reduce the initial *hypersingular* equation (49.2) and its adjoint equation (49.5) to *singular* equations in order to use the results of the theory of singular equations. To this end, apply the second regularization formula (47.6) to (49.2). Then for the derivative $\varphi'(t)$ we obtain the SIE

$$K_S\varphi' \equiv \frac{B_H(t)}{\pi i}\int_L \frac{\varphi'(\tau)}{\tau - t}\,d\tau + K\,k_0\varphi' = f(t), \tag{50.1}$$

where the operator k_0 integrates over each isolated arc according to the formula

$$k_0\varphi' \equiv \int_{a_k}^{t} \varphi'(\tau)\,d\tau = \int_L k_0(\tau,t)\varphi'(\tau)\,d\tau, \quad t \in L_k, \tag{50.2}$$

where $k_0(\tau, t) = 1$ when $\tau \in [a_k, t]$ and $k_0(\tau, t) = 0$ when $\tau \in (t, b_k]$.

Apply the third regularization formula (47.9) to the adjoint CVH-BIE (49.5). By integrating the result over an arc L_k, to which the point t belongs, from this point to end point b_k, we obtain:

$$K_S'\psi \equiv -\frac{1}{\pi i}\int_L \frac{B_H(\tau)\psi(\tau)}{\tau - t}d\tau + k_0'K'\psi = g_1(t) + C(t), \tag{50.3}$$

where

$$k_0'\psi \equiv \int_t^{b_k} \psi(\tau)d\tau = \int_L [k_0'(\tau,t)\psi(\tau)]d\tau, \quad t \in L_k,$$

k_0' is an operator adjoint to the operator k_0; for it $k_0'(\tau, t) = k_0(t, \tau)$, i.e. $k_0'(\tau, t) = 1$ when $\tau \in (t, b_k]$ and $k_0'(\tau, t) = 0$ when $\tau \in [a_k, t]$;

$$g_1(t) = k_0'g = \int_t^{b_k} g(\tau)d\tau;$$

$C(t)$ is a piece-wise constant function on isolated arcs L_k ($k = 1,\ldots, p$):

$$C(t) = C_k = -\frac{1}{\pi i}\int_L \frac{B_H(\tau)\psi(\tau)}{\tau - b_k}d\tau, \quad t \in L_k. \tag{50.4}$$

The singular equations (50.1) and (50.3) are adjoint with respect to functions of adjoint classes. We presume that SIE (50.1) is solved with respect to the derivative $\varphi'(t)$ in the widest class h_0, while the adjoint equation (50.3) is solved in the narrowest (adjoint) class h_{2p}.

We point out an important connection between solutions of hypersingular equation (49.2) and the corresponding singular equation (50.1). If a function $\varphi(t)$ is a solution of (49.2) of the class h_{2p}^*, then its derivative $\varphi'(t)$ is a solution of SIE (50.1) of the class h_0. Conversely, if $\varphi'(t)$ is a solution of SIE (50.1) of the class h_0, then the function $k_0\varphi'(t)$ is a solution of CVH-BIE (49.2) of the class h_{2p}^*, provided that the additional conditions

$$\int_{L_k} \varphi'(\tau)d\tau = 0, \quad k = 1,\ldots, p. \tag{50.5}$$

are satisfied. In this sense, CVH-BIE (49.2) is equivalent to SIE (50.1). From this it immediately follows that since the number of linearly independent solutions of SIE (50.1) is finite, the number of linearly independent solutions of CVH-BIE is also finite.

There is also a close connection between solutions of adjoint equations, hypersingular (49.5) and singular (50.3). To establish it, note that if for some set of constants C_k, SIE (50.3) has a solution $\psi(t)$, this solution certainly satisfies equations (50.4). Indeed, by taking for this solution $t = b_k$ ($k = 1,\ldots, p$) in (50.3), we have by definition of the integration operator k_0': $g_1(t) = 0$, $k_0'K\psi = 0$, and equation (50.3) turns into equation (50.4). We conclude that (50.3) is solvable only for $C(t)$ of the form (50.4). Thus, a solution of the CVH-BIE (49.5) of the class h_{2p}^* is a solution of SIE of the same class if we prescribe the constants C_k by formula (50.4). Conversely, if for some set of constants C_k the SIE (50.3) has a solution $\psi(t)$ of the class h_{2p}^*, then the function $\psi(t)$ is a solution of the CVH-BIE (49.5) of the same class. In this sense, the SVH-BIE (49.5) is equivalent to the SIE (50.3).

This equivalence between the CVH-BIE (49.2) and the SIE (50.1), and also between their adjoint CVH-BIE (49.5) and the SIE (50.3) allows us to focus on singular equations by employing Muskelishvili's theory [4].

Theorem on finite number of eigenfunctions. Consider the adjoint SIE, to which the CVH-BIE (49.2), (49.5) are reduced when their r. h. s. is zero:

$$\frac{B_H(t)}{\pi i}\int_L \frac{\varphi'(\tau)}{\tau - t}\,d\tau + K\,k_0\varphi' = 0, \tag{50.6}$$

$$-\frac{1}{\pi i}\int_L \frac{B_H(\tau)\psi(\tau)}{\tau - t}\,d\tau + k_0{}'K'\psi = C(t), \tag{50.7}$$

where the piece-wise constant function $C(t)$ is defined by (50.4). Henceforth, we consider the constants C_k entering $C(t)$ to be assembled into the column-vector C.

The index of equation (50.6) in the class h_0 may be calculated by employing the theory of Muskhelishvili [4]; it is p. The index of the adjoint equation in the adjoint class h_m is $-p$. Let q be the number of linearly independent solutions of the homogeneous ($C(t) = 0$) equation (50.7) in the class h_m. Then, the theory of SIE (Muskhelishvili [4]) implies that for solvability (50.7), the vector $C(t)$ should satisfy a system of $p + q$ equations for the p constants C_k ($k = 1,\ldots, p$):

$$\sum_{k=1}^{p} C_k \int_{a_k}^{b_k} \varphi_j'(\tau)\,d\tau = 0, \quad j = 1,\ldots, q + p, \tag{50.8}$$

where $\varphi_1',\ldots,\varphi_{q+p}'$ is a complete set of linearly independent solutions of SIE (50.6).

A homogeneous system of linear algebraic equations has a finite number of eigenvectors. (In our case, since $q \geq 0$, this number for (50.8) is zero when (50.8) has only the trivial solution $C_k = 0$ ($k = 1,\dots, p$)). Hence, the number of linearly independent vectors C, for which the system (50.7) is solvable, is finite. Consequently, the number of solutions of homogeneous (under $g(t) = 0$) CVH-BIE (49.5) is also finite. Since a similar statement holds for the CVH-BIE (49.2), we arrive at the first theorem.

Theorem 1. *The number of linearly independent solutions of the homogeneous (under $f(t) = 0$) CVH-BIE (49.2) in the class h_{2p}^* is finite. The same holds for the adjoint CVH-BIE (49.5).*

Theorem on equal number of eigenfunctions of adjoint CVH-BIEs. We now calculate the number of eigenfunctions of each of the CVH-BIE (49.2) and (49.5) in the class h_{2p}^*.We start from the second equation.

First note that, under the restrictions imposed on the kernel $K(\tau, t)$, the general theory of SIE (Muskhelishvili [4]) implies that a solution of SIE (50.7) of the class h_{2p} belongs also to the class h_{2p}^*. Then a complete set of linearly independent solutions of homogeneous (under $g(t) = 0$) CVH-BIE (49.5) consists of two parts:

1) a complete set of linearly independent solutions $\psi_j(t)$ ($j = 1,\dots, q$) of homogeneous (under $C(t) = 0$) SIE (50.7) in the class h_{2p}^*;

2) a set of linearly independent solutions of the inhomogeneous (under $C(t) \neq 0$) SIE (50.7) in the class h_{2p}^*.

The first group contains q solutions. Calculate the number of solutions in the second group. Let r be the rank of the matrix of the linear algebraic system (50.8); $0 \leq r \leq p$. Then (50.8) has $p - r$ linearly independent solutions. This implies that there are exactly $p - r$ linearly independent vectors $C^j(t) = (C_1^j,\dots, C_p^j)$ for which equation (50.7), and therefore the homogeneous CVH-BIE (49.5) has solutions. These solutions are not nul-solutions: otherwise, due to (50.4), the corresponding vector $C^j(t)$ would be zero.

We show that each of the vectors $C^j(t)$ increases the number of eigenfunctions of CVH-BIE (49.5) by exactly one. First we show that, for each $C^j(t)$, this number increases by not more than one. Indeed, suppose that it generates two solutions $\psi_1(t)$ and $\psi_2(t)$. As stated, they serve as solutions of the SIE (50.7) when $C(t) = C^j(t)$. Then the difference $\psi_1(t) - \psi_2(t)$ is a solution of homogeneous (under $C(t) = 0$) SIE (50.7). Hence, it should be a linear combination of eigenfunctions of this SIE. We conclude that one of the two solutions is a linear combination of the other and of the eigenfunctions of (50.7). Thus it is not a linearly independent eigenfunction of (49.5); this concludes the proof.

On the other hand, the eigenfunctions of (49.5) corresponding to linearly independent vectors $C^j(t)$ are linearly independent. Indeed, if these eigenfunctions were linearly dependent, then due to (50.4), the vectors $C^j(t)$ would be also linearly dependent. We conclude that each vector $C^j(t)$ generates exactly one linearly independent eigenfunction of the CVH-BIE (49.5).

We see that the number of eigenvectors belonging to the second group equals the number of linearly independent vectors $C^j(t)$. Hence, it is $p - r$.

Finally, *the total number of linearly independent solutions $\psi_j(t)$ of CVH-BIE (49.5) is equal to $q + p - r$.*

Consider now the homogeneous (under $f(t) = 0$) CVH-BIE (49.2). It is reduced to the homogeneous SIE (50.6). The latter, according to Muskhelishvili's theory, has $q + p$ linearly independent solutions $\varphi_1',\ldots,\varphi_{q+p}'$ of the class h_0. Their linear combination

$$\varphi_*' = \gamma_1\varphi_1' + \ldots + \gamma_{q+p}\varphi_{q+p}' \tag{50.9}$$

is also a solution of SIE (50.9). The function

$$\varphi_*(t) = \int_{a_j}^{t} \varphi_*'(\tau)d(\tau), \quad t \in L_j \tag{50.10}$$

will be a solution of the homogeneous (under $f(t) = 0$) CVH-BIE (49.2) in the class h_{2p}^* if the conditions (50.5) are met. They give a system of p linear algebraic equations for the $q + p$ coefficients γ_k:

$$\sum_{k=1}^{q+p} \gamma_k \int_{a_j}^{b_j} \varphi_k'(\tau)d\tau = 0, \quad j = 1,\ldots, p. \tag{50.11}$$

Comparing (50.11) with (50.8) we see that the matrix of the system (50.11) is the transposed matrix of the system (50.8). Hence, they have the same rank r ($0 \le r \le p$). Then r (basic) constants γ_k are expressed through $q + p - r$ remaining (free) constants. For definiteness, assume that the basic constants are the first r constants γ_k, while the free constants are the remaining $q + p - r$ constants.

We always can distinguish $q + p - r$ linearly independent sets of free constants $\gamma_{r+1},\ldots, \gamma_{q+p-r}$ (for instance by setting one of them equal to unity and others to zero). Then for each of such sets the system (50.11) defines corresponding basic constants. Substitution of the constants γ_k into (50.9) provides a solution φ_*', the integral of which satisfies the CVH-BIE (49.2). As a result we obtain $q + p - r$ solutions $\varphi_{*s}(t)$ ($s = 1,\ldots, q + p - r$) of the CVH-BIE (49.2).

We study these solutions. Their derivatives, by (50.9), have the form

$$\varphi_{*s}'(t) = \gamma_1^s\varphi_1' + \ldots + \gamma_r^s\varphi_r' + \tilde{\varphi}_s' \quad s = 1,\ldots, q + p - r, \tag{50.12}$$

where

$$\tilde{\varphi}_s' = \gamma_{r+1}^s\varphi_{r+1}' + \ldots + \gamma_{q+p-r}^s\varphi_{q+p}'.$$

The linear combinations $\tilde{\varphi}_s'$ of the linear independent eigenfunctions φ_k' ($k = q + p - r,\ldots, q + p$) are also eigenfunctions. They are linearly independent by the choice of the constants γ_k^s ($k, s = 1,\ldots, q + p - r$). Besides, they do not depend on the first r eigenfunctions $\varphi_1',\ldots,\varphi_r'$. This implies that the solutions of the SIE (50.6) $\varphi_{*s}'(t)$ ($s = 1,\ldots, q+p-r$) defined by (50.12) are linearly independent (otherwise, a function $\tilde{\varphi}_s'$ would be expressed through a linear combination of the functions $\varphi_1',\ldots,\varphi_r'$).

The solutions of CVH-BIE (49.2) $\varphi_{*s}(t)$, resulting from integration $\varphi_{*s}'(t)$ (s = 1,..., $q+p-r$), are also linear independent. This is a complete set of linearly independent solutions of the CVH-BIIE (49.2). Indeed, if the homogeneous (under $f(t) = 0$) CVH-BIE (49.2) has other solutions, they may be only constants, because $\varphi_1',\ldots,\varphi_{q+p}'$ is a complete set of linearly independent solutions of SIE (50.6). But since this solution should turn to zero at start points a_k, these constants are zero.

We conclude that a complete set of linearly independent solutions of CVH-BIE (49.2) consists exactly of $q+p-r$ functions $\varphi_{*s}(t)$ (s = 1,..., $q+p-r$). This number $q+p-r$ coincides with the number of linearly independent solutions of the adjoint CVH-BIE (49.5). Hence, the following theorem is true.

Theorem 2. *The adjoint CVH-BIE (49.2) and (49.5) have the same number $q+p-r$ of linearly independent solutions in the class h_{2p}^*, i.e. the index of equation (49.2) in this class is zero.*

This important property of hypersingular equations (49.2), (49.5) provides their advantage over the singular equations (50.1), (50.3), obtained by using integration by parts and/or derivatives. Often it allows us to avoid additional equations.

Theorem on solvability of CVH-BIE. It is easy to see that *necessary* conditions for solvability of CVH-BIE (49.2) are the orthogonality conditions

$$\int_L f(\tau)\psi_k(\tau)d\tau = 0, \tag{50.13}$$

where $\psi_k(t)$ is a complete set of eigenfunctions of the adjoint equation (49.5) (k = 1,..., $q+p-r$). As usual, the proof follows from the equation

$$\int_L (K_H\varphi)\psi_k\,d\tau = \int_L \varphi(K_H'\psi_k)d\tau, \tag{50.14}$$

where interchange in the integration order is justified by the regularization formulae and by similar formulae for singular integrals. Since $K_H\varphi = f$, $K_H'\psi_k = 0$, equation (50.14) reduces to (50.13).

We now prove that equations (50.13) are *sufficient* for solvability of the CVH-BIE (49.2). Suppose that the $q+p-r$ conditions (50.13) are satisfied. By convention, the first q functions ψ_k are solutions of the homogeneous (under $C(t)$ = 0) SIE (50.7). The remaining $p-r$ functions satisfy the inhomogeneous SIE (50.7):

$$K_S{}'\psi_{q+k} = C^k(t), \quad k = 1,..., p - r, \tag{50.15}$$

with the r. h. s. presenting a system of $p - r$ linearly independent vectors $C^k(t)$ ($C^k(t) = C^k_j$ when $t \in L_j, j = 1,\dots, p$).

The first q conditions (50.13) imply solvability of the inhomogeneous SIE (50.1).We show that due to the remaining $p - r$ conditions (50.13), its solution $\varphi'(t)$ satisfies (50.5). Multiply both parts of (50.1) by any of the remaining functions ψ_{q+k} ($k = 1,\dots, p - r$) and integrate over L. Using (50.13), we obtain

$$\int_L (K_S\varphi')\psi_{q+k}d\tau = \int_L f(\tau)\psi_{q+k}(\tau)d\tau = 0.$$

From properties of a singular operator we have

$$\int_L (K_S\varphi')\psi_{q+k}d\tau = \int_L \varphi'(K_S{}'\psi_{q+k})d\tau,$$

and in combination with the previous equation this yields

$$\int_L \varphi'(K_S{}'\psi_{q+k})d\tau = 0. \tag{50.16}$$

Substitution of (50.15) into (50.16) gives $p - r$ equations

$$\sum_{j=1}^{p} C^k_j \int_{a_j}^{b_j} \varphi'(\tau)d\tau = 0, \quad k = 1,..., p - r \tag{50.17}$$

with respect to p unknown integrals $\int_{a_j}^{b_j} \varphi'(\tau)\,d\tau$.

The vectors $C^k(t)$ ($k = 1,\dots, p - r$) are linearly independent. This implies that the rank of the matrix of the system (50.17) is $p - r$. As it is equal to the number of equations, the system (50.17) has only the zero solution. Hence,

$$\int_{a_j}^{b} \varphi'(\tau)d\tau = 0, \quad j = 1,..., p.$$

We conclude that SIE (50.1) has a solution satisfying the additional conditions (50.5). This implies that the CVH-BIE (49.2) is solvable in the class h_{2p}^*. We have proved the following theorem.

Theorem 3. *The necessary and sufficient conditions for solvability of inhomogeneous CVH-BIE (49.2) are*

$$\int_L f(\tau)\psi_k(\tau)d\tau = 0, \qquad (50.18)$$

where $\psi_k(t)$ is a complete set of linearly independent solutions of the adjoint CVH-BIE (49.5) in the class h_{2p}^* $(k = 1, \dots, q + p - r)$.

From theorems 2 and 3 it follows that if a homogenous CVH-BIE does not have an eigenfunction, then the inhomogeneous CVH-BIE (49.2) has a solution for any function $f(t)$ of the H_0 class, and this solution is unique.

Equations with complex conjugate unknowns. As for singular equations, the results may be extended to the CVH-BIE, which contain $\varphi(t)$ and its complex conjugate $\overline{\varphi(t)}$. Such equations are usual in plane elasticity problems. Consider such an equation:

$$K_H\varphi \equiv \frac{B_H(t)}{\pi i}\int_L \frac{\varphi(\tau)}{(\tau - t)^2}\,d\tau + K_S\varphi = f(t), \qquad (50.19)$$

where K_S is the singular operator:

$$K_S\varphi \equiv A(t)\varphi(t) + \frac{B(t)}{\pi i}\int_L \frac{\varphi(\tau)}{\tau - t}d\tau + k_1\varphi + k_2\overline{\varphi},$$

$B_H(t)$, $A(t)$, $B(t)$, $k_1(\tau, t)$, $k_2(\tau, t)$ are prescribed functions of the H_0 class on L.

The equation adjoint to (50.19) is defined by the formula

$$K_H'\psi \equiv \frac{1}{\pi i}\int_L \frac{B_H(\tau)\psi(\tau)}{(\tau - t)^2}\,d\tau + K_S'\psi = g(t), \qquad (50.20)$$

where K_S' is the singular operator adjoint to the operator K_S; it is defined by formula

$$K_S'\psi \equiv A(t)\psi(t) - \frac{1}{\pi i}\int_L \frac{B(\tau)\psi(\tau)d\tau}{\tau - t} + \int_L k_1(t,\tau)\psi(\tau)d\tau + \int_L \overline{k_2(t,\tau)}\;\overline{\psi(\tau)}d\tau$$

$g(t)$ is a prescribed function of the H_0 type; $\psi(t)$ is an unknown function of the h_{2p}^* class.

To study equations (50.19), (50.20) one can use the same approach employed earlier to transfer from hypersingular equations to equivalent singular equations. We apply the regularization formula (57.6) to the CVH-BIE (50.19) and the regularization formula (57.9) to the adjoint CVH-BIE (50.20). In this way we arrive at the adjoint SIE. Like the initial CVH-BIE, they contain both an unknown function and its complex conjugate. The theory of such equations is worked out in detail in (Mandjavidze [1, 2], Muskhelishvili [4]). Employing it in a scheme similar to that used earlier for equations (49.2), (49.5), we study equations (50.19), (50.20). Eventually, we conclude that theorems 1-3 of this paragraph stay true for the equations (50.19), (50.20) with minor changes. Specifically:

1) a linear combination of functions is understood in a *narrow sense* (only with real coefficients);

2) solvability conditions of (50.18) type take the form

$$\operatorname{Re}\int_L f(\tau)\psi_k(\tau)d\tau = 0, \tag{50.21}$$

where $\psi_k(t)$ is a complete set of linearly independent (in the narrow sense) solutions of the adjoint homogeneous CVH-BIE (50.20) in the h_{2p}^* class. As before, *the number of eigenfunctions of the homogeneous CVH-BIE is finite, and the index is zero.*

§ 51. CLOSED CONTOURS

Consider equation (49.2):

$$K_H\varphi \equiv \frac{B_H(t)}{\pi i}\int_L \frac{\varphi(\tau)}{(\tau-t)^2}d\tau + K\varphi = f(t), \tag{51.1}$$

assuming that L is a smooth closed contour while the function $\varphi(t)$ has a derivative of the class H on L. (A more general case, when the contour consists of a number of open and closed parts is discussed in the next paragraph).

By using the second regularization formula (57.6), we reduce (51.1) to the integro-differential equation (IDE):

$$\frac{B_H(t)}{\pi i}\int_L \frac{\varphi'(\tau)}{\tau-t}d\tau + K\varphi = f(t). \tag{51.2}$$

The equation adjoint to (51.1) is given by formula (49.5):

$$K'_H \psi \equiv \frac{1}{\pi i} \int_L \frac{B_H(\tau)\psi(\tau)}{(\tau - t)^2} d\tau + K'\psi = g(t), \tag{51.3}$$

where K' is the singular operator adjoint to K.

Again, by using the second regularization formula (57.6), we transform (51.3) to an IDE:

$$\frac{1}{\pi i} \int_L \frac{[B_H(\tau)\psi(\tau)]'}{\tau - t} d\tau + K'\psi = g(t). \tag{51.4}$$

It is equivalent to CVH-BIE (51.3) in the class of solutions having the derivative of the H class.

We see that studying the adjoint CVH-BIE (51.1) and (51.3) is reduced to studying IDE (51.2) and (51.4). These IDE's are special cases of more general integro-differential equations investigated by R. V. Isahanov [1, 2]:

$$\sum_{S=0}^{m} \left[A_S(t)\varphi^{(S)}(t) + \frac{1}{\pi i} \int_L \frac{R_S(\tau,t)\varphi^{(S)}(\tau)}{\tau - t} d\tau \right] = f(t), \tag{51.5}$$

where A_S are prescribed functions of the H class; $R_S(\tau, t)$ are prescribed functions having derivatives with respect to both arguments. Equation (51.5) turns into IDE (51.2) when $m = 1$, $A_1(t) = 0$, $A_0(t) = A(t)$, $R_1(\tau, t) = B_H(t)$, $R_0(\tau, t) = B(t)+k(\tau, t)(\tau\text{-}t)$.

The equation adjoint to (51.5) is defined by the formula (Isahanov [2])

$$\sum_{S=0}^{m} \left\{ (-1)^S [A_S(t)\psi(t)]^{(S)} - \frac{(-1)^S}{\pi i} \int_L \left(\frac{\partial^S}{\partial\tau^S} + \frac{\partial^S}{\partial t^S} \right) [R_S(\tau,t)\psi(\tau)] \frac{d\tau}{\tau - t} \right\} = g(t). \tag{51.6}$$

In our case the adjoint equation (51.6) takes the form of IDE (51.4). We see that IDE (51.2) and (51.4) are adjoint in the sense of the theory by Isahanov. Hence, results of this theory immediately lead to conclusions regarding (51.2), (51.4) and consequently, (51.1), (51.3).

The index of IDE (51.5) is defined as (Isahanov [2])

$$\eta = \frac{1}{2\pi} \arg \frac{A_m(t) - R_m(t,t)}{A_m(t) + R_m(t,t)} \Bigg|_L, \tag{51.7}$$

where the r. h. s. contains the increment of the argument when traveling the entire contour L counter clockwise. In our case, $A_1(t) = 0$, $R_1(\tau, t) = B_H(t)$. Then formula (51.7) gives the index $\eta = 0$.

Use now three theorems for adjoint IDE (51.5), (51.6) (Isahanov [2]).

Theorem 1. *The number of linearly independent solutions of the homogeneous (under $f(t) = 0$) IDE (51.5) is finite. The same is true also for the adjoint IDE (51.6).*

Theorem 2. *The difference between the numbers of linearly independent solutions of the homogeneous IDE (51.5) and (51.6) is equal to the index of equation (51.5).*

Theorem 3. *The necessary and sufficient conditions for solvability of the inhomogeneous IDE (51.5) are*

$$\int_L f(\tau)\psi_k(\tau)d\tau = 0, \tag{51.8}$$

where $\psi_k(t)$ is a complete set of linearly independent solutions of the adjoint homogeneous IDE (51.6).

Theorems 1-3 are true for IDE (51.2), (51.4) with the derivatives of the first order. Because of the mentioned equivalence of the last equations to the adjoint CVH-BIE (51.1), (51.3), analogous theorems hold for the CVH-BIE (51.1), which has a zero index This implies that if CVH-BIE (51.1) does not have eigenfunctions, it is solvable for an arbitrary right hand side.

So far, these theorems have been proved only for a *single* closed contour, but without reservations regarding CVH-BIE additional to those stated in § 49. In the next paragraph we will obtain results for an *arbitrary* number of open and closed contours but under additional restrictions imposed on CVH-BIE. It is important, that CVH-BIE of elasticity theory satisfy these restrictions.

§ 52. CVH-BIE OF ELASTICITY THEORY

Let the contour L of an elastic plane region S consists of a finite number q of smooth open and closed parts. For definiteness, assume that traction is prescribed on the contour L. Then the complex hypersingular equation for displacement discontinuities has the form of (29.10):

$$\frac{1}{\pi i}\int_L \frac{\Delta u(\tau)}{(\tau-t)^2}d\tau - \frac{1}{2\pi i}\int_L \Delta u(\tau)\frac{\partial}{\partial t}d\ln\frac{\tau-t}{\overline{\tau}-\overline{t}} - \frac{1}{2\pi i}\int_L \overline{\Delta u(\tau)}\frac{\partial}{\partial t}d\frac{\tau-t}{\overline{\tau}-\overline{t}} = f(t), \tag{52.1}$$

where $f(t)$ is a function defined by the prescribed traction. If for simplicity we assume that the discontinuity of traction is zero at least at crack tips, the r. h. s. of (52.1) belongs to the class H on L.

Equation (52.1) takes the form of CVH-BIE (50.19) if we set $A(t) = 0$, $B(t) = 1$ and denote

$$k_1\varphi = -\frac{1}{2\pi i}\int_L \varphi(\tau)\frac{\partial^2}{\partial t\partial\tau}\ln\frac{\tau-t}{\overline{\tau}-\overline{t}}d\tau,\quad k_2\overline{\varphi} = -\frac{1}{2\pi i}\int_L \overline{\varphi(\tau)}\frac{\partial^2}{\partial t\partial\tau}\frac{\tau-t}{\overline{\tau}-\overline{t}}d\tau.$$

An intermittent line. For an intermittent line comprised of p cuts (cracks), we can apply all the results of § 50 directly to equation (52.1). In particular, the index of CVH-BIE (52.1) is equal to zero. By considering the energy functional (see, e.g. Muskhelishvili [4, § 113]) we see that the displacement discontinuities on the cuts are defined uniquely. Then the theory of § 50 implies that equation (52.1) is solvable in the class h_{2p}^* for an arbitrary r. h. s. $f(t)$ of the class H. This solution is unique.

In the class h_{2p}^*, the displacement discontinuities tend to zero at crack tips. Thus, in contrast with singular equations for crack problems, there are no additional conditions to be satisfied.

Closed contours. Assume that the contour L is comprised of only closed contours L_k ($k = 1,\dots, q$). Then on each of these contours we have:

$$\int_{L_k} \frac{d\tau}{(\tau - t)^2} = 0, \quad \int_{L_k} \frac{\partial}{\partial t} d \ln \frac{\tau - t}{\bar{\tau} - \bar{t}} = 0, \quad \int_{L_k} \frac{\partial}{\partial t} d \frac{\tau - t}{\bar{\tau} - \bar{t}} = 0.$$

This yields an important conclusion regarding (52.1): its l. h. s. is zero for piece-wise constant on L functions $C(t)$ ($C(t) = C_k$ when $t \in L_k$, $k = 1,\dots, q$). It is clear that we may choose q such functions $C^j(t)$ ($j = 1,\dots, q$) to be linearly independent, by taking for instance, $C^j_k = 1$ when $j = k$ and $C^j_k = 0$ when $j \neq k$. Hence, equation (52.1) for closed contours has at least q eigenfunctions of the class h_{2p}^*, and these functions are constant on each particular contour L_k ($k = 1,\dots, q$).

Now consider the more general CVH-BIE (50.19) for closed contours:

$$K_H \varphi \equiv \frac{B_H(t)}{\pi i} \int_L \frac{\varphi(\tau)}{(\tau - t)^2} d\tau + K_S \varphi = f(t). \tag{52.2}$$

Assume that it also has the stated property of having at least q eigenvalues which are piece-wise constant along particular contours. The same reservation is assumed regarding the CVH-BIE (49.2), not containing the complex conjugate unknown function:

$$K_H \varphi = \frac{B_H(t)}{\pi i} \int_L \frac{\varphi(\tau)}{(\tau - t)^2} d\tau + K\varphi = f(t). \tag{52.3}$$

Recall that analysis of these equations serves as a basis for studying more general equations containing complex conjugates. Write down the equation (49.5) adjoint to it:

$$K_H{}'\psi \equiv \frac{1}{\pi i}\int_L \frac{B_H(\tau)\psi(\tau)}{(\tau - t)^2}d\tau + K'\psi = g(t). \tag{52.4}$$

Following § 50 introduce SIE of (50.1) form:

$$K_S\varphi' \equiv \frac{B_H(t)}{\pi i}\int_L \frac{\varphi'(\tau)}{\tau - t}\,d\tau + K\,k_0\varphi' = f(t) \tag{52.5}$$

and the adjoint to it, SIE (50.3):

$$K_S{}'\psi \equiv -\frac{1}{\pi i}\int_L \frac{B_H(\tau)\psi(\tau)}{\tau - t}d\tau + k_0{}'K'\psi = g_1(t) + C(t), \tag{52.6}$$

where the entire notation is that of § 50, but the contour L now is comprised of q particular closed contours.

CVH-BIE (52.3) is equivalent (in the sense of searching for solutions of the h_{2p}^*class) to SIE (52.5) under the q conditions

$$\int_L \varphi'(\tau)d\tau = 0, \quad k = 1,\ldots,p. \tag{52.7}$$

These are analogous to the conditions (50.5) for open arcs.

There is a connection between the CVH-BIE (52.4) and the SIE (52.6) similar to that stated in § 50. Specifically, a solution of the CVH-BIE (52.4) of the class h_{2p}^* is also a solution of the SIE (52.6) of the same class, if the constants C_k are defined by (50.4). Conversely, if for some set of the constants C_k, equation (52.6) has a solution $\psi(t)$ of the class h_{2p}^*, then the function $\psi(t)$ is a solution of CVH-BIE (52.4) of the same class. In this sense, the CVH-BIE (52.4) and the SIE (52.6) are equivalent.

From the general theory by Muskhelishvili ([4, § 45]) we conclude that the index of the SIE (52.3) is zero. Hence the adjoint SIE (52.5) and (52.6) have the same number m of eigenfunctions. For the equation (52.5) denote them by $\varphi_j'(t)$ ($j = 1,\ldots, m$).

We calculate the number of linearly independent solutions of homogeneous (under $g(t) = 0$) adjoint SIE (52.4). Similar to § 50, it consists of linearly independent solutions of the homogeneous (under $g_1(t) + C(t) = 0$) SIE (52.6) and some number of linearly independent solutions of the inhomogeneous (under $g_1(t) = 0$, $C(t) \neq 0$) SIE (52.6).

According to the general theory of SIE, the solvability conditions for the SIE (52.6) under $g_1(t) = 0$, $C(t) \neq 0$ are expressed by m equations

$$\sum_{j=1}^{q} C_k^j \int_{L_k} \varphi_j'(\tau)d\tau = 0, \quad j = 1,\ldots, m. \tag{52.8}$$

Let the rank of the matrix of the linear algebraic system (52.8) be r ($0 \le r \le \min(q, m)$). Then this system provides exactly $q - r$ linearly independent vectors $C^j(t)$, for which SIE (52.6) has solutions under $g_1(t) = 0$. As in § 50, we state that these solutions are linearly independent and satisfy the CVH-BIE (52.4). Hence, the inhomogeneous (under $g_1(t) = 0$, $C(t) \neq 0$) equation (52.6) provides exactly $q - r$ linearly independent solutions of the homogeneous (under $g_1(t) = 0$) CVH-BIE (52.4). Finally, a complete set of solutions of this equation contains $m + q - r$ eigenfunctions.

Turn now to the homogeneous (under $f(t) = 0$) CVH-BIE (52.3). It is reduced to the homogeneous (under $f(t) = 0$) SIE (52.5), which, as assumed, has m eigenfunctions $\varphi_j'(t)$ ($j = 1,\ldots, m$). Their linear combinations

$$\varphi_*' = \gamma_1\varphi_1' + \ldots + \gamma_m\varphi_m' \tag{52.9}$$

also serve as solutions of the SIE (52.5). We distinguish among them those which satisfy the CVH-BIE (52.3). It is necessary and sufficient for them to satisfy equations (52.7). Substitution of (52.9) into (52.7) provides a linear algebraic system for the coefficients γ_k ($k = 1,\ldots, m$):

$$\sum_{k=1}^{q} \gamma_k \int_{L_j} \varphi_k'(\tau)d\tau = 0, \quad j = 1,\ldots, p. \tag{52.10}$$

The matrix of the system (52.10) is the transposed matrix of the system (52.8). Hence, it has the same rank r. Then r constants γ_k are expressed through the $m - r$ remaining constants. Thus, the same arguments which were used in § 50 when analyzing the system (50.11), show that there exist $m - r$ linearly independent functions φ_{*s}' ($s = 1,\ldots, m - r$), which satisfy the conditions (52.7). This implies that the CVH-BIE (52.3) has linearly independent solutions obtained by integration φ_{*s}':

$$\varphi_{*s}(t) = \int_{a_j}^{t} \varphi_{*s}'(\tau)d\tau, \quad t \in L_j, \quad s = 1,\ldots,m - r, \quad j = 1,\ldots, q, \tag{52.11}$$

where a_j is an arbitrary fixed point on the contour L_j ($j = 1,\ldots, q$).

The eigenfunctions of CVH-BIE (52.3) other than given by (52.11), can be only constants on contours L_j ($j = 1,\ldots, q$). By the convention accepted above, these constants are really solutions of (52.3). As noted, the number of their linearly independent combinations is equal to the number q of the contours L_j. Hence, the number of linearly independent solutions of the homogeneous CVH-BIE (52.3) is equal to $m + q - r$.

We see that the number of linearly independent solutions of the adjoint CVH-BIE (52.3) and (52.4) is the same and equal to $m + q - r$. This means that Theorem 1 (on finite number of eigenfunctions) and Theorem 2 (on zero index), stated in § 50 for intermittent contours hold for closed contours also. Following the path of § 50, it is easy to verify that the Theorem 3 (on solvability of CVH-BIE) is also true. Extension to the CVH-BIE (52.2) containing the complex conjugate unknown function is carried out following the standard scheme mentioned in § 50. As before, the only differences are that for equation (52.2) linear independence is understood in the *narrow* sense (for combinations with *real* coefficients) and the solvability conditions (50.21) are used instead of (50.18).

We emphasize again that these results are stated only for the case when constants on particular contours L_j ($j = 1,\ldots, q$) serve as solutions of homogeneous CVH-BIE (52.2). Elasticity equation (52.1) satisfies this requirement. Hence, the three main theorems are true for (52.1) in the case of a set of closed contours. However, now unlike the case of open arcs (cracks), the CVH-BIE (52.1) has at least q eigenfunctions. Their physical interpretation is as follows. They correspond to rigid translation of interior regions (fictitious blocks) within holes and/or rigid translation of the whole region itself (for a finite region). There may be additional eigenfunctions corresponding to rigid rotation of these regions. These rigid movements may be fixed by additional conditions of (30.13), (30.14) type. The solvability conditions of (50.21) type correspond to zero total resultant force and zero moment applied to a boundary of each of the holes.

General case of open and closed contours. We can easily extend the results to the case when the contour L contains both closed and open non-intersecting contours under the assumption accepted above for closed contours. We assume again that constants on each of closed contours present solutions of the homogeneous CVH-BIE (52.3) or the more general CVH-BIE (52.2) containing the complex conjugate function. The proof of the Theorems 1-3 is obtained by joining the arguments of § 50 for intermittent contour with those of this paragraph for closed contours. The proof is lengthy, but does not contain any new elements.

From Theorems 1-3 it follows again that the CVH-BIE of elasticity theory (52.1) is solvable in the class h_{2p}^*. Each closed contour generates its eigenfunction corresponding to rigid movement of its interior region and/or of the (finite) region itself. Again, rigid movement may be fixed by additional conditions discussed.

Comment. Having in mind computational applications, we note that the theory presented in this and previous chapter concerns only *direct* values of integrals. As consistently pointed out in the previous parts of the book, this corresponds to the numerical procedure of coming to the limit *before* integration. An alternative procedure consists in coming to the limit *after* integration. It does not involve direct values of divergent integrals, i.e. principal value and finite-part integrals. Equations corresponding to this procedure contain the symbol of limit (lim) in front of singular and/or hypersingular integrals (see, e.g. § 15, 20). Solvability of such equations is an immediate consequence of solvability of equations considered above: it is sufficient to change the various limits by formulae of Sokhotski-Plemelj type. We conclude that the theory presented in this part may also serve for formal justification of numerical procedures of coming to the limit *after* integration.

PART IV

NUMERICAL SOLUTION OF COMPLEX VARIABLE BOUNDARY INTEGRAL EQUATIONS

In the introduction we mentioned advantages of the complex variables BIE over their real analogues. The main one is easy evaluation of singular and hypersingular integrals over curvilinear arcs. In real variables, this operation presents a major obstacle; it involves parameterization of the contour and numerical integration. In complex variables, as we will see, evaluation of singular and hypersingular integrals over *curvilinear* elements is as simple as their evaluation over *straight* segment of the real axis. Specifically, if a singular or hypersingular integral of a real function is evaluated analytically over a straight element, actually the same formulae are applicable for the corresponding complex function integrated over an arbitrary curvilinear element. Moreover, for a number of commonly used elements and approximations, *all* complex integrals, including regular ones, are evaluated analytically. As a result, we obtain simple quadrature rules for the complex variable boundary element method (CV-BEM).

Other computational advantages of the complex variables are also important: complex variable BIE are always more compact than their real forms; in the latter, real and imaginary parts of a complex BIE are separated, and this leads to more cumbersome expressions. Advantages of the complex variables become especially evident when employing Kelvin's singular solution. Then the number of standard integrals to be evaluated does not exceed six, often it is only three. All these integrals, as we will see, can be evaluated from just *one* integral when integration is performed over straight or circular arc elements.

Modern computers operate with complex quantities as easily as with real ones. Checking calculations and examining computer code are also simplified.

These advantages of the CV-BIE are very attractive for numerical implementation. They excite interest in complex equations and stimulated the writing of this book. For these reasons, in this last part, we present the two most popular methods for solving CV-BIE, the complex variable boundary element method (CV-BEM) and the complex variable method of mechanical quadratures (CV-MMQ).

Chapter 12

COMPLEX VARIABLE BONDARY ELEMENT METHOD (CV-BEM)

§ 53. GENERAL STAGES OF BEM

The boundary element method (BEM) is connected with elements of the *boundary* of a region. *Boundary* integral equations (BIE) serve as its basis. The essence of the method lies in a specific approach to solving such equations. The BEM uses additivity of integrals: an integral over a whole boundary (a surface in 3D or a contour in 2D) is equal to the sum of integrals over elements of which the boundary is composed or into which it is divided. So, first of all, the boundary is represented by a set of such elements termed *boundary elements*. This procedure is called *discretization* of a boundary.

We may approximate functions within each element by a combination (usually linear) of some basis functions. This comprises the *approximation* stage.

Substitution of approximated functions into integrals over boundary elements transforms a BIE into an approximate equation. In it the integrals are taken only from basis functions over elements. Evaluation of the integrals from basis functions comprises the stage of *integration* over elements.

Next, we require the approximate equation to be satisfied exactly at a finite number of collocation points, or in the mean over an element with a finite number of weight functions. (In the latter case, one needs to perform iterated integration). As a result we arrive at an algebraic system with respect to unknown coefficients in the approximations to the unknown functions. Its *solution* provides the unknown coefficients of functions on boundary elements and, consequently, approximates unknown functions.

These approximate expressions are used in the final stage to calculate values within the region. We use integral representation of fields within the region by integrals over its boundary. To evaluate the integrals at this stage, we again use boundary elements and approximations over them, actually reproducing the first two stages. This normally allows us to employ the former procedures but for field points, not boundary points. Now all the coefficients of the approximations are known.

We see that the boundary element method, both real and complex, includes the following five stages:

discretization of a boundary into elements;

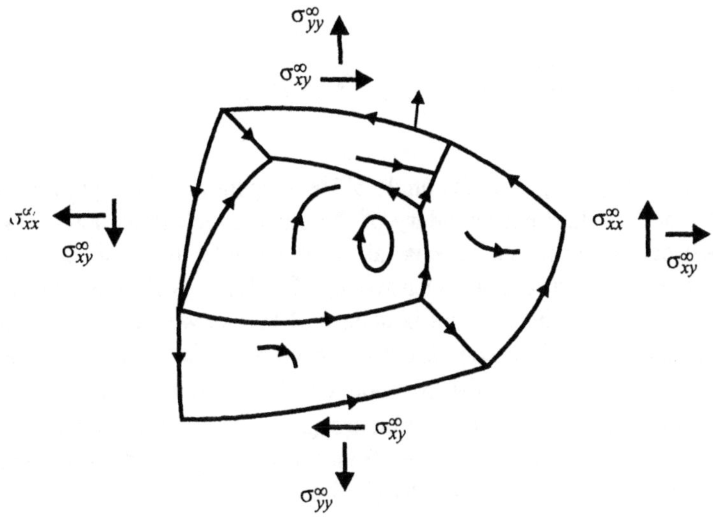

Fig. 41.

approximation of functions within each element;

evaluation of integrals from basis functions over elements;

formation and solution of an algebraic system of equations;

calculation of values within the region; if necessary, previously unknown values, such as tangential stresses, are calculated on the boundary as well.

For the CV-BEM, the listed stages, except the fourth, are performed in complex variables. In the fourth stage it is convenient to separate real and complex parts. This is because the BIE of elasticity theory contain not only an unknown function but also its complex conjugate. Certainly, one might take the conjugate values as additional complex unknowns and continue calculations in complex variables to the end. However, this doubles the number of complex unknowns.

We illustrate these stages of the CV-BEM by using the hypersingular equation (23.4) for blocky structures with continuous traction through the contacts of blocks (Figure 41):

$$\frac{1}{2\pi i}\left[2\int_L \frac{\Delta u(\tau)}{(\tau-t)^2}d\tau - \int_L \Delta u(\tau)\frac{\partial}{\partial t}dk_1(\tau,t) - \int_L \overline{\Delta u(\tau)}\frac{\partial}{\partial t}dk_2(\tau,t) + \int_L (2a_1 - a_3)\frac{\sigma(\tau)}{\tau - t}d\tau + \right.$$

$$\left. + \int_L (a_1 - a_3)\sigma(\tau)\frac{\partial}{\partial t}k_1(\tau,t)d\tau + \int_L a_1\overline{\sigma(\tau)}\frac{\partial}{\partial t}k_2(\tau,t)d\bar{\tau}\right] =$$

$$= \frac{1}{2}a_2\sigma(t) + F(t), \quad t \in L. \tag{53.1}$$

Here, we have accounted for the stresses s_{xx}, s_{yy}, s_{xy} at infinity when the blocks are imbedded in an infinite matrix with elasticity parameters χ_∞, μ_∞ in accordance with (30.18)-(30.20):

$$F(t) = -\frac{\chi_\infty + 1}{2\mu_\infty}\left[s_{xx} + s_{yy} + (s_{yy} - s_{xx} - 2i\, s_{xy})\exp(-2i\alpha_0)\right].$$

For a finite region, $F(t) = 0$; α_0 is the angle of the tangent at a field point t with the axis x. Recall that L is the total boundary of the blocks, inclusions and cracks, and the contact between adjacent surfaces is taken as a single line with the normal **n** directed to the right to the direction **t** of traveling. The latter is chosen arbitrarily for each element of L, but for the external boundary of blocks we, as usual, presume it is taken counter-clockwise. The complex traction vector σ is taken in the local co-ordinates (n, t): $\sigma = \sigma_{nn} + i\sigma_{nt}$, while the complex displacement discontinuity vector $\Delta u = u^+ - u^-$ is taken in the global co-ordinates (x, y): $\Delta u = \Delta u_x + i\, \Delta u_y$. The sign "plus" ("minus") refers to the limit values from the left (right) side of the traveling path. The parameters a_1, a_2, a_3 are given by (30.12):

$$a_1 = \frac{1}{2\mu^+} - \frac{1}{2\mu^-}, \quad a_2 = \frac{\chi^+ + 1}{2\mu^+} + \frac{\chi^- + 1}{2\mu^-}, \quad a_3 = \frac{\chi^+ + 1}{2\mu^+} - \frac{\chi^- + 1}{2\mu^-}.$$

The kernels $k_1(\tau, t)$ and $k_2(\tau, t)$ are defined by (12.18): $k_1(\tau, t) = \ln[(\tau - t)/(\bar{\tau} - \bar{t})]$, $k_2(\tau, t) = (\tau - t)/(\bar{\tau} - \bar{t})$. If the region is finite and traction is prescribed on its entire boundary, then an arbitrary rigid displacement satisfies the CVH-BIE (53.1). To exclude this eigenfunction, we can use additional equations (30.13), (30.14); if the traction is continuous across the contour:

$$\int_L \frac{\Delta u + a_1 f}{\tau - z_{01}} d\tau = 0, \quad \mathrm{Re}\int_L \frac{\Delta u + a_1 f}{(\tau - z_{02})^2} d\tau = 0.$$

Here, z_{01}, z_{02} are fixed points in some blocks (if the origin is within a block, we may for instance set $z_{01} = z_{02} = 0$). If there are holes within the blocks, we add similar equations with fixed points in each hole.

Equation (53.1) is true also for points off the boundary: within blocks and in the surrounding matrix if the blocks are embedded in a matrix. When applying it to these points, we set $t = z$, $\partial/\partial t = \partial/\partial z$ and in the definition of a_2 we assume $\mu^+ = \mu^- = \mu^j$, $\chi^+ = \chi^- = \chi^j$, where μ^j, χ^j is the shear modulus and the Muskhelishvili parameter at the point where the stresses are calculated. (Recall § 11, that by taking the direction ∂z along each of the co-ordinate axes we obtain all the components of the stress tensor in the global system).

We need to solve (53.1) under prescribed conditions on the contour L. They may include the following: prescribed traction on some parts of L (for instance, on surfaces of cracks); stresses at infinity (for infinite regions); prescribed dependence between displacement discontinuity and traction on contacts between blocks, inclusions and/or on interacting surfaces of cracks; prescribed displacements on some parts of L (in particular, $\Delta u = 0$ on ideal contacts). After solving CVH-BIE (53.1), this formula can be used to find stresses within the region. If needed, displacements are found by employing (30.25).

We discuss the stages of the CVH-BIE (53.1).

1. ***Discretization of the boundary.*** This stage concerns the geometry of the problem. We divide the boundary L into a finite number Q of boundary elements L_q ($q = 1,\dots, Q$). The elements may be arbitrary smooth arcs. They may coincide with parts of L, or one may represent a contour approximately by standard straight, circular arc, or elliptic elements. This allows us to evaluate analytically not only singular and hypersingular integrals but also ordinary (proper) integrals. To avoid ill-conditioning at stage 4, it is advisable to use elements with similar sizes. Smaller elements may be used in local parts of a boundary in repeated (adaptive) calculations, by employing the results of preliminary calculations with large elements.

After the discretization of the boundary, equation (53.1) is approximated by

$$\frac{1}{2\pi i}\sum_{q=1}^{Q}\left[2\int_{L_q}\frac{\Delta u(\tau)}{(\tau-t)^2}d\tau - \int_{L_q}\Delta u(\tau)\frac{\partial}{\partial t}dk_1(\tau,t) - \int_{L_q}\overline{\Delta u(\tau)}\frac{\partial}{\partial t}dk_2(\tau,t)\right] +$$

$$+\sum_{q=1}^{Q}\left[(2a_1 - a_3)\int_{L_q}\frac{\sigma(\tau)}{\tau - t}d\tau + (a_1 - a_3)\int_{L_q}\sigma(\tau)\frac{\partial}{\partial t}k_1(\tau,t)d\tau + a_1\int_{L_q}\overline{\sigma(\tau)}\frac{\partial}{\partial t}k_2(\tau,t)d\overline{\tau}\right] =$$

$$= \frac{1}{2}a_2\sigma(t) + F(t), \quad t \in L. \tag{53.2}$$

2. ***Approximation of functions within elements.*** Equations (53.1) and (53.2) contain known or unknown displacement discontinuities and tractions. Approximate them in the q-th element by linear combinations of basis functions:

$$\Delta u^q(t) \approx \sum_{m=1}^{M_q} u^{qm} g^{qm}(t), \quad \sigma^q(t) \approx \sum_{m=1}^{M_q} \sigma^{qm} g^{qm}(t), \tag{53.3}$$

where $g^{qm}(t)$ is a chosen system of M_q complex basis functions ($m = 1,\dots, M_q$) on the element q; assume for simplicity, that they are taken the same both for Δu^q and σ^q, albeit in general the systems for Δu^q and σ^q may be different; u^{qm} and σ^{qm} are complex approximation coefficients; they are known for prescribed functions and unknown for the functions to be found on the element q. For a particular choice of basis functions, the coefficient u^{qm} may express the value Δu^{qm} of the displacement discontinuity at a fixed point (node) t^m of the element. Such basis functions are sometimes termed form-functions. We will discuss an appropriate choice of basis functions in the next paragraph.

Substitution of (53.3) into (53.2) yields the approximate equation

$$\frac{1}{2\pi i}\sum_{q=1}^{Q}\sum_{m=1}^{M_q}\left[2u^{qm}\int_{L_q}\frac{g^{qm}(\tau)}{(\tau-t)^2}d\tau-u^{qm}\int_{L_q}g^{qm}(\tau)\frac{\partial}{\partial t}dk_1(\tau,t)-\overline{u^{qm}}\int_{L_q}\overline{g^{qm}(\tau)}\frac{\partial}{\partial t}dk_2(\tau,t)+\right]+$$

$$+\sum_{q=1}^{Q}\sum_{m=1}^{M_q}\left[(2a_1-a_3)\sigma^{qm}\int_{L_q}\frac{g^{qm}(\tau)}{\tau-t}d\tau+(a_1-a_3)\sigma^{qm}\int_{L_q}g^{qm}(\tau)\frac{\partial}{\partial t}k_1(\tau,t)d\tau+\right.$$

$$\left.+a_1\overline{\sigma^{qm}}\int_{L_q}\overline{g^{qm}(\tau)}\frac{\partial}{\partial t}k_2(\tau,t)d\overline{\tau}\right]\approx\frac{1}{2}a_2\sum_{p=1}^{M_p}\sigma^{pm}g^{pm}(t)+F(t),\quad t\in L_p\subset L. \tag{53.4}$$

On parts of L with prescribed dependence $\sigma(\Delta u)$, there are obvious changes. For such parts, the first of (53.3) is substituted into the dependence $\sigma(\Delta u)$ and the result is inserted into (53.2) as σ. Analogous operations are performed for a reverse dependence $\Delta u = \Delta u(\sigma)$.

3. ***Evaluation of integrals from basis functions.*** The approximate equation (53.4) contains integrals of known basis functions. In this case, the number of these integrals is six. Two of them are the integrals having a singularity when $\tau = t$. These are the *hypersingular* (finite-part) integral

$$I_g^{qm}(t)=\int_{L_q}\frac{g^{qm}(\tau)}{(\tau-t)^2}d\tau \tag{53.5}$$

and the *singular* (principal value) integral

$$S_g^{qm}(t)=\int_{L_q}\frac{g^{qm}(\tau)}{\tau-t}d\tau. \tag{53.6}$$

Four remaining integrals are *proper* integrals:

$$I_{1g}^{qm}(t)=\int_{L_q}g^{qm}(\tau)\frac{\partial}{\partial t}dk_1(\tau,t), \tag{53.7}$$

$$I_{2g}^{qm}(t)=\int_{L_q}g^{qm}(\tau)\frac{\partial}{\partial \bar{t}}d\overline{k_2(\tau,t)}, \tag{53.8}$$

$$S_{1g}^{qm}(t)=\int_{L_q}g^{qm}(\tau)\frac{\partial}{\partial t}k_1(\tau,t)d\tau, \tag{53.9}$$

$$S_{2g}^{qm}(t)=\int_{L_q}g^{qm}(\tau)\frac{\partial}{\partial \bar{t}}\overline{k_2(\tau,t)}d\tau. \tag{53.10}$$

In (53.8) and (53.10), for uniformity, we have written conjugate values of integrals to have basis functions without the conjugation symbol. Naturally, this does not influence methods of integration.

We point out that the integrals denoted with the symbol I are easily expressed through the corresponding integrals denoted with the symbol S by means of differentiation or integration by parts. In particular,

$$I_g^{qm}(t) = \frac{dS_g^{qm}(t)}{dt}. \tag{53.11}$$

Hence, if there are available analytical expressions for the integrals denoted by the symbol S, the remaining three integrals immediately follow in analytical form as well.

When using the real variables, evaluation of singular and hypersingular integrals presents the main obstacle. In complex variables, these integrals are incomparably simpler. Below in § 55 we will discuss formulae for evaluating integrals (53.5)-(53.10) from basis functions for the most important applications. For the moment, we will assume that for any t within or outside an element the needed integration can be performed. Then the approximate equation (53.4) takes the form

$$\frac{1}{2\pi i}\sum_{q=1}^{Q}\sum_{m=1}^{M_q}\Big\{u^{qm}\left[I_g^{qm}(t) - I_{1g}^{qm}(t)\right] - \overline{u^{qm}}\;\overline{I_{2g}^{qm}(t)} +$$
$$+\sigma^{qm}\left[(2a_1 - a_3)S_g^{qm}(t) + (a_1 - a_3)S_{1g}^{qm}(t)\right] + a_1\overline{\sigma^{qm}}\;\overline{S_{2g}^{qm}(t)}\Big\} \approx$$
$$\approx \frac{1}{2}a_2\sum_{p=1}^{M_p}\sigma^{pm}g^{pm}(t) + F(t), \quad t \in L_p \subset L. \tag{53.12}$$

There are M unknown complex coefficients, $2M$ real ones.

4. ***Formation of the algebraic system of equations.*** To find the M unknown complex coefficients, we need to replace the approximate equations (53.12) by M exact complex equalities. In the *collocation* method we satisfy (53.12) exactly at M *collocation* points t_k (k = 1,..., M). The points t_k may or may not be taken as the nodes of the approximation.

As a result, we obtain an algebraic system of M complex equations for the M unknown coefficients:

$$\frac{1}{2\pi i}\sum_{q=1}^{Q}\sum_{m=1}^{M_q}\Big\{u^{qm}\left[I_g^{qm}(t_k) - I_{1g}^{qm}(t_k)\right] - \overline{u^{qm}}\;\overline{I_{2g}^{qm}(t_k)} +$$

$$+\sigma^{qm}\left[(2a_1 - a_3)S_g^{qm}(t_k) + (a_1 - a_3)S_{1g}^{qm}(t_k)\right] + a_1\overline{\sigma^{qm}}\;\overline{S_{2g}^{qm}(t_k)}\Big\} =$$

$$=\frac{1}{2}a_2\sum_{p=1}^{M_p}\sigma^{pm}g^{pm}(t_k)+F(t_k),\quad k=1,\ldots,M,\quad t_k\in L_p\subset L. \quad (53.13)$$

We write (53.13) in compact form:

$$\sum_{q=1}^{Q}\sum_{m=1}^{M_q}\left\{a_k^{qm}u^{qm}+b_k^{qm}\overline{u^{qm}}+c_k^{qm}\sigma^{qm}+d_k^{qm}\overline{\sigma^{qm}}\right\}+$$

$$+\sum_{p=1}^{M_p}e_k^{pm}\sigma^{pm}=F_k,\ k=1,\ldots,M, \quad (53.14)$$

where we denote $a_k^{qm}=\frac{1}{2\pi i}\left[I_g^{qm}(t_k)-I_{1g}^{qm}(t_k)\right]$, $b_k^{qm}=-\frac{1}{2\pi i}I_{2g}^{qm}(t_k)$,

$c_k^{qm}=\frac{1}{2\pi i}\left[(2a_1-a_3)S_g^{qm}(t_k)+(a_1-a_3)\,S_{1g}^{qm}(t_k)\right]$, $d_k^{qm}=\frac{1}{2\pi i}a_1\,\overline{S_{2g}^{qm}(t_k)}$,

$e_k^{qm}=-\frac{1}{4\pi i}a_2\,g^{pm}(t_k)$, $F_k=F(t_k)$.

The system (53.14) has the form

$$\sum_{j=1}^{M}\left(a_{kj}w_j+b_{kj}\overline{w_j}\right)=f_k,\quad k=1,\ldots,M. \quad (53.15)$$

Separating real and imaginary parts, we obtain $2M$ real equations:

$$\sum_{j=1}^{M}\left(a'_{kj}+b'_{kj}\right)w'_j+\left(-a''_{kj}+b''_{kj}\right)w''_j=f'_k$$

$$\sum_{j=1}^{M}\left(a''_{kj}+b''_{kj}\right)w'_j+\left(a'_{kj}-b'_{kj}\right)w''_j=f''_k,\quad k=1,\ldots,M.$$

This is how the algebraic system is obtained by the collocation method. A more general approach consists in satisfying (53.12) *in the mean*. For this, both parts of (53.12) are multiplied by complex weight functions $h_k(t)$ (k = 1,…, M) and integrated over L. As a result, we arrive at a system of M complex equations with respect to M complex unknowns.

Usually each of the weight functions is chosen so that it is nonzero only on one element. For each element, the number of weight functions is taken equal to the number of unknown coefficients on it. By taking *weight* functions equal to conjugate *basis* functions and by properly using them, we arrive at a *symmetric* CV-

BEM (Dobroskok and Linkov [1, 2]). It retains all the well-known advantages of the real variable symmetric BEM (see, e.g. Maier and Polizotto [1], Bonnet [1]) and also, in contrast to the real variable symmetric BEM, it simplifies the integrations for commonly used basis functions and boundary elements.

5. ***Calculation of values at internal points.*** After the algebraic system is solved, all the coefficients of approximations are known. Then the approximation formulae like (53.3) provide approximate values of displacement discontinuities and traction for points on the contour. By using potential theory or employing integral representations of K-M functions, we can find stresses and displacements *within* a region. Often the integrals evaluated at the matrix formation stage allow us to find values at points off the contour. This is so when calculating stresses at internal points by means of CVH-BIE (53.1). Indeed, according to (24.1), for these points we have formula analogous to (53.1):

$$\frac{\chi^j+1}{2\mu^j}\sigma(z)=\frac{1}{2\pi i}\left\{2\int_L\frac{\Delta u(\tau)}{(\tau-z)^2}d\tau-\int_L\Delta u(\tau)\frac{\partial}{\partial t}dk_1(\tau,z)-\right.$$

$$-\int_L\overline{\Delta u(\tau)}\frac{\partial}{\partial t}dk_2(\tau,z)+\int_L(2a_1-a_3)\frac{\sigma(\tau)}{\tau-z}d\tau+\int_L(a_1-a_3)\sigma(\tau)\frac{\partial}{\partial t}k_1(\tau,z)d\tau+$$

$$\left.+\int_L a_1\overline{\sigma(\tau)}\frac{\partial}{\partial t}k_2(\tau,z)d\overline{\tau}\right\}-F(z),$$

where μ^j and χ^j are elastic constants of a medium at its point z.

The approximation (53.3) and integrals (53.5)-(53.10) now yield:

$$\frac{\chi^j+1}{2\mu^j}\sigma(z)\approx\frac{1}{2\pi i}\sum_{q=1}^{Q}\sum_{m=1}^{M_q}\left\{u^{qm}\left[I_g^{qm}(z)-I_{1g}^{qm}(z)\right]-\right.$$

$$-\overline{u^{qm}}\;\overline{I_{2g}^{qm}(z)}+\sigma^{qm}\left[(2a_1-a_3)S_g^{qm}(z)+(a_1-a_3)S_{1g}^{qm}(z)\right]+$$

$$\left.+a_1\overline{\sigma^{qm}}\;\overline{S_{2g}^{qm}(z)}\right\}-F(z). \tag{53.16}$$

Comparison (53.16) with (53.12) shows that we really can calculate the stresses at internal points by using the same procedures to evaluate the integrals (53.5)-(53.10) in the matrix of the algebraic system.

Strains are found from stresses by using Hooke's law (1.9); displacements are found from formula (30.25).

Comment 1. Sometimes, it is necessary to have the left and right limiting values of displacements. Formula (30.21) gives these. Assume that the normal at the point where we are finding the displacement is outward with respect to the region from which we take the limit; equation (30.21) gives

$$\frac{\chi^j+1}{2\mu^j}\varphi_j^+(t)=\frac{1}{2}\left[\Delta u(t)+a_1 f(t)\right]+\frac{1}{2\pi i}\int_L\frac{\Delta u(\tau)+a_1 f(\tau)}{\tau-t}d\tau+\frac{\chi_\infty+1}{2\mu_\infty}A_\varphi t,$$

where $f(t)$ is the resultant force obtained by integration (25.3), A_φ is defined by (30.20).

Now use formula (25.8) which expresses the K-M function in terms of physical values. Then for the needed limiting value $u^+(t)$ we obtain:

$$u^+(t)=\frac{1}{2}\Delta u(t)+\left(\frac{a_1}{2}-\frac{1}{2\mu^j}\right)f(t)+$$

$$+\frac{1}{2\pi i}\int_L\left[\frac{\Delta u(\tau)+a_1 f(\tau)}{\tau-t}\right]d\tau+\frac{\chi_\infty+1}{2\mu_\infty}A_\varphi t. \qquad (53.17)$$

To find the limiting values of tangential stresses $\sigma_{tt}^{\pm}$ we need derivatives of displacements. By differentiating (53.17), we obtain:

$$u'^+(t)=\frac{1}{2}\Delta u'(t)+\left(\frac{1}{2a_1}-\frac{1}{2\mu^j}\right)\sigma(t)+\frac{1}{2\pi i}\int_L\left[\frac{\Delta u(\tau)}{(\tau-t)^2}+\frac{a_1\sigma(\tau)}{\tau-t}\right]d\tau+\frac{\chi_\infty+1}{2\mu_\infty}A_\varphi.$$

The derivative $\Delta u'(t)$ on the r. h. s. may be found either by differentiation of the approximations (53.3), or by solving the BIE (30.16). Having u'^+, we find the tangential strain ε_{tt}^+:

$$\varepsilon_{tt}^+=\mathrm{Im}\left[iu'^+\exp(-i\alpha)\right],$$

where α is the angle of the tangent to the contour at the point t and the x-axis. Then from Hooke's law (1.9) we obtain the formula

$$\sigma_{tt}^+=\frac{1}{\chi^j+1}\left[8\mu^j\varepsilon_{tt}^+ +\left(3-\chi^j\right)\sigma_{nn}\right]$$

for the tangential stress in terms of the tangential strain ε_{tt} and the normal stress σ_{nn}, which is either prescribed, or known from the BIE.

Comment 2. Inspection of the CV-BIE derived in the second part of the book, shows that singular and Fredholm equations contain only three integrals other than (53.5)-(53.10). Meanwhile, the derivatives of these integrals with respect to t lead to integrals (53.6), (53.9), (53.10). These integrals allow us to evaluate integrals (53.5), (53.7), (53.8). Thus, only three independent quadratures are necessary for the evaluation of all the integrals. Below we will see that for a straight or circular arc boundary element there is further simplification: we need only *one* quadrature in an analytical form; all the others can be expressed through it.

Formulae for control. For any *closed* contour traveled counter clockwise, we have the following simple formulae:

$$\frac{1}{2\pi i}\int_L \frac{d\tau}{\tau - z} = \begin{cases} 1, & z \in D, \\ 1/2, & z \in L, \\ 0 & z \notin D + L, \end{cases} \qquad \int_L \frac{d\tau}{(\tau - z)^2} = 0,$$

$$\frac{1}{2\pi i}\int_L dk_1(\tau, z) = \begin{cases} 1, & z \in D, \\ 1/2, & z \in L, \\ 0 & z \notin D + L, \end{cases} \qquad \int_L dk_2(\tau, z) = 0,$$

where D is the finite region encircled by L. They are useful for checking intermediate results when developing computer programs based on the CV-BIE. Actually, they correspond to the case of the constant (unit) density. It is important that they allow us to check separately each of the integrals entering the CV-BIE. There is no such an opportunity for real variables: one may check only some combination of integrals corresponding to rigid movement.

We have additional options for checking the correctness and accuracy of calculations for singular and hypersingular integrals. For any function $\varphi(z)$ holomorphic in D we have:

$$\frac{1}{2\pi i}\int_L \frac{\varphi d\tau}{\tau - z} = \begin{cases} \varphi(z), & z \in D, \\ 1/2\varphi(t), & z = t \in L, \\ 0, & z \notin D + L, \end{cases} \qquad \frac{1}{2\pi i}\int_L \frac{\varphi d\tau}{(\tau - z)^2} = \begin{cases} \varphi'(z), & z \in D, \\ 1/2\varphi'(t), & z = t \in L, \\ 0, & z \notin D + L, \end{cases}$$

We may take a polynomial of an arbitrary order as $\varphi(z)$. This allows us to control both the correctness and accuracy of calculations for the usual polynomial approximations, in particular, when employing Lagrange's polynomials on boundary elements.

For curvilinear *open* contours, the holomorphicity theorems also provide an opportunity to control singular and hypersingular integrals. For instance, for the density $X(t) = [(t - a)(t - b)]^{1/2}$ prescribed on an open contour (a, b) with start point a and end point b, we have:

$$\frac{1}{\pi i}\int_a^b \frac{X(\tau)d\tau}{\tau - z} = \begin{cases} X(z) - z + (a + b)/2, & z \notin L, \\ -t + (a + b)/2, & z = t \in L. \end{cases}$$

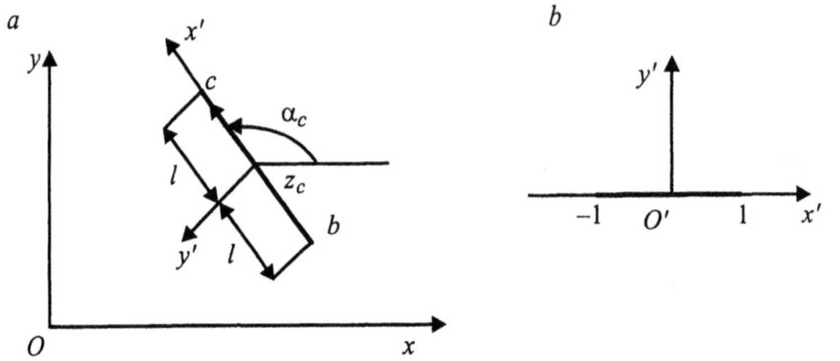

Fig. 42.

For definiteness, we assume that $X(t)$ represents the values of the function $X(z) = [(z - a)(z - b)]^{1/2}$ on the *left* side of the cut (a, b) and that $X(z)$ behaves as $+ z + \ldots$ at infinity.

§ 54. CHOICE OF APPROXIMATING FUNCTIONS

Ordinary (not tip) elements. Consider a part of the boundary not adjoining a tip of a cut, or a point where boundary conditions change, or a corner point. An element belonging to such a boundary part is called an *ordinary* (not tip) element.

We see that the physical values on the l. h. s. of (25.1), (25.2) are expressed in terms of holomorphic functions and their *conjugates*. Holomorphic functions may be represented by convergent power series, which may be approximated by polynomials. Thus a proper approximation of physical values should include not only polynomials but also conjugate polynomials (Linkov, Mogilevskaya, Napier [1], Linkov and Mogilevskaya [5]).).

Many contours may be approximated by using straight lines and circular arcs.

Straight element. Consider a straight element with length l and subtending an angle α_c with the x-axis. (Figure 42, *a*). The center of the element has co-ordinate z_c. We assume that the tangent is directed in the direction of traveling, and the normal to the right. Introduce a new co-ordinate system (x', y'), in which the element is located along the real axis x', its center is at the origin and its half-length is unity. The co-ordinates $z' = x' + i y'$, $z = x + iy$ are related by

$$z = z_C + lz' \exp(i\alpha_c). \tag{54.1}$$

On the interval [-1, 1], corresponding to the straight boundary element, the co-ordinate z' is real, so that the displacements are represented by some polynomial $P(\tau')$ with complex coefficients and *real* argument τ':

$$u(\tau') \approx P(\tau').$$

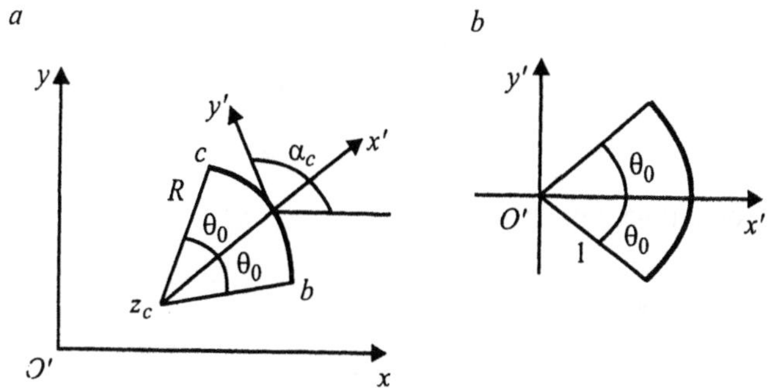

Fig. 43.

Displacement discontinuities, resultant force and traction have similar representations in the new co-ordinates. Since polynomials form a complete set, we can use them to obtain any required accuracy. In practice, it is convenient to employ Lagrangian interpolation polynomials.

2. ***Circular-arc element.*** Consider an arc element of a circle of radius R and center z_c (Figure 43, a); the angle of the element is $2\theta_0$. Introduce a local Cartesian system (x' , y') with the origin at the center of the circle and with the axis x' along a radius through the midpoint of the circular arc. The y' - axis is directed to the left of x' to provide a right Cartesian system. Thus, the y' - axis has the direction of the tangent at the midpoint, its angle with the x - axis of the global system is α_c when traveling the element *counter*-clockwise. (Below we will comment on changes needed when traveling the circumference clockwise).

The co-ordinates $z' = x' + i\,y'$, $z = x + iy$ are linked by

$$z = z_c - i\,Rz'\exp(i\alpha_c). \tag{54.2}$$

In the new co-ordinate system, the element becomes a circular arc of unit radius, shown in Figure 43, b. Its points have complex co-ordinates

$$\tau' = \exp(i\theta),$$

where θ is the polar angle counted from the axis x'. Now all functions are represented by Fourier series approximations:

$$u(\tau') = \sum_{k=-m}^{m} a_k \exp(ik\theta) = \sum_{k=-m}^{m} a_k \tau'^k. \tag{54.3}$$

Again such trigonometric functions form a complete set. Formula (54.3) accounts for conjugate polynomials because $1/\tau' = \overline{\tau'}$.

As for a straight element, we can choose trigonometric polynomials so that their coefficients will give the values of approximating functions at given nodes:

$$u(\tau')\approx\sum_{k=1}^{2m+1}u_k\widetilde{P}^k(\tau'), \tag{54.4}$$

where now the form-functions $\widetilde{P}^k(\tau')$ are

$$\widetilde{P}^k(\tau')=\sum_{s=-m}^{m}c_{ks}\tau'^s=\begin{cases}1, & \tau'=\tau'_k,\\ 0, & \tau'=\tau'_j, \quad j\neq k,\end{cases}\qquad k=1,...,m. \tag{54.5}$$

The form-functions (54.5) entering the approximation (54.4) are not Lagrangian polynomials with the nodes at points τ'_k, but they may be expressed through these polynomials. Indeed, by multiplying (54.3) by τ'^m we have:

$$\tau'^m u(\tau')\approx\sum_{k=0}^{2m}a_{k-m}\tau'^k.$$

The r. h. s. is an ordinary polynomial of power $2m$. It naturally may be represented by usual combination of Lagrangian polynomials. Then

$$\tau'^m u(\tau')\approx\sum_{k=1}^{2m+1}\tau_k'^m u_k P^k(\tau'),$$

where as before $u_k = u(\tau'_k)$ is the value of the function $u(\tau)$ at the node τ'_k, and $P^k(\tau')$ is the Lagrangian polynomial of the even power $2m$:

$$P^k(\tau')=\prod_{\substack{i=1\\ i\neq k}}^{2m}\frac{\tau'-\tau'_i}{\tau'_k-\tau'_i}=\begin{cases}1, & \tau'=\tau'_k,\\ 0, & \tau'=\tau'_j, \quad j\neq k,\end{cases}\qquad k=1,...,2m+1.$$

Thus,

$$u(\tau')\approx\sum_{k=1}^{2m+1}u_k\left(\frac{\tau'_k}{\tau'}\right)^m P^k(\tau')$$

and comparing with (54.4) gives the simple expression for the form functions with arbitrarily located nodes:

$$\widetilde{P}^k(\tau')\approx\left(\frac{\tau_k'}{\tau'}\right)^m P^k(\tau'). \tag{54.6}$$

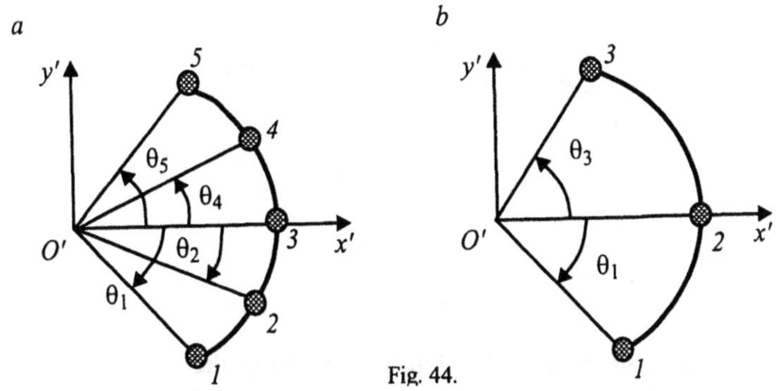

Fig. 44.

We write down explicit formulae for the coefficients c_{ks} in (54.5) when there are 5 (Figure 44, *a*) or 3 (Figure 44, *b*) nodes located symmetrically.

For five nodes, we have $\theta_5 = -\theta_1$, $\theta_4 = -\theta_2$. The coefficients c_{ks} ($k = 1,\ldots, 5$) are

$$c_{k0} = \frac{a_{0k}}{\Delta_A}, \quad c_{k1} = \overline{c_{k(-1)}} = \frac{1}{2}\left(\frac{a_{1k}}{\Delta_A} - i\frac{b_{1k}}{\Delta_B}\right), \quad c_{k2} = \overline{c_{k(-2)}} = \frac{1}{2}\left(\frac{a_{2k}}{\Delta_A} - i\frac{b_{2k}}{\Delta_B}\right)$$

where $a_{01} = a_{05} = -(a_{11} + a_{21})$, $a_{02} = a_{04} = -(a_{12} + a_{22})$, $a_{03} = \Delta_A - (a_{13} + a_{23})$, $a_{11} = a_{15} = -\sin^2\theta_2$, $a_{12} = a_{14} = \sin^2\theta_1$, $a_{13} = 2(\sin^2\theta_2 - \sin^2\theta_1)$, $a_{21} = a_{25} = \sin^2(\theta_2/2)$, $a_{22} = a_{24} = -\sin^2(\theta_1/2)$, $a_{23} = 2[\sin^2(\theta_1/2) - \sin^2(\theta_2/2)]$, $b_{11} = -b_{15} = 1/2\sin(2\theta_2)$, $b_{12} = -b_{14} = -1/2\sin(2\theta_1)$, $b_{13} = 0$, $b_{21} = -b_{25} = -1/2\sin\theta_2$, $b_{22} = -b_{24} = 1/2\sin\theta_1$, $b_{23} = 0$, $\Delta_A = 4[\sin^2(\theta_1/2)\sin^2\theta_2 - \sin^2(\theta_2/2)\sin^2\theta_1]$, $\Delta_B = \sin\theta_1\sin(2\theta_2) - \sin\theta_2\sin(2\theta_1)$.

For three nodes (Figure 44, *b*) we have $\theta_3 = -\theta_1$. The coefficients c_{ks} ($k = 1, 2, 3$) are

$$c_{10} = c_{30} = c_{21} = \overline{c_{2(-1)}} = \frac{1}{4\sin^2(\theta_1/2)}, \; c_{20} = 1 - \frac{1}{2\sin^2(\theta_1/2)},$$

$$c_{11} = \overline{c_{1(-1)}} = -c_{3(-1)} = -\overline{c_{31}} = -\frac{\cos(\theta_1/2) + i\sin(\theta_1/2)}{8\sin^2(\theta_1/2)\cos(\theta_1/2)}.$$

If an element is traveled *clockwise* one may make the following changes. Assume, as before, that for 5 nodes, θ_1 and θ_2 are negative, and for 3 nodes, θ_1 is negative. Then it is sufficient to change the signs of the coefficients c_{ks} corresponding to the odd powers of τ'^s in (54.5). As a result, both for five and three nodes, only the signs of c_{k1} and $c_{k(-1)}$ are changed ($k = 1,\ldots, 5$ or $k = 1, 2, 3$).

Tip elements. To increase accuracy of calculations and to find stress intensity factors, it is useful to account for asymptotic behavior of functions near singular points. Such points are tips of cracks, common vertices of blocks (grains), points where boundary conditions change. The asymptotic behavior of functions and integrals may be studied on the basis of Muskhelishvili's theory [4]. Here we consider only tips of cracks. A similar scheme serves for singular elements in other cases as well.

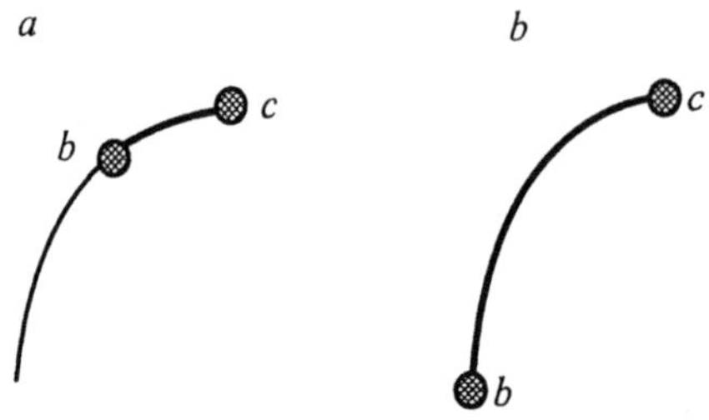

Fig. 45.

Consider a tip element, that is an element adjoining the end c of an open arc (Figure 45, a). From the general theory (Muskhelishvili [4,5]) it follows that if traction is prescribed in the vicinity of the point c, then the function $\varphi(z)$ near this point has a representation

$$\varphi(z) \approx \sqrt{z-c}\, P(z),$$

where $P(z)$ is a polynomial in z. For the displacement discontinuities from (25.8) we have $2\mu\Delta u(\tau) = \Delta\varphi(\tau)(\chi + 1) - \Delta f(\tau)$. Hence, if there is no traction discontinuity at the crack tip ($\Delta f(c) = 0$), then the displacement discontinuity is approximated as

$$\Delta u(\tau) \approx \sqrt{\tau - c}\, P(\tau). \tag{54.7}$$

When $\Delta f(c) \neq 0$, we include a term approximating the traction discontinuity on the r.h.s.

Sometimes it is convenient to approximate an entire crack by a single element (Figure 45, b). Then both start b and end c points are singular. In this case, we account for asymptotic behavior at both these points:

$$\Delta u(\tau) \approx \sqrt{(\tau - b)(\tau - c)}\, P(\tau). \tag{54.8}$$

The approximations (54.7), (54.8) are very convenient when calculating singular and hypersingular integrals because these integrals can be evaluated analytically (see § 55).

Note that instead of a polynomial $P(z)$ we can use other functions holomorphic in the vicinity of a tip; for a circular-arc element we can use form-functions (54.6). We will consider this important case below.

Straight element. Let the straight element shown in Figure 42 be a tip element. Assume that the traveling direction leads to the end point c. In the new (local) coordinates defined by (54.1), we have:

$$\sqrt{\tau - c} = \sqrt{l}\exp\frac{i\alpha}{2}\sqrt{\tau' - 1}. \tag{54.9}$$

Then the representation (54.7) becomes

$$\Delta u(\tau') \approx \sqrt{l}\exp\frac{i\alpha}{2}\sqrt{\tau' - 1}P(\tau'),$$

where $P(\tau')$ is a polynomial of the real coordinate τ' on the segment [-1, 1]. We may take this polynomial as a linear combination of Lagrangian polynomials. Then we obtain:

$$\Delta u(\tau') \approx \sqrt{l}\exp\frac{i\alpha}{2}\sqrt{\tau' - 1}. \tag{54.10}$$

Circular-arc tip element. Consider a circular-arc element (Figure 43) assuming that the traveling direction leads to end point c. In the new (local) co-ordinates defined by (54.2), we have

$$\sqrt{\tau - c} = \sqrt{R}\exp\left[i\left(\frac{\alpha_c}{2} - \frac{\pi}{4}\right)\right]\sqrt{\tau' - c'},$$

where $c' = \exp(i\theta_0)$ is the new co-ordinate of the crack tip. This leads to approximations in the new coordinates.

$$\Delta u(\tau') \approx \sqrt{R}\exp\left[i\left(\frac{\alpha_c}{2} - \frac{\pi}{4}\right)\right]\sqrt{\tau' - c'}\sum_{k=1}^{2m+1} f_k \widetilde{P}^k(\tau'), \tag{54.11}$$

where we use the form-functions (54.6).

Calculation of stress intensity factors. By distinguishing square-root asymptotic in formulae for tip elements (54.7), (54.8), we obtain an easy means to find SIF. Use (31.19):

$$k_1 - ik_{11} = -i\sqrt{\frac{\pi}{2}}\frac{2\mu}{\chi + 1}\exp(-i\alpha_*)\lim\frac{\Delta u(\tau)}{\sqrt{s}}, \tag{54.12}$$

where $s = |c - \tau|$ is a distance from the tip c to a point τ on the tip element, α_* is the angle of the tangent (in the direction of traveling to the tip) with the x – axis. Substitution of (54.7) into (54.12) and coming to the limit yield:

$$k_1 - ik_{11} = \sqrt{\frac{\pi}{2}}\frac{2\mu}{\chi + 1}\exp(-i\alpha_*)P(c). \tag{54.13}$$

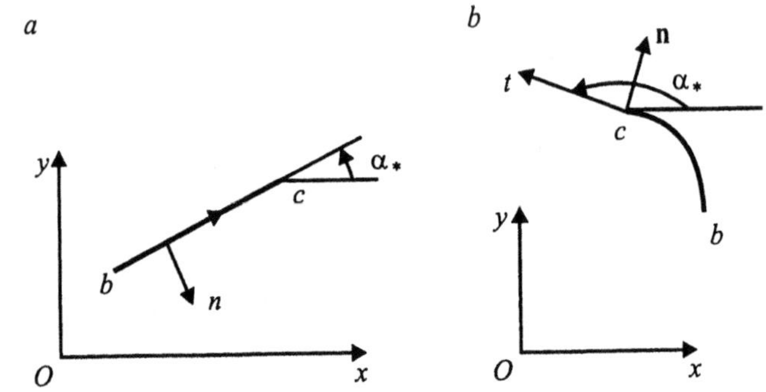

Fig. 46.

When employing form-functions, we represent the polynomial on the r. h. s. of (54.13) via boundary values $\Delta u_k / \sqrt{\tau_k - c}$ at the nodes. We give the final formulae for straight and circular-arc tip elements.

SIF for straight tip element (Figure 46, *a*). Substitution of (54.10) into (54.12) and coming to the limit yields

$$k_1 - ik_{11} = \sqrt{\frac{\pi}{2}}\frac{2\mu}{\chi+1}\exp\left(-\frac{i\alpha_*}{2}\right)\sum_{k=1}^{m} f_k P^k(1) \tag{54.14}$$

In § 31 we mentioned that not to make a mistake in the sign of SIF, it is reasonable to use the displacement discontinuity Δv in the local co-ordinates (**n**, **t**). Then

$$\Delta u(t_k) = i\exp(-i\alpha_*)\Delta v(t_k),$$

where α_* is the angle of the tangent to the element with the x –axis (for a straight element this angle is constant). Assuming that the BIE is solved with respect to $\Delta v(t)$, we may employ an approximation for Δv analogous to (54.10):

$$\Delta v(\tau) \approx \sqrt{\tau - c}\sum_{k=1}^{m} g_k P^k(\tau'),$$

(Note that $g_k = i\exp(-i\alpha_*)f_k$). The formula (31.16), which expresses SIF via $\Delta v(\tau)$ yields:

$$k_1 - ik_{11} = -i\sqrt{\frac{\pi}{2}}\frac{2\mu}{\chi+1}\exp\left(-i\frac{\alpha_*}{2}\right)\sum_{k=1}^{m} g_k P^k(1) \tag{54.15}$$

SIF for circular arc tip element (Figure 46, *b*). For this element from (54.11) and (54.12) we obtain the formula analogous to (54.14):

$$k_1 - ik_{11} = \sqrt{\frac{\pi}{2}}\frac{2\mu}{\chi + 1}\exp\left(-i\frac{\alpha_*}{2}\right)\sum_{k=1}^{2m+1} f_k \tilde{P}^k(c')$$

Alternatively, by using the displacement discontinuity Δv in the local co-ordinates (**n**, **t**), we obtain the formula analogous to (54.15):

$$k_1 - ik_{11} = -i\sqrt{\frac{\pi}{2}}\frac{2\mu}{\chi + 1}\exp\left(i\frac{\alpha_*}{2}\right)\sum_{k=1}^{2m+1} g_k \tilde{P}^k(c').$$

The coefficients of approximation in the global f_k and local g_k co-ordinates are connected by the dependence $g_k = i\exp(-i\alpha_k)f_k$, where α_k is the angle, now not constant, between the tangent at the node t_k and the axis x ($k = 1,\ldots, 2m + 1$).

§ 55. EVALUATION OF SINGULAR AND HYPERSINGULAR INTEGRALS

Singular and hypersingular integrals present the main difficulty when employing the real BEM. In complex variables they may be evaluated easily for an arbitrary curvilinear element L. A singular integral has the form

$$S_g(t) = \int_L \frac{g(\tau)}{\tau - t} d\tau \tag{55.1}$$

and a hypersingular integral is

$$I_g(t) = \frac{dS_g(t)}{dt} = \int_L \frac{g(\tau)}{(\tau - t)^2} d\tau. \tag{55.2}$$

The variables τ and t may be global or local co-ordinates..

Consider first those basis functions, which do not contain conjugate values but rather are represented by products of the form

$$g(\tau) = g_s(\tau) = K(\tau)\tau^s, \tag{55.3}$$

where s is integer (positive, negative or zero), and $K(\tau)$ is a function reflecting peculiarities of an approximated function. For ordinary elements $K(\tau) = 1$; for one-sided tip elements, $K(\tau) = \sqrt{\tau - c}$; for two-sided elements, $K(\tau) = \sqrt{(\tau - b)(\tau - c)}$

We can evaluate the integrals (55.1) and (55.2) by means of recurrence formulae for any power s when they are known for one value of this power.

Recurrence formulae for singular and hypersingular integrals. By substituting (55.3) into (55.1), (55.2) we have the singular integral

$$S_s(t)=\int_L \frac{K(\tau)\tau^s}{\tau - t}d\tau \tag{55.4}$$

and the hypersingular integral

$$I_s(t)=\frac{dS_s(t)}{dt}=\int_L \frac{K(\tau)\tau^s}{(\tau - t)^2}d\tau, \tag{55.5}$$

We use the identity $\tau^s = \tau^{s-1}(\tau - t) + t\tau^{s-1}$ to give the recurrence formula:

$$S_s(t)=c_{s-1}+tS_{s-1}(t), \quad s=\ldots,-1,0,1,\ldots, \tag{55.6}$$

where

$$c_{s-1}=\int_L K(\tau)\tau^{s-1}\,d\tau.$$

The quantities c_{s-1} do not depend on t. They are obtained by simple integration for many functions $K(\tau)$ used in the CV-BEM (see below). We will consider these quantities to be known.

We may re-write (55.6) in the form

$$S_{s-1}(t)=\frac{S_s(t)-c_{s-1}}{t}. \tag{55.7}$$

Given $S_0(t)$, we may obtain all the integrals for positive exponents by using(55.6), and for negative exponents by using (55.7).

The recurrence formulae for the hypersingular integral (55.5) are:

$$I_s(t)=S_{s-1}+tI_{s-1}(t), \quad s=\ldots,-1,0,1,\ldots$$

and

$$I_{s-1}(t)=\frac{I_s(t)-S_{s-1}}{t}.$$

In theory, we may use

$$I_s(t)=\frac{dS_s(t)}{dt},$$

but in practice it is not convenient, except for $s = 0$.

For some calculations we need hypersingular integrals of the third order:

$$\frac{dI_s(t)}{dt} = 2\int_L \frac{K(\tau)\tau^s}{(\tau - t)^3} d\tau.$$

These may be obtained in a similar way from that for $s = 0$.

Conclusion. The *only* analytical expression that is needed is $S_0(t)$.

We write explicit expressions for the quantities c_s and for the starting integrals $S_0(t)$, $I_0(t)$ and $dI_0(t)/dt$ for ordinary and tip elements (one- and two-sided).

Ordinary (not tip) element. For such an element (b, c) we have $K(\tau) = 1$, and

$$c_{-1} = \ln\frac{c}{b} \quad \text{when } s = -1; \quad c_s = \int_b^c \tau^s d\tau = \frac{c^{s+1} - b^{s+1}}{s+1} \quad \text{when } s \neq -1,$$

$$S_0(t) = \ln\frac{c - t}{b - t} + i\pi\delta_t, \quad I_0(t) = \frac{1}{b - t} - \frac{1}{c - t}, \quad \frac{dI_0(t)}{dt} = \frac{1}{(b - t)^2} - \frac{1}{(c - t)^2}.$$

The symbol δ_t denotes 1 if $t \in (b, c)$, or 0 if $t \notin (b, c)$.

One-sided tip element (Figure 45, *a*). Now $K(\tau) = \sqrt{\tau - c}$, and

$$c_s = \frac{2csc_{s-1} - 2b^s(b - c)\sqrt{b - c}}{1 + 2(s + 1)}, \quad c_{s-1} = \frac{(2s + 3)c_s + 2b^s(b - c)\sqrt{b - c}}{2cs},$$

and $c_0 = 2/3\sqrt{b - c}\,(c - b)$.

For the starting ($s = 0$) expressions for integrals we have:

$$S_0(t) = -2\sqrt{b - c} + \sqrt{t - c}\left\{\ln\frac{\sqrt{b - c} + \sqrt{t - c}}{\sqrt{b - c} - \sqrt{t - c}} + i\pi(1 - \delta_t)\right\},$$

$$I_0(t) = \frac{1}{2}\frac{S_0(t) + 2\sqrt{b - c}}{t - c} + \frac{\sqrt{b - c}}{b - t},$$

$$\frac{dI_0(t)}{dt} = -\frac{1}{2}\frac{S_0(t) + 2\sqrt{b - c}}{(t - c)^2} + \frac{1}{2}\frac{I_0(t)}{t - c} + \frac{\sqrt{b - c}}{(b - t)^2}.$$

Two-sided tip element (Figure 45, *b*). Now $K(\tau) = \sqrt{(\tau - b)(\tau - c)}$. Singular integrals (55.4) for $s \geq 0$ are easily evaluated by employing the general theory (see, e. g. Muskhelishvili [5, § 110]):

$$S_s(t)=\int_b^c \sqrt{(\tau-b)(\tau-c)}\frac{\tau^s}{\tau-t}d\tau=-\pi i\left[G_\infty(t)-(1-\delta_t)t^s\sqrt{(t-b)(t-c)}\right] \quad (55.8)$$

Here,

$$G_\infty(z)=a_0+a_1z+a_2z^2+\ldots+a_{s+1}z^{s+1}$$

is the principal part of the function $\sqrt{(z-b)(z-c)}z^s$ at infinity. The coefficients a_k are defined by formulae

$$a_{s+1-k}=\frac{1}{k!}\frac{d^k}{d\zeta^k}\sqrt{(1-c\zeta)(1-b\zeta)}\bigg|_{\zeta=0};$$

specifically, $a_{s+1}=1$, $a_s=-(b+c)/2$, $a_{s-1}=-(b+c)^2/8+bc/2$, $a_{s-2}=-(b+c)^3/16+bc(b+c)/4$.

For $s=0$ we have $a_0=-(b+c)/2$, $a_1=1$. Then (55.8) gives the starting integrals:

$$S_0(t)=-\pi i\left[-\frac{1}{2}(b+c)+t-(1-\delta_t)\sqrt{(t-c)(t-b)}\right],$$

$$I_0(t)=-\pi i\left[1-\frac{1}{2}\frac{(1-\delta_t)(2t-b-c)}{\sqrt{(t-c)(t-b)}}\right],\quad \frac{dI_0(t)}{dt}=-\frac{\pi i}{4}\frac{(1-\delta_t)(b-c)^2}{[(t-b)(t-c)]^{3/2}}.$$

Singular and hypersingular integrals from conjugate functions. Approximation formulae contain conjugate functions. Consider an integral from the conjugate function $\overline{Q(\tau)}$:

$$S_{\overline{Q}}(t)=\int_b^c \overline{Q(\tau)}\frac{d\tau}{\tau-t}.$$

Take into account that $\dfrac{d\overline{\tau}}{\overline{\tau}-\overline{t}}=\dfrac{d\tau}{\tau-t}-dk_1$, where as usual $k_1(\tau, t)$ is defined by (12.18): $k_1(\tau, t)=\ln\dfrac{\tau-t}{\overline{\tau}-\overline{t}}$. Then we obtain:

$$S_{\overline{Q}}(t)=\overline{S_Q(t)}-\overline{M_{1Q}(t)}, \quad (55.9)$$

where the proper integral $M_{1Q}(t)$ is defined as

$$M_{1Q}(t)=\int_b^c Q(\tau)dk_1(\tau,t) \tag{55.10}$$

A hypersingular integral from the conjugate function

$$I_{\overline{Q}}(t)=\int_b^c \overline{Q(\tau)}\frac{d\tau}{(\tau-t)^2}$$

is obtained by differentiation (55.9) with respect to t:

$$I_{\overline{Q}}(t)=\left[\overline{I_Q(t)}-\overline{I_{1Q}(t)}\right]\exp(-2i\alpha_0), \tag{55.11}$$

where α_0 is the angle of the tangent (in the direction of traveling) with the x – axis. The regular integral $I_{1Q}(t)$ is defined by (53.7). We discuss methods to evaluate these integrals in the next paragraph.

Singular and hypersingular integrals from functions $t\overline{P'(t)}$. The function of the form $t\overline{P'(t)}$ is also present in the approximations of displacements. For a singular integral from this function we have:

$$S_{\tau\overline{P'}}(t)=c_{\overline{P'}}+tS_{\overline{P'}}(t), \tag{55.12}$$

where $c_{\overline{P'}}=\int_b^c \overline{P'(\tau)}d\tau$; the function $S_{\overline{P'}}(t)$ is defined via a singular integral from $P'(t)$ and a proper integral $M_{1P'}(t)$ in accordance with (55.9) where now we use $P'(t)$ as $Q(t)$.

We obtain the hypersingular integral from $t\overline{P'(t)}$ by differentiation of (55.12):

$$I_{\tau\overline{P'}}(t)=S_{\overline{P'}}(t)+tI_{\overline{P'}}(t).$$

Here, the integral $I_{\overline{P'}}(t)$ is defined in accordance with (55.11), where now we use $P'(t)$ as $Q(t)$.

We see that the evaluation of singular and hypersingular integrals from a conjugate functions $\overline{Q(\tau)}$ and from the functions of the form $t\overline{P'(t)}$ is reduced to evaluation of (i) integrals from functions $Q(t)$ and $P'(t)$ without conjugation and (ii) standard proper integrals. Hence, if we take $Q(t)$ and $P'(t)$ in the form (55.3), then the crucial step, evaluation of singular and hypersingular integrals, is again carried out by employing the recurrence formulae given above. Again, the single integral $S_0(t)$ is sufficient to make this crucial step.

Comment. The recurrence formulae of this paragraph are based on the trivial identity $\tau^s = \tau^{s-1}(\tau - t) + t\tau^{s-1}$, which decreases the exponent in singular and hypersingular integrals, as suggested in (Linkov, Mogilevskaya and Napier [1]). An alternative less convenient approach uses the identity: $\tau^s = [(\tau - t) + t]^s$ (Linkov and Mogilevskaya [2, 3], Mogilevskaya [1, 2]).

§ 56. EVALUATION OF REMAINING (PROPER) INTEGRALS

The remaining integrals in CV-BIE are common proper integrals which may be evaluated with any prescribed accuracy over curvilinear elements by using quadrature rules. The simplest, rectangular or trapezoid formulac, do not even require parameterization of an element. It is necessary to be cautious when evaluating integrals containing the kernel $k_1(\tau, t) = \ln[(\tau - t)/(\bar{\tau} - \bar{t})]$ because the complex logarithm is a multi-valued function. The limiting values of these integrals from the left and right are not the same and differ also from the direct value used in the CV-BIE. One needs to fix a proper branch of the logarithm, for instance, by employing comments given in § 63 and 98 of Muskhelishvili's monograph [5].

Meanwhile, we can avoid numerical integration of proper integrals if the boundary elements are straight or circular-arc elements. For them all the integrals (53.7)-(53.10) are expressed in terms of the singular integral $S_g(t)$ and its derivatives $I_g(t)$ and $dI_g(t)/dt$, and all these integrals may be evaluated recurrently from $S_0(t)$.

Straight element. Integrals for a straight element (Figure 42) are evaluated in the new (local) system, connected with the global co-ordinates by (54.1). As a result of rather lengthy calculations we obtain the following formulae for the proper integrals (53.7)-(53.10), (55.10) over a straight element:

$$I_{1g}(t) = \int_L g(\tau)\frac{\partial}{\partial t}dk_1(\tau,t) = \left[I_1(t') - \lambda_1 I_g\left(\overline{t'}\right)\right]\frac{\lambda_2}{l}, \tag{56.1}$$

$$I_{2g}(t) = \int_L g(\tau)\frac{\partial}{\partial \bar{t}}d\overline{k_2(\tau,t)} = \left[\frac{\lambda_1 - 1}{\lambda_1}I_g(t') - \frac{t' - \overline{t'}}{\lambda_1}\frac{dI_g(t')}{dt'}\right]\frac{\lambda_2}{l}, \tag{56.2}$$

$$S_{1g}(t) = \int_L g(\tau)\frac{\partial}{\partial t}k_1(\tau,t)d\tau = -S_g(t') + \lambda_1 S_g\left(\overline{t'}\right), \tag{56.3}$$

$$S_{2g}(t) = \int_L g(\tau)\frac{\partial}{\partial \bar{t}}\overline{k_2(\tau,t)}d\tau = \left(\frac{1}{\lambda_1} - 1\right)S_g(t') + \frac{1}{\lambda_1}\left(t' - \overline{t'}\right)I_g(t'), \tag{56.4}$$

$$M_{1g}(t') = \int_L g(\tau)dk_1(\tau,t) = S_g(t') - S_g\left(\overline{t'}\right), \tag{56.5}$$

where $\lambda_1 = \exp[2i(\alpha_c - \alpha_0)]$, $\lambda_2 = \exp(-i\alpha_c)$, α_0 is the angle between the tangent at the field point t and the global axis x (commonly, for a field point out of the considered element, this angle differs from the angle α_c of inclination of the element).

The argument on the l. h. s. of (56.1)-(56.5) is the *global* co-ordinate of the field point t. The r. h. s. contains its new co-ordinate t' defined in the *local* system of the considered element. Accordingly to (54.1), it is $t' = (t - z_c)\lambda_2/l$. The singular and hypersingular integrals on the r. h. s. are evaluated in the *local* co-ordinates:

$$S_g(t') = \int_{-1}^{1} \frac{g(\tau')}{\tau' - t'} d\tau', \quad I_g(t') = \int_{-1}^{1} \frac{g(\tau')}{(\tau' - t')^2} d\tau'.$$

Note that formula (56.2) contains the derivative of hypersingular integral, i. e. a hypersingular integral of the third order. Its application is justified by the theory presented in Chapter 10.

Circular-arc element. We obtain the following formulae for all the proper integrals used in the CV-BEM:

$$I_{1g}(t) = \int_L g(\tau) \frac{\partial}{\partial t} dk_1(\tau, t) = \left[I_g(t') - \lambda_1 \left(\frac{1}{\bar{t}'} \right)^2 I_g\left(\frac{1}{\bar{t}'} \right) \right] i \frac{\lambda_2}{R}, \tag{56.6}$$

$$I_{2g}(t) = \int_L g(t) \frac{\partial}{\partial \bar{t}} d\,\overline{k_2(\tau, t)} =$$

$$= \left\{ I_g(t') - \frac{1}{\lambda_1}\left(\frac{1}{t'} \right)^2 \left[I_g(t') - I_g(0') \right] + \frac{1}{\lambda_1}\left(\frac{1}{t'} - \bar{t}' \right) \frac{dI_g(t')}{dt'} \right\} i \frac{\lambda_2}{R}, \tag{56.7}$$

$$S_{1g}(t) = \int_L g(\tau) \frac{\partial}{\partial t} k_1(\tau, t) d\tau = -S_g(t') + \lambda_1 \left(\frac{1}{\bar{t}'} \right)^2 S_g\left(\frac{1}{\bar{t}'} \right) + \lambda_1 \frac{c_0}{\bar{t}'}, \tag{56.8}$$

$$S_{2g}(t) = \int_L g(\tau) \frac{\partial}{\partial \bar{t}} \overline{k_2(\tau, t)} d\tau = -S_g(t') + \frac{1}{\lambda_1}\left(\frac{1}{t'} \right)^2 \left[S_g(t') - S_g(0') \right] - $$

$$- \frac{1}{\lambda_1}\left(\frac{1}{t'} - \bar{t}' \right) I_g(t'), \tag{56.9}$$

$$M_{1g}(t) = \int_L g(\tau) dk_1(\tau, t) = S_g(t') - S_g\left(\frac{1}{\bar{t}'} \right) + S_g(0'), \tag{56.10}$$

where $\lambda_1 = \exp[2i(\alpha_c - \alpha_0)]$, $\lambda_2 = \exp(-i\alpha_c)$, α_0 is the angle between the tangent at the field point t and the global axis x. In accordance with (54.2), the new co-ordinate t' on the r. h. s. of (56.6)-(56.10) is defined as $t' = i(t - z_c)\lambda_2/R$. The coefficient c_0', the singular and hypersingular integral are evaluated in the local system:

$$c_0' = \int_{b'}^{c'} g(\tau')d\tau', \quad S_g(t') = \int_{b'}^{c'} \frac{g(\tau')}{\tau' - t'} d\tau', \quad I_g(t') = \int_{b'}^{c'} \frac{g(\tau')}{(\tau' - t')^2} d\tau',$$

where b', c' are the new co-ordinates of the start b and end c point of an element; when traveling along the element *counter*-clockwise, we have $b' = \exp(-i\theta_0)$, $c' = \exp(i\theta_0)$, where $2\theta_0$ is the angle of a circular-arc element shown in Figure 43; when traveling *clockwise* we have $b' = \exp(i\theta_0)$, $c' = \exp(-i\theta_0)$. The prime near 0 in (56.9), (56.10) denoted the point $t' = 0$. This is a nonsingular point because the terms containing t' in their denominator mutually cancel when $t' \to 0$.

Such simple formulae do not exist in real variables. To see this, it is sufficient to separate the real and imaginary parts in any of the formulae: the results look formidable. The simplicity of the quadrature formulae is the most attractive feature of the complex variable method.

Chapter 13

NUMERICAL EXPERIMENTS USING CV-BEM

In this chapter we present numerical results that show the accuracy and capability of the CV-BEM. In numerical examples we reveal the importance of using a complete set of basis functions and accounting for the asymptotic behavior of functions at singular points, in particular, by employing tip elements. Under a reasonably chosen approximation, a solution for displacements and SIF normally has at least three correct significant digits even for a rather small number of boundary elements. It is sufficient to follow the simple general rules discussed below.

The CV-BEM allows us to solve problems for finite, infinite and semi-infinite problems concerning regions with holes, cracks (curvilinear, branching, intersecting, growing), inclusions, blocky systems and problems of homogenization. Examples may be found in papers by Zoubkov [1], Linkov, Zoubkov and Mogilevskaya [1], Mogilevskaya [1-3], Linkov and Koshelev [1, 2], Dobroskok et al., Wang et al. [1] (see also the review by Linkov and Mogilevskaya [5]).

§ 57. ROLE OF CONJUGATE POLYNOMIALS AND TIP ELEMENTS

We consider the accuracy of solutions obtained by the CV-BEM. We first use two simple problems having analytical solutions: one for a circular inclusion and another for a straight crack.

A circular inclusion in a plane under given stresses at infinity does not contain singular points on the contour; all elements are ordinary. This allows us to reveal effects not connected with asymptotic behavior of functions near singular points. In particular, we may study the role of conjugate polynomials for proper approximation of a solution, and the importance of accounting for the curvature of elements. The problem for a straight crack allows us to ignore these factors and to reveal the drastic influence of accounting for asymptotic behavior near singular points.

Role of conjugate polynomials. In § 54 we mentioned that for curvilinear elements, polynomials are not sufficient for approximating functions entering the CV-BIE. One has to combine them with conjugate polynomials and products of the form $\tau\overline{P(\tau)}$, where $P(\tau)$ is an ordinary polynomial.

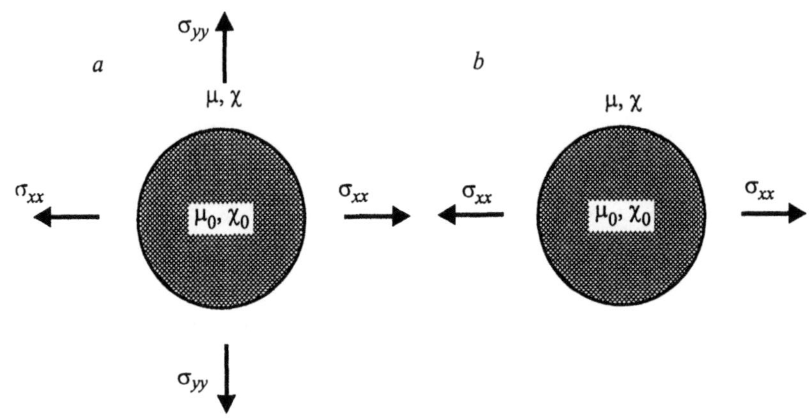

Fig. 47.

Problem for circular inclusion (Figure 47)[1]. Consider a circular inclusion with elastic constants μ_0 and χ_0 bonded with an infinite matrix with elastic constants μ and χ. At infinity we have the stresses σ_{xx}^{∞}, σ_{yy}^{∞} and σ_{xy}^{∞}.

This problem has a simple analytical solution (see, e. g. Muskhelishvili [5, § 58]). The traction on the contour of the inclusion depends linearly on both τ and $\bar{\tau}$. The conjugate term vanishes only when $\sigma_{xx}^{\infty} = \sigma_{yy}^{\infty}$, $\sigma_{xy}^{\infty} = 0$; any approximation that does not use conjugate polynomials may be satisfactory in this particular case, but cannot be good in the general case when $\sigma_{xx}^{\infty} \neq \sigma_{yy}^{\infty}$.

Numerical experiments for this and some other problems were carried out by applying the CV-BEM to the complex equation (53.1) and, when necessary, additional equations given in § 53 to exclude rigid movement. In the considered case, there is no displacement discontinuity on the contour ($\Delta u = 0$). Hence, the hypersingular term in (53.1) vanishes and the equation is singular. Calculations were performed with circular-arc elements coinciding with the boundary of the inclusion. Each of the elements contained three nodes: one at its center and two others at the distance $a/6$ from the ends of an element (a is the length of an element). In all cases we used the collocation method to obtain an algebraic system. Collocation points were taken at the nodes.

Part of the calculation was performed with the usual polynomials only; in others we added conjugate polynomials. To compare results, we took the same number of elements both times, fifteen; for three-point nodes on each element and two components of traction that gave 15x3x2 = 90 unknowns. All integrals were evaluated exactly by using formulae from the previous paragraph. The number of unknowns (90) being small, the algebraic system of equations was not ill-conditioned. Thus errors 2-3 orders of magnitude greater than rounding errors could arise only due an unsatisfactory approximation.

Evaluation of rounding-off errors when calculating with ordinary (not double) precision showed that the computer being used gave the number π with six correct

[1] Calculations and analysis for this example were carried out by S. G. Mogilevskaya (Linkov and Mogilevskaya [5]). The author participated in formulation of the problem and in analysis of the results.

digits. Accumulation of errors in arithmetic operations used to obtain the matrix and to solve the linear algebraic system could be expected to yield a numerical solution with not less than five correct digits. If the accuracy were worse than this, the errors must be due to inappropriate approximation.

Numerical experiments gave the following quantitative estimation of errors arising when the approximation did not involve conjugate polynomials. For uniform tension at infinity ($\sigma_{xx}^{\infty} = \sigma_{yy}^{\infty}$; Figure 47, *a*) when conjugate terms are not present in the analytical solution, the approximation without conjugate polynomials provided a solution with six correct digits. However, if $\sigma_{xx}^{\infty} \neq \sigma_{yy}^{\infty}$, the accuracy of this same approximation drastically decreased. In particular, for uniaxial tension (($\sigma_{xx}^{\infty} = 1$, $\sigma_{yy}^{\infty} = 0$; Figure 47, *b*) the solution had only three correct digits. Adding conjugate terms immediately improved results: the solution again had six correct digits.

Quite similar results were obtained for other locations of nodes, and other numbers of boundary elements. From this example we see that for curvilinear elements it is necessary to use both ordinary and conjugate polynomials. Moreover, we may conclude that the quality of the approximation is more important than the number of elements and the positions of nodes.

Crack along a circular arc (Figure 48)[1]. If the crack surfaces are loaded by equal and opposite tractions, the discontinuity $\Delta\varphi(\tau)$ of K-M function $\varphi(z)$ is proportional to the displacement discontinuity $\Delta u(\tau)$. Thus, it might appear that due to the holomorphicity of $\varphi(z)$, common polynomials are sufficient for approximating the displacement discontinuity on elements along the crack, except tip elements. However, this is not so. The reason is that the function $1/z$ entering the analytical solution for a crack along a circular-arc with the center at the origin, is also holomorphic in the vicinity of the arc. Naturally, it approximates the behavior of $\Delta\varphi(\tau)$ on the contour much better than the second and third powers of τ. Moreover, as follows from § 54, usual polynomials taken on a circular arc too strongly connect the real Δu_x and imaginary Δu_y parts of the displacement discontinuity Δu. By including negative powers of τ we avoid this shortcoming. Indeed, in § 54 we could see that under symmetric distribution of nodes with respect to the midpoint of an arc, the inclusion of the negative powers of polynomials corresponds to approximation by a complete set of trigonometric polynomials. Recalling that on a unit circle $1/\tau = \bar{\tau}$, we see that accounting for negative powers of τ is equivalent to introducing conjugate polynomials. Hence, these polynomials are important also for crack problems.

We illustrate improvements resulting from accounting for conjugate polynomials (negative powers of τ) by an example of a semi-circular crack of unit radius (Figure 48). At infinity we have uniform tension ($\sigma_{xx}^{\infty} = \sigma_{yy}^{\infty} = p$, $\sigma_{xy}^{\infty} = 0$). In this case the analytical solution by Muskhelishvili (see, e. g. [5, § 124a]) gives:

---

[1] Calculations and analysis for this example were carried out by J. A. L. Napier (Linkov, Mogilevskaya and Napier [1]). The author participated in working out a computer program, formulation of the problem and in the analysis of the results.

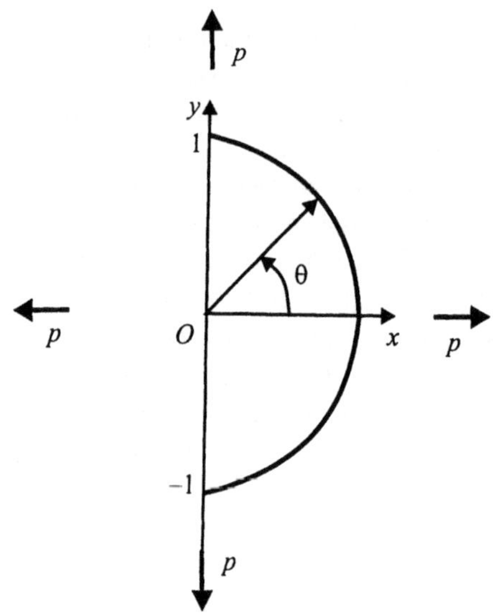

Fig. 48.

$$\Delta u_x = \frac{\sqrt{2}}{3}\frac{p(\chi+1)}{\mu}\sqrt{\cos\theta}\cos\left(\frac{\theta}{2}\right), \quad \Delta u_y = \frac{\sqrt{2}}{3}\frac{p(\chi+1)}{\mu}\sqrt{\cos\theta}\sin\left(\frac{\theta}{2}\right).$$

We may see from Table 1 that for points apart from the tips, the error in the component Δu_x becomes notably (1.4 – 2 fold) less if we account for the conjugate terms. For the component Δu_y the improvement is even more significant, being five fold near the midpoint. Only near the crack tips where both approximations are unsatisfactory, do the errors become similar. In these regions, as pointed out, we have to use special tip elements accounting for asymptotic behavior of unknown functions. We will see how drastically such elements increase accuracy later. In the meantime, we continue discussion of ordinary elements.

TABLE 1. Relative error (in %) of displacement discontinuity calculation for a semi-circular crack

θ, degrees	Δu_x			Δu_y		
	Ordinary polynomial	+conjugate polynomial	DIGS	Ordinary polynomial	+conjugate polynomial	DIGS
7.5	0.41	0.20	10.9	2.38	0.41	27.0
22.5	0.50	0.31	12.1	1.81	0.50	3.7
37.5	0.71	0.50	12.1	1.98	0.71	13.7
52.5	1.61	0.97	17.0	3.99	1.61	11.0
67.5	4.5	2.92	19.7	5.8	4.5	18.1
82.5	12..9	13.22	50.6	12.8	12.9	35.6

Role of contour approximation. To illustrate the role of contour approximation, two columns of Table 1 contain results obtained by using the program DIGS (Napier [1]). This program does not account for the curvature of a curve: elements are taken as straight segments. The displacement discontinuities are represented by linear functions. To compare the results, we represented the semicircle by six straight elements and took collocation nodes at the same nodes, which served us for the mentioned four circular arc elements. The number of unknowns was the same as before, 24.

The results presented in the table show that this approximation of the contour and displacement discontinuities is rather unsatisfactory. The relative error, as a rule, exceeds 10 %, it is 1-2 orders greater than that arising when the contour is approximated by circular-arc elements. Additional calculations show that this error reduces to the level provided by circular-arc elements only when the number of elements was increased to 48 (96 nodes, 192 unknown displacement discontinuities). In other words, using improved approximation and basis functions allows us to have 8 fold decrease in the order of the algebraic system. This gives us more than 50 fold decrease in time and memory.

There are many other examples showing that it is very important to account for contour curvature. We conclude that when solving the CV-BIE for curvilinear contours it is unreasonable to use only straight elements; we must use *circular-arc* elements. This is especially useful and important because, as we saw in the previous paragraph, complex variable calculations for circular-arc elements are no more difficult than those for straight elements.

Role of tip elements. We have just seen that accuracy provided by ordinary elements, even when employing conjugate polynomials, dramatically decreases near crack tips. The reason is quite clear: usual elements do not account for asymptotic behavior of a sought function near tips. It is impossible to drastically improve accuracy without accounting for it.

Problem for a straight crack (Figure 49). To evaluate the role of tip elements, consider a straight crack. There is no influence of conjugate polynomials and curvature in this case, and we can distinguish the effect of tip elements. The crack has length $2a$ along the real axis in an infinite plane with uniaxial stress $\sigma_{yy}^{\infty} = p$ at infinity (Figure 49, *a*). The exact analytical expression for the displacement discontinuity Δu_y, following from the classical solution by Muskhelishvili [5, § 120] is:

$$\Delta u_y(x) = pa\frac{\chi+1}{2\mu}\sqrt{1-\left(\frac{x}{a}\right)^2}.$$

The stress intensity factor (SIF) k_I^0 is given by the formula (see, e. g. Murakami [1])

$$k_I^0 = p\sqrt{\pi a}.$$

In numerical experiments it is convenient to normalize co-ordinates by a, displacements by the multiplier $[pa(\chi + 1)/(2\mu)]$ and SIF by $p\sqrt{\pi a}$. Now the half-length and load are unit (Figure 49, *b*), while the displacements and SIF are given by formula

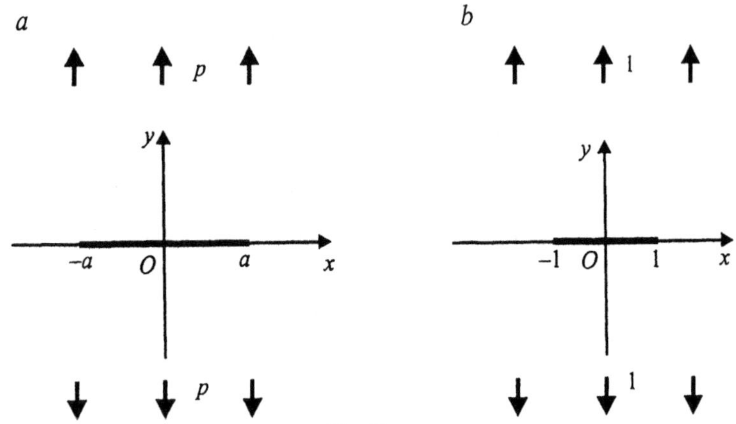

Fig. 49.

$$\Delta u_y(x)=\sqrt{1-x^2}, \quad k_1^0=1.$$

The length of boundary elements is now measured in fractions of the unit half-length. Use the approximation by Lagrangian polynomials on ordinary elements and the approximation of the type (54.7) for tip elements. For *ordinary* elements we use quadratics, corresponding to three-node elements. One of the nodes we locate at the center of an element while two others will normally be placed 1/6 of the length from the ends. The polynomial in the approximation (54.7) for *tip* elements is either constant or quadratic. In the first case we have an *one-node* tip element, in the second a *three-node* tip element. In both cases, in accordance with (54.7) a tip element accounts for a square-root asymptotic.

Influence of one-node tip elements.[1] Table 2 contains the results of calculations without tip elements and with two (the simplest, one-node) elements. The number of nodes was almost the same (15 and 14). All the elements had the same length. The table contains the relative error of the displacement discontinuity Δu_y at various points of the crack.

We see that the simplest tip elements increase accuracy by at least an order, even at points distant from the tips. For points near the tips, the advantage of accounting for the asymptotic becomes more and more evident, and at the distance 0.1 of the crack half-length the accuracy becomes two orders higher.

TABLE 2. Relative error (in %) of displacement discontinuity calculation for a straight crack

x	0	0.1	0.3	0.4	0.5	0.7	0.8	0.9
Without tip elements	1.8	1.8	1.9	2.1	2.4	3.5	6.5	14.7
With tip elements	0.04	0.04	0.04	0.13	0.26	0.23	0.19	0.16

[1] Calculations and analysis for this and all following examples were carried out by V. F. Koshelev. The author participated in working out a computer program, formulation of the problem and in the analysis of the results.

SIF calculation. Naturally, employing tip elements is extremely useful also for calculation of SIF: the latter are defined by the very same square-root asymptotic. In a problem for a straight crack, Mogilevskaya [1] obtained the SIF with four correct digits even when employing only four elements: two three-node ordinary elements and two *three*-node tip elements[1]. When applying *three*-node tip elements we also achieved accuracy close to that described by Mogilevskaya in a wide range of the sizes of elements and the location of nodes.

Three-node elements. Choice of sizes. The corresponding results are presented in Table 3. It shows the relative error of the displacement discontinuity Δu_y at the center of a straight crack for various combinations of the length $2l_t$ of three-node tip elements and the length $2l$ of three-node ordinary elements. The number of ordinary elements was taken such that the midpoint of the central element was at the origin. Recall, that with the accepted normalizing of values, the error 0.001% means one unit in the sixth digit after the decimal point. This error is on the threshold of the computer accuracy for calculations performed with ordinary (not double) precision.

The table shows the high accuracy of the calculated displacement discontinuity Δu_y. However, we see that the accuracy deviates from the threshold of six correct digits for values underlined in the table. This occurs in two cases. First, the accuracy decreases if for a small ($2l_t \le 0.2$) tip element, the length of an ordinary element is too large ($2l \ge 0.5$). Secondly, it decreases when a tip element is too large ($2l_t \ge 0.5$).

Two recommendations follow from this observation. To maintain accuracy, it is reasonable: (i) to take the length of a tip element less than 1/5 of a crack half-length and (ii) to take the length of ordinary elements adjacent to tip elements not exceeding the length of the tip elements.

TABLE 3. Relative error (in %) of displacement discontinuity calculation in crack midpoint for various sizes of three-nodes tip $2l_t$ and ordinary $2l$ elements

$2l_t$	0.1	0.2			0.4				0.5			0.6		
$2l$	0.094	0.094	0.0(17)	0.32	0.092	0.17	0.24	0.4	0.0(90)	0.2	0.(3)	0.0(8)	0.16	0.2(6)
Error, %	0.003	0.001	0.003	0.055	0.003	0.001	0.00001	0.002	0.011	0.008	0.011	0.027	0.026	0.027

TABLE 4. Relative error (in %) of SIF calculation for various sizes of three-nodes tip $2l_t$ and ordinary 2l elements

$2l_t$	0.1	0.2			0.4			
$2l$	0.094	0.094	0.0(17)	0.32	0.092	0.17	0.24	0.4
Error, %	0.07	0.006	0.006	0.03	0.001	0.004	0.02	0.09

[1] In the Russian edition of the book, the results by Mogilevskaya [1] were wrongly attributed to using one-node tip elements. The author thanks S. G. Mogilevskaya identifying this misunderstanding and apologizes for the misstatement.

0.5			0.6		
0.0(90)	0.2	0.(3)	0.0(8)	0.16	0.2(6)
0.002	0.004	0.02	0.006	0.006	0.008

One may of course increase the sizes of the elements and/or decrease their number if very high accuracy is not needed.

We obtain similar accuracy when calculating SIF. For the same problem, the relative error in SIF is presented in Table 4.

Now, in contrast with the displacement discontinuity, even large ($2l_t \geq 0.5$) three-node tip elements provide high accuracy. This difference with the results for the displacement discontinuity is to be expected; tip elements reflect behavior near the tips of a crack, while ordinary elements relate to its middle. However, for large tip elements, it is better not to use ordinary elements of the same length. As the table shows, in such cases, best accuracy comes from taking the length of ordinary elements 2 or 3 fold less than the length of a tip element.

Similar numerical results for the displacement discontinuity and the SIF are obtained for a circular-arc crack along the arc ($-\pi/2$, $\pi/2$) of a circumference. We will not present these results because they do not contain new information as compared with tables 3 and 4. All the conclusions on the limiting precision remain; however, the three-node approximation of ordinary and tip elements must include the conjugate terms discussed in § 54.

Influence of node location. Quite similar results are obtained for various locations of two sets of three nodes on each element (ordinary and tip). The calculated SIF do not change notably when the distance of the mid-nodes from the crack tips changes from 1/6 to 1/24 of the element size. This observation is in contrast with that of Mogilevskaya [1] who reported significant (from four to six correct digits) increase in accuracy when changing the distance from 1/6 to 1/13 of the element size (the tip elements were one-node elements). The difference may be due to the fact that three-node tip elements used in our experiments provided high accuracy by themselves. Nor did we manage to reproduce Mogilevskaya's results for one-node elements. The reason of this disagreement is not clear.

Fan of cracks. It is interesting to compare the conclusions on the accuracy of SIF calculations with the results for a fan of cracks with equal size (Figure 50). Uniform tension is applied at infinity: $\sigma_{xx}^{\infty} = \sigma_{yy}^{\infty} = p$, $\sigma_{xy}^{\infty} = 0$. Table 5 contains data on relative error in SIF for a fan with four (Figure 50, *a*) and twelve (Figure 50, *b*) cracks. The error is found by comparing the results with the exact values given in the handbook by Murakami [1]. These exact values are $\dfrac{k_1}{p\sqrt{\pi a}} = 0.86354$ for four cracks and $\dfrac{k_1}{p\sqrt{\pi a}} = 0.55106$ for twelve cracks. The difference between these two cases is due to the fact that twelve cracks, being closer to each other, exhibit a stronger shielding effect.

a *b*

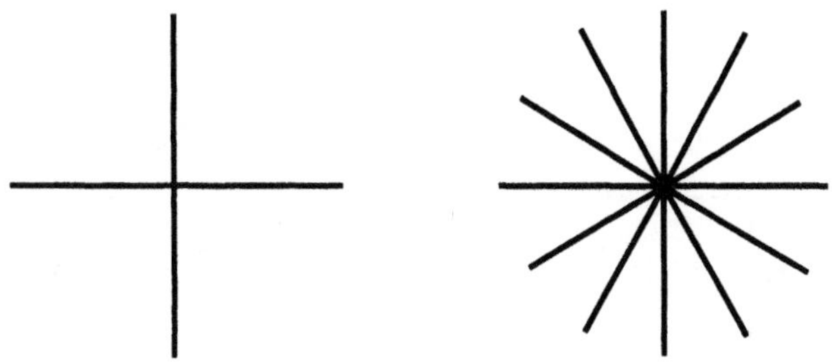

Fig. 50.

TABLE 5. Relative error (in %) of SIF calculation for various sizes of three-nodes tip $2l_t$ and ordinary $2l$ elements

$2l_t/a$	0.1	0.2			0.4			
$2l/a$	0.1	0.1	0.2	0.4	0.1	0.2	0.3	0.6
4cracks, error	0.06	0.01	0.09	0.45	0.01	0.04	0.04	0.25
12cracks, error	0.09	0.09	0.18	0.7	0.9	0.9	1.1	1.2

$2l_t/a$	0.5			0.6		
$2l/a$	0.1	0.25	0.5	0.1	0.2	0.4
4cracks, error	0.04	0.04	0.06	0.10	0.08	0.09
12cracks, error	1.7	0.2	1.8	1.7	0.7	0.7

From the table we see that for four cracks the results are largely similar to those presented in the previous table for a single crack. Meanwhile, for strong interaction in the case of twelve cracks, the accuracy markedly decreases. Later, when discussing periodic problems we will see that this decrease in accuracy is a common effect for strongly interacting cracks. One can diminish its influence by using *adaptive* calculations, which allow us to account for actual behavior of a solution on a contour.

Adaptive calculation. Accuracy decreases when adjacent contours strongly interact because of local perturbations induced by the interaction. To overcome this difficulty, we first reveal the nature and the region of perturbation by means of preliminary calculation employing quite arbitrary elements. Then at the next stage of calculations we can concentrate elements in zones of rapid change of the sought function, and space them out in zones where the function changes slowly. With such *adaptive* two-stage calculations the accuracy may be significantly increased without increasing the number of elements.

Consider for example the same fan of cracks (Figure 50). Preliminary calculations used seven three-nodes elements on each crack: one of them is a tip element with length $2l_t/a = 0.4$ and the other six are ordinary elements with length $2l/a = 0.1$. From Table 5 we see that for twelve cracks, the relative error for this choice was 0.9%.

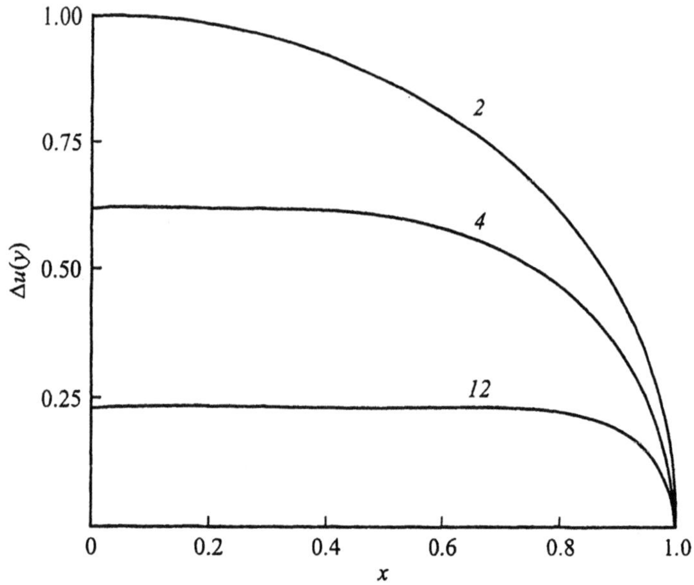

Fig. 51.

The distribution of the normal displacement discontinuity normalized by its value $\Delta u_0 = pa(\chi + 1)/(2\mu)$ in the midpoint of a single crack is shown in Figure 51 for two, four and twelve cracks. (Two cracks correspond to a single crack of length $2a$). From the graph for the strongly interacting twelve cracks, we see that the displacement discontinuity is nearly constant along the major part of a crack; the changes occur only at a distance $0.2a$ from the crack tip. Thus, at the second stage of calculations we take (i) the length of the tip element $2l_t/a = 0.15$ to account for the square-root behavior and (ii) the length of two adjacent ordinary elements $2l/a = 0.05$ to account for rapid changes in the transitional zone. The remaining ordinary elements are in the zone of slow changes of the displacement discontinuity. We take their length $2l_t/a = 0.1875$ to have the same number (seven) of elements and the same number of unknowns on a crack 7x3x2=42.

The second stage of calculations performed for this choice of elements gives $\dfrac{k_1}{p\sqrt{\pi a}} = 0.5515$, which is close to the exact value 0.55106; the error is 0.04 %. This is more than an order less than the error 0.9% at the first (preliminary) stage.

§ 58. PERIODIC PROBLEMS

A computer program worked out for a finite number of contours may serve also to solve periodic problems. Indeed, take *finite* but sufficiently large number of strips with a prescribed period. Then we may expect that values in the central strip will be close to those which correspond to an *infinite* system of the periods. Take into account that modern computers allow us to solve systems of equations with hundreds and even thousands unknowns. Hence, one may try to solve a problem for a large but finite system of strips directly, not developing a special algorithm for an initial infinite system.

There may be two approaches when employing this suggestion. The first of them do not require solving a system with unknowns on all congruent contours. We consider the unknowns to be equal to those on the contour in the main strip. In fact, we sum a finite number of terms in series for the kernels entering the BIE for a periodic problem. The number of the terms in the truncated series is equal to the number of strips used to represent an infinite system. Actually, we solve the BIE (35.2) for a finite number of strips.

The number of the strips does not influence the order of an algebraic system. This allows us to examine accuracy by comparing results for increasing numbers of the strips (actually, terms of the expansions). When the difference decreases to an accepted level, we may consider that the convergence is reached. Now the accuracy within the accepted level depends solely on the quality of function approximation. If necessary we may improve the latter.

The second approach is based on the possibility to solve large algebraic systems. Employing it, we consider unknown values on congruent contours to be independent. As compared with the first approach, the number of unknowns increases proportionally to the number of strips. Thus the second approach certainly looks unattractive. However, it gives some impression on the accuracy with which periodicity conditions are satisfied. Indeed, after solving the system we can compare calculated values, for instance the displacement discontinuities or SIF, in the central strip and in the adjacent strips. Comparing these values we may see the number of coinciding digits and estimate the accuracy with which the periodicity conditions are approximated.

Numerical experiments discussed below show that for the same number of strips, both approaches provide approximately the same accuracy. Hence, the first is significantly more efficient because it allows us to account for hundreds and even thousands of strips. Meanwhile, the second, being independent, serves as an additional tool to control accuracy of results.

We consider examples illustrating accuracy of the approaches.

Periodic system of collinear cracks. Consider cracks of length $2a$ located with period 2ω along the x-axis (Figure 52). The distance between cracks is $d = 2\omega - 2a$. This problem has a simple analytical solution presented in § 34. In a particular case of the tension $\sigma_{yy} = p$ at infinity, the SIF is given by formula (see, e. g. Murakami [1])

$$\frac{k_1}{k_1^0} = \sqrt{\frac{2\omega}{\pi a}\tan\frac{\pi a}{2\omega}}, \tag{58.1}$$

where k_I^0 is the SIF for a single crack of the same length: $k_I^0 = p\sqrt{\pi a}$. The SIF tends to k_I^0 when the period 2ω tends to infinity.

For both approaches we employed a CV-BEM program for the hypersingular BIE (53.1). Each crack was divided into 11 three-nodes elements, two of them

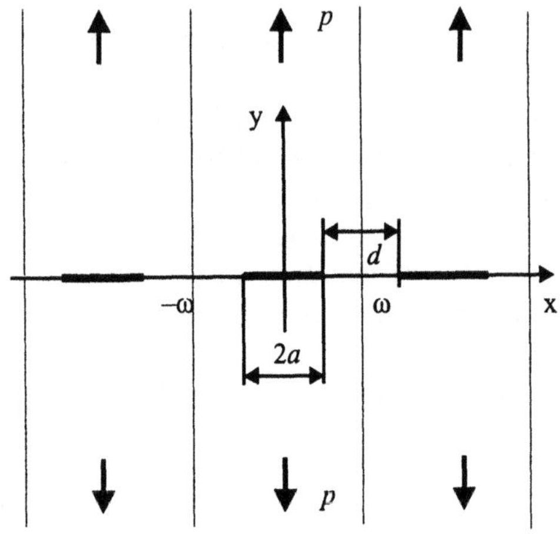

Fig. 52.

being three-nodes tip elements with the length $2l_t/a = 0.15$; the remaining nine elements were ordinary three-nodes elements of length $2l/a = 0.2$.

When using the first approach, we increased the number of strips to have convergence with five significant digits. For this, depending on the relative distance $d/(2\omega)$ between cracks, it was necessary to take from 20 (for $d/(2a) \geq 1$) to 100 (for $1 > d/(2a) \geq 0.5$) strips. For $0.25 > d/(2a) \geq 0.01$, the number of strips increased to 300-500. To be sure of results, some experiments were reproduced for 2000 strips.

For the second approach, computer resources restricted the number of strips. In the experiments we increased the number from three to seventeen. For 17 strips, the SIF for the central crack had three digits coinciding with the SIF for cracks in two adjacent strips. Thus, the SIF also coincided within the same accuracy with the SIF obtained in the first approach when using the same number of strips. Henceforth, we employed only the first approach

Table 6 presents relative errors in SIF with respect to the exact values given by (58.1)

TABLE 6. Relative error (in %) of SIF calculation by CV-BEM for a periodic system of collinear cracks

$d/(2a)$	1.5	1.0	0.5	0.25	0.05	0.025	0.01
Error, %	0.01	0.04	0.04	0.06	0.08	0.3	2.5

The errors depend now solely on the quality of the approximation. For the chosen elements it is not great (less than half-percent) when the relative distance between crack tips exceeds 0.05. The accuracy drastically decreases when the distance becomes less than 0.01. This is a result of strong interaction between cracks with attendant asymptotics generated by small necks between the tips of adjacent cracks. To solve the problem accurately when the necks tend to zero, we need to distinguish the new asymptotic explicitly and to apply asymptotic methods (see, e. g. Nazarov et al [1]). We cannot dwell on such methods. Note only that in

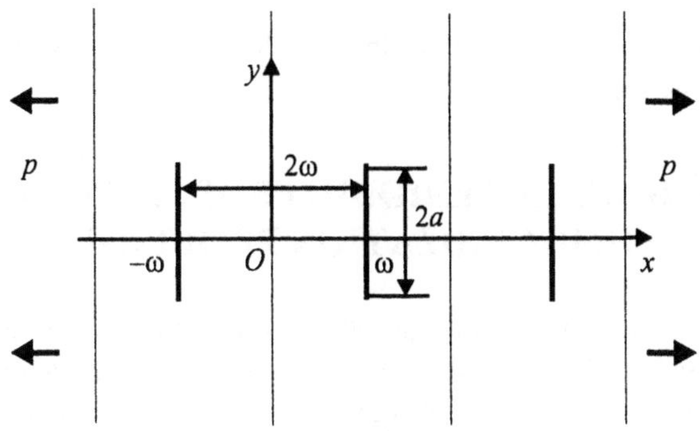

Fig. 53.

this problem the small parameter is d/a; the new asymptotic corresponds to a problem for two *semi-infinite* cracks separated by a neck.

Periodic system of parallel cracks. Consider a row of cracks of length $2a$ parallel to the y-axis and located with period 2ω along the x-axis (Figure 53). The calculated SIF's are compared with numerical results given in the handbook by Murakami [1] (the relative error of these results is near 0.1 %). For sufficiently small distances between cracks ($\omega/a < 0.25$), the SIF is given with error less than 0.1 % by the asymptotic formula

$$\frac{k_1}{k_1^0} = \sqrt{\frac{\omega}{\pi a}\,\mathrm{th}\left(\pi\frac{a}{\omega}\right)}.$$

In this case we carried out calculations by employing adaptive boundary elements. As for the fan with twelve cracks, displacement discontinuities are nearly constant along the major portion of a crack, while changes are localized on edge portions of length approximately $0.2a$. In accordance with this observation, we took the tip elements with length $2l_t/a = 0.15$ and, nearest to them, ordinary elements of length $2l/a = 0.05$. The remaining ordinary elements were rather large: their length was $2l/a = 0.47$. All the elements were three-node elements. We augmented the number of strips as in the previous example. Comparing the two approaches leads to conclusions similar to those for collinear cracks. Again we carried out the major part of calculations by the more efficient first approach. The difference as compared with collinear cracks was that for close cracks ($\omega/a < 0.5$) we needed to use more than 1000 strips and to calculate with double precision, because the large number of terms led to round-off errors in single precision. Table 7 contains data on the relative errors.

TABLE 7. Relative error (in %) of SIF calculation by CV-BEM for a periodic system of parallel cracks

ω/a	2.00	1.00	0.50	0.25	0.10
Error, %	0.1	0.1	0.3	0.2	0.5

From the table we conclude that this method provides a quite acceptable accuracy, 0.5% up to $\omega/a = 0.1$.

§ 59. DOUBLY PERIODIC PROBLEMS AND HOMOGENIZATION PROBLEM

We are interested in doubly periodic problems in the study of media with two or more structural levels. We noted in § 39 that one can easily adjust a CV-BEM program for a *finite* system of cracks, blocks, holes and/or inclusions in the main cell to solve problems for *doubly periodic* systems. As the CV-BEM for finite systems has proved to be efficient, this suggests high potential of the form (39.9).

The following examples illustrate efficiency of this approach. A program of the CVH-BEM, primarily developed for finite systems of blocks, inclusions, holes and cracks, is adjusted for doubly periodic problems following the line presented. As usual, three-nodes ordinary and tip elements are employed. All integrals for straight and circular-arc elements are evaluated exactly by using basis functions and quadrature formulae given in § 55, 56. This provides accurate results (four-five correct digits) for the displacement discontinuities and the SIF even for a moderate number of boundary elements, which normally does not exceed twenty. But to preserve accuracy when using (39.9) we need also to choose appropriate parameters:

(i) the numbers N_1 and N_2 in truncated series, and

(ii) the number of iterations if we do not change the algebraic matrix by accounting for integrals in the constants A_α, A_β, A, B but find the latter with successive steps starting from their zero values.

These numbers are found from numerical experiments. Take the same number $N_1=N_2=N$ along each of the periods $2\omega_1$, $2\omega_2$. In this case whole "chains" of cells enclosing previous ones with growing number N embrace the main cell. The total number of cells and consequently the number of terms in the truncated series is $(2N+1)^2$. Numerical experiments show that even for strongly interacting cracks, it is sufficient to take $N=10$ to reproduce results with five digits. The experiments show also that ten successive steps are always enough to reproduce five digits in the constants A_α, A_β, A, B even for cracks with length $2a$ close to the length 2ω of the smallest period ($a/\omega=0.9$). Below we use these numbers: *ten* for "chains" embracing the main cell and *ten* for iterations improving the constants.

To compare the results with those by other authors, we consider *straight* cracks of length $2a$ in rectilinear, square and triangle lattices. The length of tip elements is taken to be $0.15a$; three neighboring ordinary elements have length $0.05a$; the remaining part of a crack is represented with eleven uniform ordinary elements of length $1.4a/11$. Thus, the total number of boundary elements is 19. From calculations for an isolated crack, we could see that such a choice of elements always provided from four to six correct digits. This accuracy is confirmed by calculations with a doubled number of elements and double precision. The particular case of periodic collinear cracks, for which an analytical solution is

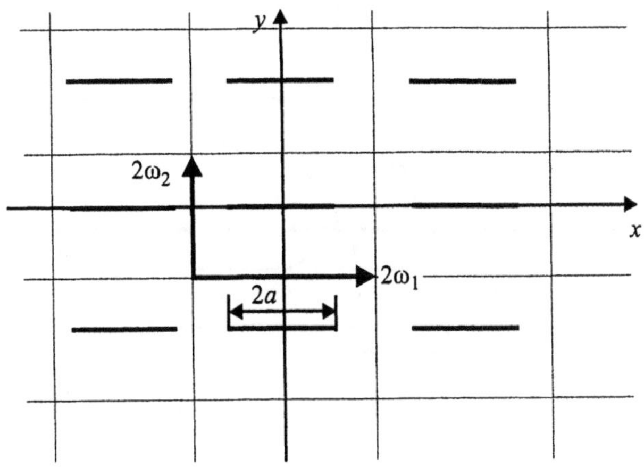

Fig. 54.

available, was studied in detail to examine accuracy for closely located cracks. We conclude that the numerical data obtained by the CV-BEM, and presented in the following tables, have at least four correct significant digits.

For definiteness and to compare with the results of other authors, we consider plane stress. The SIF is normalized by the value $p\sqrt{\pi a}$ corresponding to a single crack with the length $2a$, the effective elastic modulus E_{ef} is normalized by the modulus E of a plate without cracks, the effective shear modulus μ_{ef} by the shear modulus μ of a plate without cracks.

Rectangular lattice of cracks. Consider a doubly periodic system of cracks in rectangular cells with sides $2\omega_1$, $2\omega_2$ along the axes x and y respectively (Figure 54). For this problem the handbook (Murakami [1]) contains graphs of SIF k_I for tension normal to the cracks. This data refer to moderate crack density: the distance between cracks along the y-axis is greater than the crack half-length, the distance along the x-axis exceeds a quarter of the length. The accuracy of the results presented in the handbook is about 1.5%.

Table 8 contains values of k_I for the lattice with $|\omega_1/\omega_2|$ = 0.4. This case is rather interesting. It corresponds to the most significant changes in the dependence of k_I on the crack concentration in a range of $|\omega_1/\omega_2|$ represented by graphs in the handbook. The table shows satisfactory agreement with the published results.

Square lattice of cracks. The results for a square lattice ($|\omega_1/\omega_2|=1$) are given in Tables 9 and 10. The first contains values of k_I for extension normal to the cracks; the second contains k_{II} for shear along the cracks. For comparison, the tables include data by Panasiuk et al. [1]. For normal tension, Table 9 contains data from the handbook by Murakami [1].

The values obtained by various methods agree within the accuracy provided by these methods. Note that the CV-BEM provides accurate values of SIF up to a concentration of cracks corresponding a/ω_1 = 0.9. This concentration exceeds those reached by other methods.

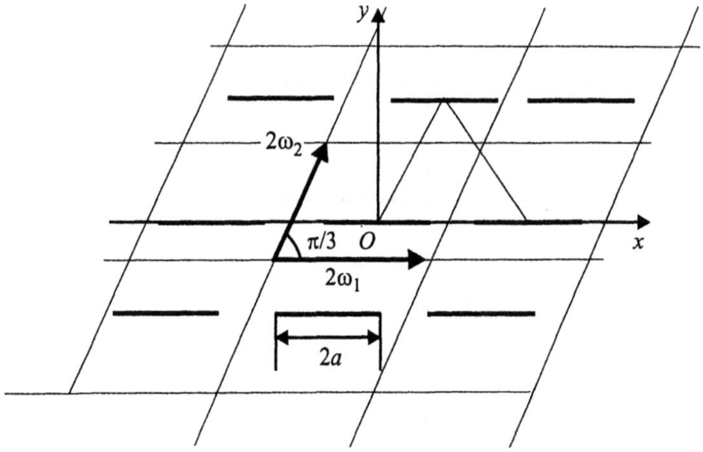

Fig. 55.

Right triangle lattice of cracks. Consider a triangular lattice of cracks (Figure 55). In this case $\omega_2/\omega_1 = \exp(i\pi/3)$ and the points z, $z + 2\omega_1$ and $z + 2\omega_2$ are located at vertices of a right triangle. Accordingly, by connecting crack midpoints located at these points we have a right triangle.

Table 11 contains the results of calculations of k_I for normal tension; Table 12 contains k_{II} for shear. For comparison, these tables include also data by Filshtinski [1] and Panasiuk et al. [1].

TABLE 8. Normalized values of SIF k_I for a rectangular lattice of cracks ($|\omega_1/\omega_2| = 0.4$)

a/ω_1	0.3	0.4	0.5	0.6	0.7	0.75
CV-BEM	0.8532	0.8381	0.8864	0.9771	1.141	1.263
Murakami [1]	0.85	0.84	0.89	0.98	1.14	1.27

TABLE 9. Normalized values of SIF k_I for a square lattice of cracks

a/ω_1	0.3	0.4	0.5	0.6	0.7	0.8	0.9
CV-BEM	1.0304	1.0624	1.1136	1.1941	1.325	1.558	2.112
Murakami [1]	1.03	1.06	1.11	1.19	1.30		
Panasiuk et al. [1]	1.03	1.063	1.12	1.19	1.32		

TABLE 10. Normalized values of SIF k_{II} for a square lattice of cracks

a/ω_1	0.3	0.4	0.5	0.6	0.7	0.8	0.9
CV-BEM	1.0468	1.0853	1.1399	1.2197	1.3451	1.571	2.116
Panasiuk et al. [1]	1.0461	1.0827	1.131	1.193	1.27	1.37	

TABLE 11. Normalized values of SIF k_I for a right triangular lattice of cracks ($|\omega_1/\omega_2| = 0.4$)

a/ω_1	0.3	0.4	0.5	0.6	0.7	0.8	0.9
CV-BEM	1.061	1.105	1.168	1.253	1.385	1.63	2.27
Panasiuk et al. [1]	1.059	1.103	1.158	1.228	1.319	1.445	1.63
Filshtinski [1]	1.06	1.10	1.16	1.23	1.36	1.58	1.90

TABLE 12. Normalized values of SIF k_{II} for a right triangular lattice of cracks

a/ω_1	0.3	0.4	0.5	0.6	0.7	0.8	0.9
CV-BEM	1.0273	1.0576	1.109	1.192	1.329	1.580	2.20
Panasiuk et al. [1]	1.027	1.057	1.106	1.18	1.28	1.42	1.6
Filshtinski [1]	1.03	1.06	1.12	1.18	1.31	1.57	2.05

TABLE 13. Normalized values of elasticity moduli for a square lattice of cracks

a/ω_1	0.3	0.4	0.5	0.6	0.7	0.8	0.9
E_{eff}/E, CV-BEM	0.873	0.790	0.698	0.602	0.504	0.405	0.300
E_{eff}/E, Filshtinski [1]	0.86	0.80	0.71	0.61	0.50	0.41	0.29
μ_{eff}/μ, CV-BEM	0.9461	0.905	0.853	0.791	0.718	0.632	0.521
μ_{eff}/μ, Filshtinski [1]	0.93	0.90	0.85	0.80	0.72	0.63	0.50

TABLE 14. Normalized values of elasticity moduli for a right triangular lattice of cracks

a/ω_1	0.3	0.4	0.5	0.6	0.7	0.8	0.9
E_{eff}/E, CV-BEM	0.859	0.757	0.652	0.548	0.447	0.350	0.250
E_{eff}/E, Filshtinski [1]	0.84	0.77	0.65	0.56	0.45	0.35	0.24
μ_{eff}/μ, CV-BEM	0.939	0.895	0.840	0.774	0.696	0.603	0.482
μ_{eff}/μ, Filshtinski [1]	0.92	0.89	0.84	0.78	0.69	0.60	0.47

We see that for high crack density there are differences between the results of other authors. Meanwhile, both for k_I and k_{II}, the results by Filshtinski are closer to our results obtained by the CV-BEM. The error in the calculations by Panasiuk et al. [1] increases with the ratio a/ω_1. This could be expected, because these authors used expansions with respect to powers of a/ω_1.

Homogenization problem. As we could see in § 40, equations (40.5) or (40.7) allow us to solve homogenization problems after we have found unknown values on boundaries of blocks, inclusions, cracks, holes within the main cell. Therefore, for the examples discussed above, we can easily obtain effective moduli as well. Table 13 contains the results for a square lattice, Table 14 for a right triangle lattice. To compare with the results by Filshtinski [1], calculations are performed for a Poisson's ratio of 0.3.

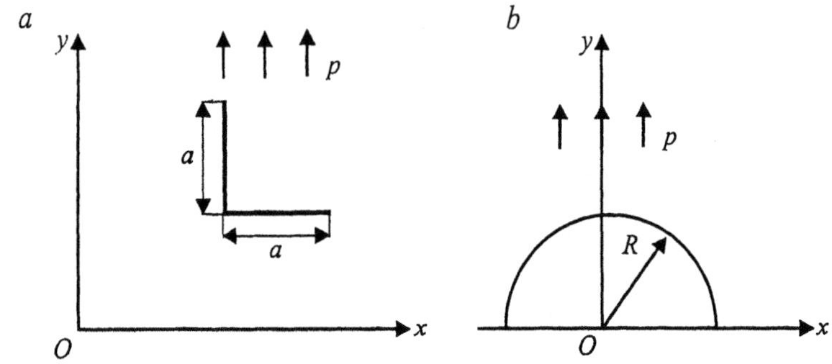

Fig. 56.

We again see that the results agree within the accuracy of the cited paper. Note however that calculations by the CV-BEM are actually carried out by employing a *universal* program for a finite system of contours. One may consider arbitrary curvilinear, broken, branching contours and systems of blocks with internal structure in doubly periodic problems.

Two examples for cracks are given in Figure 56.

Square lattices with angular and semi-circular cracks. Results for angular (Figure 56, *a*) and semi-circular (Figure 56, *b*) systems of cracks in a square lattice with periods $2\omega_1=2\omega$, $2\omega_2=i\cdot 2\omega$ are given in Tables 15, 16. The additional compliances $b_{11}^a=b_{22}^a$ and b_{33}^a are normalized by the values respectively $b_{110}=b_{220}=(\chi+1)/(8\mu)=1/E$ and $b_{330}=1/\mu$, corresponding to a plane without cracks.

TABLE 15. Normalized values of the additional compliances b_{11}^a, b_{33a} for a square lattice with kinked cracks

a/ω	0.3	0.4	0.5	0.6	0.7	0.8
$b_{11a} = b_{22}^a$	0.2110	0.3895	0.6457	1.0157	1.581	2.552
b_{33}^a	0.2416	0.4782	0.8451	1.3645	2.084	3.136

TABLE 16. Additional compliances b_{11}^a, b_{33}^a for a square lattice with semi-circular cracks

R/ω	0.3	0.4	0.5	0.6	0.7	0.8
b_{22}^a	0.1627	0.2965	0.4828	0.7432	1.126	1.756
b_{33}^a	0.1987	0.3887	0.6921	1.190	2.073	3.914

We present two more examples; these are abstract, but illustrate the ability to account for contact interaction of blocks and internal cracks (Linkov and Koshelev [2]).

Masonry type structure with and without cracks. Consider a specific type of doubly periodic structure that resembles brick masonry. A brick has a thickness equal to ¼ of its length. The period $2\omega_1$ and the x-axis are taken along the brick length ($\omega_1 = |\omega_1|$); the period $2\omega_2$, being orthogonal to $2\omega_1$, is directed along the y-axis ($\omega_2 = i\omega_1/2$). Schemes without cracks (*a*), a crack within a brick (*b*) and a crack terminating at cement (*c*) are shown in Figure 57. In cases *b*) and *c*) the angle between cracks and the x-axis is $\pi/9$, crack length is $\omega_1/2$.

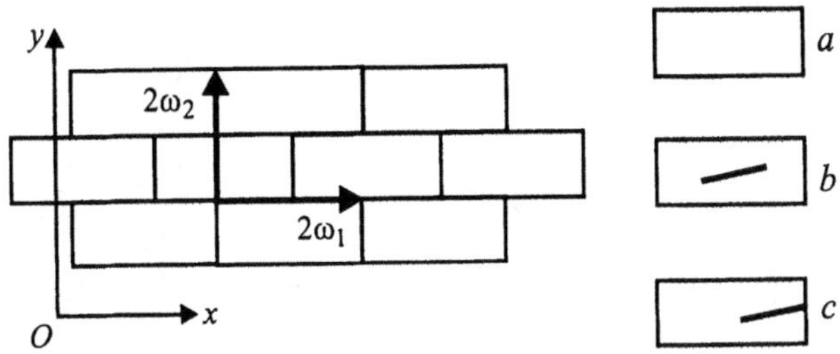

Fig. 57.

Interaction on the boundaries of bricks is prescribed by the equations $\sigma_{nn} = a_{nn}\Delta u_n$, $\sigma_{nt} = a_{nt}\Delta u_t$ accounting for the properties of cement. It is assumed that $a_{nn} = 2a_{\tau\tau}$ (the normal rigidity of contacts is two fold greater then the shear rigidity). The value $8\mu_0/(\chi_0 + 1)$ is equal (for plane strain) or close (for plane stress) to the Young's modulus where μ_0 and χ_0 are the shear modulus and Muskhelishvili's parameter of brick material. The ratio $[8\mu_0/(\chi_0 + 1)]/[2\omega_1 a_{nn}]$ is a dimensionless parameter which characterizes relative input of bricks and cement into rigidity of the whole system.

Table 17 illustrates the influence of this parameter, and also of cracks. Additional compliances $[8\mu_0/(\chi_0+1)]b_{11}^a$, $[8\mu_0/(\chi_0+1)]b_{22}^a$, $[8\mu_0/(\chi_0+1)]b_{33}^a$ are given for three values of this ratio: 0.25, 0.50 and 1.25.

TABLE 17. Normalized effective compliances for bricks interacting on boundaries: (a) bricks without cracks, (b) bricks with internal cracks, (c) bricks with cracks terminating on boundaries

$[8\mu_0/(\chi_0+1)]/$	(a)			(b)			(c)		
$(2\omega_1 a_{nn})$	0.25	0.50	1.25	0.25	0.50	1.25	0.25	0.50	1.25
$[8\mu_0/(\chi_0+1)]b_{11}^a$	0.198	0.372	0.875	0.280	0.466	0.998	0.360	0.555	1.115
$[8\mu_0/(\chi_0+1)]b_{22}^a$	0.998	1.997	4.981	2.174	3.650	8.012	2.176	3.718	8.391
$[8\mu_0/(\chi_0+1)]b_{33}^a$	1.168	2.277	5.550	1.877	3.156	6.883	1.942	3.246	7.128

As it could be expected, the growth of the cement rigidity increases the effective rigidity of the system. The presence of cracks naturally decreases the effective rigidity. The table shows that cracks terminating at brick edges reduce rigidity more severely than internal cracks. Note that two hierarchical levels are accounted for in this example.

Hexagonal structure without and with cracks. Consider a hexagonal structure shown in Figure 58. Interaction on boundaries is the same as in the previous example. Again we consider three schemes: (*a*) without cracks, (*b*) a horizontal crack within a cell and (*c*) an inclined crack.

The results of calculations normalized as in the previous example are presented in Table 18.

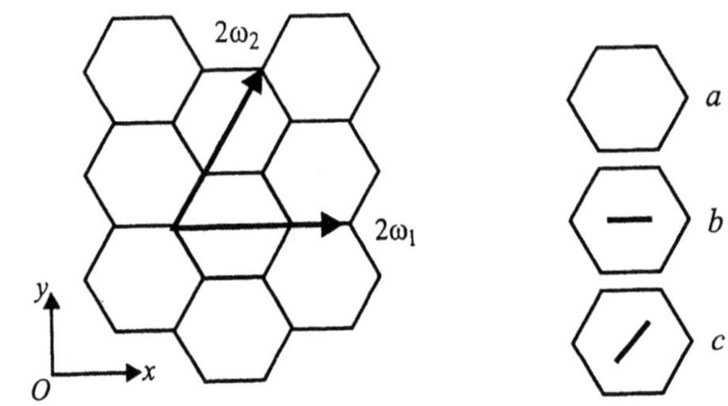

Fig. 58.

TABLE 18. Normalized effective compliances for hexagonal cells interacting on boundaries: *a*) cells without cracks, *b*) cells with horizontal cracks, *c*) cells with inclined cracks

$[8\mu_0/(\chi_0+1)]$ / $(2\omega_1 a_{nn})$	(a)			(b)			(c)		
	0.25	0.50	1.25	0.25	0.50	1.25	0.25	0.50	1.25
$[8\mu_0/(\chi_0+1)]b_{11}^a$	0.704	1.350	3.606	0.660	1.283	3.030	0.776	1.520	3.240
$[8\mu_0/(\chi_0+1)]b_{22}^a$	0.795	1.600	3.920	1.050	1.800	4.451	0.776	1.513	4.96
$[8\mu_0/(\chi_0+1)]b_{33}^a$	1.607	3.242	8.200	1.958	3.700	10.22	1.720	3.720	10.18

On the whole, the results are similar to those for the brick structure. But now we see the much stronger influence of cell interaction on the horizontal compliance b_{11a}. This could be expected because now we have inclined zigzag lines of interaction uninterrupted by intact material.

We could present numerous other examples. A code based on the CVH-BEM allows us to consider arbitrary lattices with arbitrary contours represented or approximated with a set of straight and circular-arc elements. With a user-friendly graphic interface, the program may serve as a specific handbook to find SIF and effective compliances for numerous configurations which cannot be listed in a conventional handbook.

In conclusion we will comment on one application, which looks very promising for obtaining constitutive equations for systems with at least two structural levels.

Numerical modeling of constitutive equations. The homogenization problem is a classical topic of physics, elasticity, material and fracture. Physical experiments used to solve it are restricted in number, difficult and expensive. Thus, having in view the progress in computers and informatics, it looks reasonable to complement or replace physical experiments by numerical ones. A new approach to modeling constitutive equations for a number of structural levels employs solutions for *regular* structures to describe properties of arbitrary, in general *irregular*, media (Linkov and Koshelev [2], Dobroskok et al. [1]).

The approach has its roots in the usual concepts of continuum mechanics. We assume that (i) there exists a representative volume of a finite size that may be treated as a mathematical point, and (i) there is infinite multiplicity (continuum) of such volumes (points) in the vicinity of each of them. As a pattern, we may take a representative volume in the form of a principal cell, and reproduce the multiplicity of surrounding volumes by infinitely repeating the principal cell in 2 directions (in 2D) or 3 (in 3D). In this way we arrive at a homogenization problem for a regular system: doubly periodic in 2D or triply periodic in 3D.

Recall that this approach has three virtues:

(i) cyclic constants of displacements provide *rigorous* (not approximate) connections between average strains and average stresses for whatever geometries and interaction of structural elements within the principal cell;

(ii) calculations involve only a restricted area represented by the principal cell;

(iii) we can account for at least two distinct hierarchical levels of the structure: the first, external, includes whole cells, which may have displacement discontinuities on their interacting boundaries; the second, internal with respect to a cell, includes its internal structure.

In 2D, the approach is strongly supported by the CV-BEM for doubly periodic blocky systems with cracks, inclusions and cavities presented in this chapter. By employing it as explained, one obtains dependence between average strains and average stresses for a prescribed loading path. Naturally, for irreversible contact deformations and for crack growth, *increments* of values are used. In this case, calculated macroscopic strains depend on the loading path, as discussed in the papers by Dobroskok [2] and Dobroskok et al. [1].

Chapter 14

COMPLEX VARIABLE METHOD OF MECHANICAL QUADRATURES (CV-MMQ)

The complex variable method of mechanical quadratures is a popular method for solving complex variable *singular* equations. There is a vast literature on this method as applied to elasticity problems (see, e.g. Theocaris and Ioakimidis [1-4], Ioakimidis and Theocaris [1-5], Panasiuk, Savruk and Datsishin [1], Savruk [4], Savruk, Osiv and Prokopchuk [1], Grigoliuk and Filshtinski [2] and reviews in the last four monographs. The L'vov school of solid mechanics systematically employed this method. Since the results have been comprehensively presented in the cited monographs by Savruk, we will discuss only the essence of the method and the differences between it and the CV-BEM.

§ 60. GENERAL STAGES OF CV-MMQ

In elasticity problems, the CV-MMQ is employed primarily for singular BIE; it has restricted applicability for hypersingular equations. We will follow its general stages by applying it to the singular integral equation (29.9). For the case of prescribed tractions, write this equation in the form

$$\int_L \left[K_s(\tau,t)\Delta u'\, d\tau + K_0(\tau,t)\overline{\Delta u'\, d\tau}\right] = h(t), \tag{60.1}$$

where

$$K_S(\tau,t) = \frac{1}{2\pi i}\left[\frac{2}{\tau - t} + \frac{\partial k_1(\tau,t)}{\partial t}\right],\ K_0(\tau,t) = \frac{1}{2\pi i}\frac{\partial k_2(\tau,t)}{\partial t},\ h(t) = \frac{\chi+1}{2\mu}\frac{\sigma^+ + \sigma^-}{2},$$

$K_S(\tau, t)$ is a singular kernel; $K_0(\tau, t)$ is a regular kernel; $h(t)$ is a given sufficiently smooth function; $\Delta u'$ is the derivative of the displacement discontinuity $\Delta u = u^+ - u^-$ (recall that for closed contours, we apply $u^+ = u$, $u^- = 0$ when the traveling path leaves the region on the left). We consider the contour L to be comprised of m closed and p open parts L_k ($k = 1,\dots, m + p$). The contour L may include boundaries of holes, cracks terminating on these boundaries, internal cracks, and an external contour enclosing all internal contours (for a finite region, Figure 59). For isolated cracks and also for cracks terminating on boundaries of holes, it is necessary to use

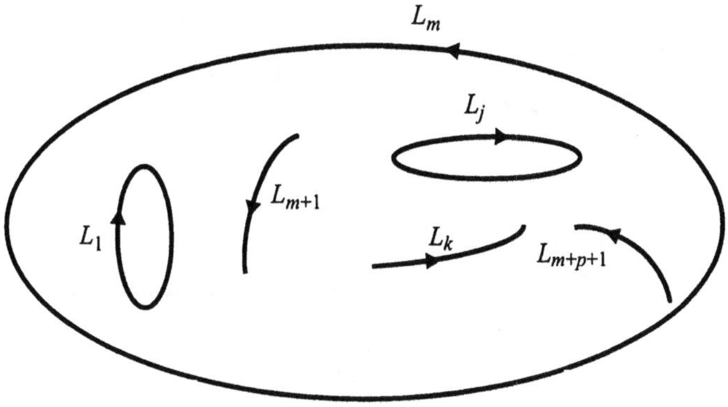

Fig. 59.

the condition that displacements are single-valued functions, because equation (60.1) contains the derivative $\Delta u'$. In particular, for isolated cracks, these conditions have the form

$$\int_{L_k} \Delta u' \, d\tau = 0. \tag{60.2}$$

In plane elasticity the CV-MMQ has six stages:

1) divide contours into closed and open parts;

2) approximate each closed and open contour by employing canonical parametrization;

3) approximate unknown functions;

4) evaluate integrals over particular contours;

5) form and solve algebraic system of equations;

6) calculate values within a considered region; if necessary, previously unknown values, such as tangential stresses, are calculated on the boundary as well.

On the whole, the CV-MMQ contains the same stages as the CV-BEM (cf. § 53). It differs primarily in the methods used in these stages. We consider the stages, to see the peculiarities of the CV-MMQ.

1. ***Divide contours into sets of closed and open parts.*** In essence, the initial division of the contour L into m closed and p open parts is the first stage. (Note that for branching cracks, each branch is assumed to be a separate open arc). As a result, the integral on the l. h. s. of (60.1) becomes a sum of integrals over closed and open parts:

$$\sum_{j=1}^{m} \int_{L_j} \left[K_s(\tau,t)\Delta u' d\tau + K_0(\tau,t)\overline{\Delta u' d\tau}\right] +$$

$$+ \sum_{k=m+1}^{m+p} \int_{L_k} \left[K_s(\tau,t)\Delta u' d\tau + K_0(\tau,t)\overline{\Delta u' d\tau}\right] = h(t). \tag{60.3}$$

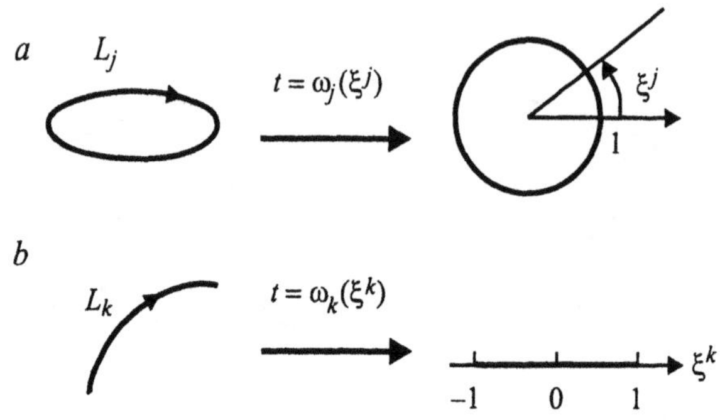

Fig. 60.

In contrast to the CV-BEM, each closed contour is now taken as a whole, without division with smaller parts. The same holds for open arcs representing cracks. If a crack does not contain corners, it is taken as a whole. If there are corner points, the parts between them are also taken as wholes. We see that the CV-MMQ operates with strictly fixed and larger elements of the boundary than the CV-BEM.

2. ***Approximate particular contours.*** Each of the separated parts is mapped onto one of two canonical contours: 1) a *circumference* of unit radius for *closed* parts or 2) an *interval* [-1, 1] of the real axis for *open* arcs. Afterwards, this serves for the standard approximation of functions on the separate "super-elements", and for simple integration over them.

The mapping employs a *real* parameter. For a circumference, it is a polar angle ξ, which changes from 0 to 2π. In general, the mapping of a *closed* contour L_j onto a unit circumference is fulfilled approximately by choosing an appropriate function $\omega_j(\xi^j)$ (Figure 60, *a*):

$$t = \omega_j\left(\xi^j\right), \quad 0 \le \xi^j \le 2\pi, \quad j = 1,\ldots,m. \tag{60.4}$$

The mapping of an *open* arc L_k onto an interval [-1, 1] in general is also approximate, it employs an appropriate function $\omega_k(\xi^k)$ (Figure 60, *b*):

$$t = \omega_k\left(\xi^k\right), \quad -1 \le \xi^k \le 1, \quad k = p+1,\ldots,m+p. \tag{60.5}$$

Certainly, an infinity of functions may serve for the parametrization; we choose simple, appropriate ones; they are taken sufficiently smooth; normally they have a Holder continuous derivative.

For approximate mapping, real curvilinear contours are always distorted to some extent, and actually we solve a problem for an approximation of the contour L. The CV-BIE (60.3) is replaced by the approximate equation on canonical contours:

$$\sum_{j=1}^{m}\int_{0}^{2\pi}\left[M_s^j\left(\eta^j,\xi^l\right)\varphi^j\left(\eta^j\right)+M_0^j\left(\eta^j,\xi^l\right)\overline{\varphi^j\left(\eta^j\right)}\right]d\eta^j +$$

$$+\sum_{k=m+1}^{m+p}\int_{-1}^{1}\left[M_s^k\left(\eta^k,\xi^l\right)\varphi^k\left(\eta^k\right)+M_0^k\left(\eta^k,\xi^l\right)\overline{\varphi^k\left(\eta^k\right)}\right]d\eta^k \approx h^l\left(\xi^l\right),$$

$$0\le\xi^l\le 2\pi,\quad l=1,\dots,m;\quad -1\le\xi^l\le 1,\quad l=m+1,\dots,m+p, \tag{60.6}$$

where

$$\varphi^j\left(\eta^j\right)=\Delta u'\left(\omega_j\left(\eta^j\right)\right)\omega_j'\left(\eta^j\right),$$

$$M_s^j\left(\eta^j,\xi^l\right)=K_s\left(\omega_j\left(\eta^j\right),\omega_l\left(\xi^l\right)\right),\quad M_0^j\left(\eta^j,\xi^l\right)=K_0\left(\omega_j\left(\eta^j\right),\omega_l\left(\xi^l\right)\right),$$

$$j=1,\dots,m+p,\quad h^l\left(\xi^l\right)=h\left(\omega_l\left(\xi^l\right)\right), \tag{60.7}$$

the kernels $M_s^j(\eta^j, \xi^j)$ are singular, $M_0^j(\eta^j, \xi^j)$ regular.

Note that each integral is integrated with respect to a real variable. For this reason, we placed the differential symbol behind the square brackets. An analogous transformation is performed for the additional conditions (60.2).

3. ***Approximate functions.*** For functions on the canonical contours, one normally employs approximations that are *canonical.* They differ for functions on closed contours transformed into unit circumferences, and for functions on open arcs, which become intervals [-1, 1].

Closed contours (Figure 60, *a*). A closed contour does not have singular points. We approximate a function on it by a trigonometric polynomial with an even or odd number of uniformly spaced nodes. Such a polynomial corresponds to a combination of usual and complex conjugate polynomials of the complex argument.

An interpolation formula with N nodes for a 2π-periodic function $\varphi(\eta)$ has the form

$$\varphi(\eta)\approx\sum_{k=1}^{N}\varphi_k P_{nk}(\eta), \tag{60.8}$$

where $\varphi_k = \varphi(\eta_k)$ is the value of the function $\varphi(\eta)$ at the node η_k; P_{nk} is a trigonometric form-function; this is a trigonometric polynomial of a degree not exceeding $N/2$.

For an *even* number of nodes $N = 2n$ the interpolating polynomial P_{nk} is

$$P_{nk}(\eta)=\frac{1}{N}\sin\frac{N(\eta-\eta_k)}{2}\cot\frac{\eta-\eta_k}{2}. \tag{60.9}$$

With $P_{nk}(\eta)$ defined by (60.9), formula (60.8) is exact for any trigonometric polynomial of a degree not exceeding $n - 1$.

For an odd number of nodes $N = 2n + 1$ the interpolating polynomial P_{nk} is

$$P_{nk}(\eta)=\frac{1}{N}\sin\frac{N(\eta_k-\eta)}{2}\operatorname{cosec}\frac{\eta_k-\eta}{2}. \tag{60.10}$$

With $P_{nk}(\eta)$ defined by (60.10), formula (60.8) is exact for any trigonometric polynomial of a degree not exceeding n.

Open contours (Figure 60, *b*). For open contours we account for asymptotic behavior of a function near tips. For the displacement discontinuities Δu, the approximation is taken in accordance with formula (54.8) for a two-sided tip element. Now the element is a standard interval [-1, 1].

If the initial CV-BIE is formulated with respect to *displacement discontinuity* itself, the integrals after parametrization contain the product $\Delta u(\omega(\eta))\,\omega'(\eta)$, which we denote by $\psi(\eta)$. For this function, (54.8) gives

$$\psi(\eta)=\Delta u(\omega(\eta))\omega'(\eta)\approx\sqrt{1-\eta^2}\,P(\eta),$$

where $P(\eta)$ is a polynomial of the real argument η. This polynomial can be represented by a combination of form-functions that are common Lagrangian polynomials $P^k(\eta)$. For N nodes at the points η_k ($k = 1,\ldots, N$), and the polynomial $P_N(\eta)$, we have for the function $r(\eta) = \Delta u\,\omega'(\eta)/\sqrt{1-\eta^2}$:

$$r(\eta)=\frac{\Delta u\omega'}{\sqrt{1-\eta^2}}\approx P_N(\eta)=\sum_{k=1}^{N} r_k P^k(\eta). \tag{60.11}$$

Here, $r_k = \Delta u_k\,\omega_k'/\sqrt{1-\eta_k^{\,2}}$ is the value of $r(\eta)$ at the node η_k. Now we can express the product $\Delta u\,\omega'(\eta)$ through $r(\eta)$ as

$$\Delta u(\omega(\eta))\,\omega'(\eta)=\psi(\eta)=r(\eta)\sqrt{1-\eta^2}. \tag{60.12}$$

and solve the transformed BIE containing Δu itself with respect to the function $r(\eta)$.

If the initial CV-BIE contains the *derivative* $\Delta u'$ of the displacement discontinuity, as in equation (60.1), the integrals after parametrization contain the product $\Delta u'(\omega(\eta))\,\omega'(\eta)$, which was called $\varphi(\eta)$ in (60.6). Its approximation should account for the square-root singularity. Thus

$$\varphi(\eta)=\Delta u'(\omega(\eta))\omega'(\eta)\approx\frac{P(\eta)}{\sqrt{1-\eta^2}}.$$

Again use Lagrangian polynomials as form-functions, now of an order N for the non-singular function $g(\eta) = \varphi(\eta)\sqrt{1-\eta^2}$:

$$g(\eta) = \varphi(\eta)\sqrt{1-\eta^2} \approx \sum_{k=1}^{N} g_k P^k(\eta), \tag{60.13}$$

where $g_k = \varphi(\eta_k)\sqrt{1-\eta_k{}^2} = \Delta u_k{}'\,\omega_k{}'\sqrt{1-\eta_k{}^2}$; as above, $P^k(\eta)$ are Lagrangian polynomials. We can take $g(\eta)$ as a new unknown for open arcs in the transformed equation (60.6) by setting

$$\varphi(\eta) = \frac{g(\eta)}{\sqrt{1-\eta^2}}. \tag{60.14}$$

The standard interval [-1, 1] has an important advantage: it allows us to use the roots of Chebyshev polynomials as the nodes of approximation. This allows us to use quadrature rules of the highest algebraic accuracy in the integration step.

Consider first the substitution (60.12) $\psi(\eta) = r(\eta)\sqrt{1-\eta^2}$ used in the CV-BIE containing the *displacement discontinuities* themselves. For it we employ the roots of Chebyshev polynomial $U_N(\eta)$ of the *second* kind. This polynomial is defined by formula

$$U_N(\eta) = \frac{\sin[(N+1)\theta]}{\sin\theta},$$

where $\eta = \cos\theta$. Its roots used as nodes for approximation of $r(\eta)$ are

$$\eta_k = \cos\frac{k\pi}{N+1}, \quad k = 1,\ldots,N.$$

For this choice of the nodes, the Lagrangian polynomials $P^k(\eta)$ of order $N-1$ are expressed through Chebyshev polynomial $U_N(\eta)$ of order N by the formula

$$P^k(\eta) = (-1)^{k+1}\frac{1}{N+1}\frac{U_N(\eta)(1-\eta_k^2)}{\eta-\eta_k}.$$

Now consider the substitution (60.14) $\varphi(\eta) = g(\eta)/\sqrt{1-\eta^2}$ used in the CV-BIE (60.1) containing the *derivative* of the displacement discontinuities. For it we employ the roots of Chebyshev polynomial of the *first* kind $T_N(\eta)$. This polynomial is defined by formula

$$T_N(\eta) = \cos(N\theta),$$

where $\eta = \cos\theta$. Its roots used as nodes for approximation of $g(\eta)$ are

$$\eta_k = \cos\frac{\pi(2k-1)}{2N}, \quad k = 1,\ldots,N.$$

For this choice of nodes, the Lagrangian polynomial $P^k(\eta)$ of order $N-1$, is

$$P^k(\eta) = \frac{1}{N} T_{N\text{-}1}(\eta_k) \frac{T_N(\eta)}{\eta - \eta_k}.$$

Sometimes, approximation of the derivative of the displacement discontinuities is taken in a form with two main terms reflecting major input from the asymptotic behavior near tips (Savruk, Osiv and Prokopchuk [1]):

$$\varphi(\eta) \approx A\sqrt{\frac{1-\eta}{1+\eta}} + B\sqrt{\frac{1+\eta}{1-\eta}} + \varphi_0(\eta), \tag{60.15}$$

where A and B are constants to be found, and $\varphi_0(\eta)$ is a new unknown function. Its behavior is prescribed by a formula of (60.12) type.

One form of this is

$$\Delta u'(t) = A\sqrt{\frac{t-c}{t\text{-}b}} + B\sqrt{\frac{t\text{-}b}{t-c}} + u_0(t)$$

where asymptotic terms are distinguished *before* parametrization of a contour. In this case, the approximation of (60.12) type is applied after parametrization to the new unknown function $u_0(\omega(\eta))$. Savruk et al. [1] note that in their numerical examples this approach provided more accurate results than the approach based on (60.15), while the latter was better than employing (60.14).

Comment 1. It is not quite correct to use these approximations for a crack terminating on a contour because the square-root asymptotic occurs only at the opposite tip. Nevertheless, numerical results obtained by Savruk et al. [1] by employing the substitution (60.14) with approximation (60.13) for cracks terminating on a contour of a circle appeared to be satisfactory. In our view, in such cases it is more consistent to use "one-sided" approximations following from (54.7) (see also comment 2 below).

4. ***Integration.*** After inserting approximations into an equation we need to integrate each form-function. The intervals of integration are real, and the form-functions are standard. This suggests employing universal quadrature rules well known for real functions. They are available both for singular and regular integrals.

For *singular* integrals the quadrature rules have the highest algebraic accuracy. This is due to the special choice of approximating functions, nodes, and collocation points. Here we will present the proper choice of nodes. Note that the transition to real intervals does not provide the best way of evaluating singular integrals in the complex plane. Also, we will see that employing analogous formulae for *hypersingular* integrals is a much more difficult problem.

Closed contours. When calculating a singular integral over a closed contour, we transform it into a circumference. Assume that the singularity is of Hilbert kernel type: $\cot[(\xi - \eta)/2]$. We can always reduce it to this by using a standard 'subtract and add' procedure. For form-functions in (60.8) with $N = 2n$, the approximation (60.8), (60.9) provides a quadrature rule of the highest accuracy (Chawla, Ramakrishnan [2])

$$S(\xi_j) = \int_0^{2\pi} \varphi(\eta)\cot\frac{\eta - \xi_j}{2}d\eta \approx \frac{\pi}{N/2}\sum_{k=1}^{N}\varphi_k \cot\frac{\eta_k - \xi_j}{2}. \tag{60.16}$$

if we choose the points ξ_j as the roots of the function $\cot(N\xi/2)$:

$$\xi_j = \frac{\pi(2j-1)}{N}, \quad j = 1,\ldots,N. \tag{60.17}$$

Recall that η_k are nodes uniformly distributed on the unit circumference: $= 2\pi k/N$ ($k = 1,\ldots, N$), $\varphi_k = \varphi(\eta_k)$ is the value of an integrated function at the node η_k. The quadrature rule turns into an identity for any trigonometric polynomial up to order N.

Note that the formula (60.16) for the *singular* integral is precisely the same as the formula of the highest accuracy for *regular* integrals of 2π-periodic functions:

$$\int_0^{2\pi} f(\eta,\xi)d\eta \approx \frac{\pi}{N/2}\sum_{k=1}^{N} f(\eta_k,\xi). \tag{60.18}$$

Indeed, we may consider (60.16) as a special case of (60.18) when $f(\eta, \xi) = \varphi(\eta)\mathrm{ctg}(\xi - \eta)/2]$ while the values of ξ are taken according to (60.17). Note that this complete agreement occurs only for the specific choice of $\xi = \xi_j$.

From (60.16), (60.18) it follows that there is no need to distinguish the Hilbert kernel in a singular term. It is sufficient to use a common quadrature rule with one reservation: we have to take collocation points at the points defined by (60.17).

Open arcs. Quite similar results follow for open arcs. Consider first the substitution (60.12) $\psi(\eta) = r(\eta)\sqrt{1-\eta^2}$ used in the CV-BIE containing the *displacement discontinuities* themselves. The approximation (60.11) yields (Korneychuk [1]):

$$S(\xi_j)=\int_{-1}^{1}\frac{\sqrt{1-\eta^2}\,r(\eta)}{\eta-\xi_j}d\eta\approx\frac{\pi}{N+1}\sum_{k=1}^{N}\frac{\left(1-\eta_k^2\right)r_k}{\eta_k-\xi_j}. \qquad (60.19)$$

Here, ξ_j are roots of Chebyshev polynomial of the *first* kind $T_{N+1}(\xi)$:

$$\xi_j=\cos\frac{\pi(2j-1)}{2(N+1)},\quad j=1,\ldots,N+1. \qquad (60.20)$$

Recall that the nodes η_k are chosen as the roots of the Chebyshev polynomial of the *second* kind $U_N(\eta)$: $\eta_k=\cos[k\pi/(N+1)]$, $k=1,\ldots,N$. For this choice of the points ξ_j, the quadrature rule (60.19) has the highest algebraic accuracy: it becomes an identity for any polynomial of order up to $2N-1$.

Again the formula (60.19) for the *singular* integral is precisely the same as the formula of the highest accuracy for *regular* integrals with the weight function $\sqrt{1-\eta^2}$:

$$\int_{-1}^{1}\sqrt{1-\eta^2}\,f(\eta,\xi)d\eta\approx\frac{\pi}{N+1}\sum_{k=1}^{N}\left(1-\eta_k^2\right)f(\eta_k,\xi). \qquad (60.21)$$

This formula is exact for polynomials of the order up to $2N-1$ for arbitrary $\xi\notin[-1,1]$. Hence, we may consider (60.19) as a special case of (60.21) when $f(\eta,\xi)=r(\eta)/(\eta-\xi)$, while the values of ξ are taken according to (60.20). Note that this complete agreement occurs only for the specific choice of $\xi=\xi_j$. From this we conclude that it is sufficient to use a common quadrature rule (60.21) with one reservation: we have to take collocation points given by (60.20).

Now consider the substitution (60.14) $\varphi(\eta)=g(\eta)/\sqrt{1-\eta^2}$ used in the CV-BIE (60.1) containing the *derivative* of the displacement discontinuities. The approximation (60.13) yields (Chawla and Ramakrishnan [1]):

$$S(\xi_j)=\int_{-1}^{1}\frac{g(\eta)}{\sqrt{1-\eta^2}\left(\eta-\xi_j\right)}d\eta\approx\frac{\pi}{N}\sum_{k=1}^{N}\frac{g_k}{\eta_k-\xi_j}. \qquad (60.22)$$

Here, ξ_j are roots of the Chebyshev polynomial of the *second* kind $U_{N-1}(\xi)$:

$$\xi_j=\cos\frac{\pi j}{N},\quad j=1,\ldots,N-1. \qquad (60.23)$$

Recall that the nodes η_k are chosen as the roots of the Chebyshev polynomial of the *first* kind $T_N(\eta)$: $\eta_k=\cos[(\pi/2)(2k-1)/N]$, $k=1,\ldots,N$. For this choice of the points ξ_j the quadrature rule (60.22) has the highest algebraic accuracy: it becomes an identity for any polynomial of an order up to $2N-1$.

Again the formula (60.22) for the *singular* integral is precisely the same as the formula of the highest accuracy for *regular* integrals with the weight function $\dfrac{1}{\sqrt{1-\eta^2}}$:

$$\int_{-1}^{1}\frac{f(\eta,\xi)d\eta}{\sqrt{1-\eta^2}}\approx\frac{\pi}{N}\sum_{k=1}^{N}f(\eta_k,\xi). \tag{60.24}$$

This formula is exact for polynomials of the order up to $2N-1$ for arbitrary $\xi\notin[-1,1]$. Hence, we may consider (60.22) as a special case of (60.24) when $f(\eta,\xi) = g(\eta)/(\eta-\xi)$, while the values of ξ are taken according to (60.23). This implies that it is sufficient to use a common quadrature rule (60.24) with one reservation: we have to take the collocation points given by (60.23).

It is easy to see that the number of collocation points in (60.19) is greater by one, while in (60.22) one less, than the number of nodes in approximations. This reflects peculiarities of the corresponding BIE. In the first case, each "redundant" equation serves to find an unknown constant expressing, for instance, the value of the resultant force at the crack start point. In the second case, a "missing" equation is replaced by an additional condition of the (60.2) type, stating that displacements are single valued. In mathematical terms, these peculiarities of singular equations arise due their non-zero index.

***Comment** 2.* For cracks terminating on a contour, as mentioned, the approximation (60.13), (60.14) is not correct. The additional condition (60.2) also cannot be used as a condition providing single valued displacements: it does not account for crack opening at the termination point. Hence, when trying to apply (60.13), (60.14) for cracks terminating on a contour, one should apply some other condition to have a complete set of equations instead of (60.2). As such a condition, Savruk, Osiv and Prokopchuk [1] used the condition of zero derivative of the displacement discontinuity at the termination point. Although it is difficult to justify this condition, its application provided quite satisfactory numerical results.

Comment 3. After the mapping onto canonical contours, the classical CV-MMQ employs quadrature rules of the highest accuracy (of Gauss type). Such integration is especially favorable for evaluation of singular integrals. However, we need to pay for this advantage. The price is this: in regular integrals, approximation involves not only densities but also the kernels of integrals.

Comment 4. In the quadrature rules of the highest accuracy (60.16), (60.19), (60.22) for *singular* integrals, the approximation nodes η_k do not coincide with the collocation points ξ_j; meanwhile, both the nodes and the collocation points belong to an integration interval. The matter becomes trickier for *hypersingular* integrals, in particular for the integrals

$$\int_{-1}^{1}\frac{\sqrt{1-\eta^2}\,r(\eta)}{(\eta-\xi)^2}d\eta,\quad \int_{-1}^{1}\frac{g(\eta)d\eta}{\sqrt{1-\eta^2}(\eta-\xi)^2}.$$

There is a quadrature formula of Gauss type for each particular value of ξ (Tsamasphyros and Dimou [1]). But now the location of the nodes η_k depends on the point ξ. Sometimes, the nodes become complex, i.e. they do not belong to the integration interval; this makes them and consequently the CV-MMQ inconvenient for approximation.

Comment 5. Certainly, one may integrate without using quadrature rules of the highest accuracy, and locate the nodes η_k as convenient. Then there is no need map to the canonical contours: in this case hypersingular integrals are simply evaluated over arbitrary curvilinear contours by employing the recurrence formulae of § 55. The property of the CV-MMQ of being a *very special* variety of the CV-BEM vanishes. The CV-MMQ becomes a *common* variety of the CV-BEM corresponding to large elements. The same holds for singular equations if one does not employ quadrature rules of the highest accuracy. For this reason we do not discuss such approaches separately: they are covered by the content of chapter 12.

5. ***Forming and solving an algebraic system of equations.*** In the CV-MMQ, the field points ξ_j are fixed by the requirement of using quadrature rules of the highest accuracy for integration. Thus, in formation of an algebraic system we are confined to the collocation method. For open contours, the substitution (60.14), the quadrature rule (60.24) and collocation points (60.23), gives, via (60.6) an algebraic system for the approximation coefficients φ_k^i, g_k^i:

$$\sum_{i=1}^{m}\frac{\pi}{2N_i}\sum_{k=1}^{N_i}\left[M_s^i\left(\eta_k^i,\xi_j^l\right)\varphi_k^i+M_0^i\left(\eta_k^i,\xi_j^i\right)\overline{\varphi_k^i}\right]+$$

$$+\sum_{i=m+1}^{m+p}\frac{\pi}{N_i}\sum_{k=1}^{N_i}\left[M_s^i\left(\eta_k^i,\xi_j^l\right)g_k^i+M_0^i\left(\eta_k^i,\xi_j^i\right)\overline{g_k^i}\right]=h^l\left(\xi_j^l\right),$$

$$\xi_j^l=\frac{\pi(2j-1)}{N_l},\quad j=1,\ldots,N_l \text{ for } l=1,\ldots,m,$$

$$\xi_j^l=\cos\frac{\pi j}{N_l},\quad j=1,\ldots,N_l-1 \text{ for } l=m+1,\ldots,m+p, \tag{60.25}$$

where N_i is the number of nodes on the i-th canonical contour ($i = 1,\ldots, m+p$). The additional conditions (60.2) are transformed analogously.

The algebraic system contains conjugated values of unknowns. Hence, as for the CV-BEM, it is better to separate the real and imaginary parts of unknowns and equations.

6. ***Calculation of values within a considered region.*** By solving the algebraic system, we find values of unknown functions at the approximation nodes. Using

them in the approximating formulae we obtain values at other points of the contour. If we need to find values within a region, then we employ integral representations to express these values via contour integrals and follow the scheme of points 1 – 4. Now all the integrals are regular and hence the quadrature rules of the highest accuracy are available for any $t = z$.

Stress intensity factors. SIF at crack tips, as usual, are found by means of some of the formulae (31.16)-(31.20). In particular, when employing singular equation (60.1), notation (60.7), (60.13) and (60.14) we obtain from (31.20):

$$k_1 - ik_{11} = \pm i \frac{2\mu}{\chi + 1} \omega(\pm 1)\sqrt{\pi|\omega'(\pm 1)|} \sum_{k=1}^{N} g_k P^k(\pm 1),$$

where the upper (lower) sign refers to end (start) point of a crack. In this case,

$$P^k(+1) = -\frac{1}{N}(-1)^k \cot\frac{\pi(2k-1)}{4N}, \quad P^k(-1) = \frac{1}{N}(-1)^{k+N} \tan\frac{\pi(2k-1)}{4N}.$$

We see that, as in the CV-BEM, accounting for asymptotic behavior of a solution by appropriate approximation allows us to evaluate SIF easily at the final stage of calculations.

INDEX

Numbers in bold type refer to text sections

REFERENCES

Ang W. T., Park Y. S.

[1] CVBEM for a system of second-order elliptic partial differential equations // *Engineering Analysis with Boundary Elements*, 1998, v. 21, p. 179-184.

Ang W. T., Clements D. L., Cooke T.

[1] A complex variable boundary element method for a class of boundary value problems in anisotropic thermoelasticity // *Intern. J. Computer Mathematics*, 1999, v. 70, p. 571-586.

Bardzokas D.Ya., Parton V. Z., Theocaris P. S.

[1] Integral equations of elasticity theory for a multiconnected domain with inclusions. // *Journal of Applied Mathematics and Mechanics*, 1989, v. 53, p. 375-384.

Benerjee P. K.

[1] *The Boundary Element Methods in Engineering*. London-New York: McGraw-Hill, 1994.

Benerjee P. K., Butterfield R

[1] *Boundary Element Methods in Engineering Science*. London: McGraw-Hill, 1981.

Bonnet M.

[1] *Boundary Integral Equations Methods in Solids and Fluids*. London: John Willey and Sons, 1999.

Brebbia, C. A., Walker, S.

[1] *The Boundary Element Techniques in Engineering*. London: Newnes-Butterworths, 1980.

Carrier G. F., Krook M., Pearson C. E.

[1] *Functions of a Complex Variable*. London: McGrow-Hill, 1966.

Chawla M. M., Ramakrishnan T. R.

[1] Modified Gauss-Jacoby quadrature formulas for the numerical evaluation of Cauchy type singular integrals // *BIT* (Sver.), 1974, v. 14, p. 14-21.

[2] Numerical evaluation of integrals of periodic functions with Cauchy and Poisson type kernels // *Numer. Math.*, 1974, v. 22, p. 317-323.

Chen J.-T.

[1] *On Hadamard Principle Value and Boundary Integral Formulation of Fracture Mechanics* // M. Sc. Thesis, Institute of Applied Mechanics. National Taiwan University, 1986.

Chen Y. Z.

[1] Solutions of multiple crack problems of a circular plate or an infinite plate containing a circular hole by using Fredholm integral equations approach // *Int. J. Fracture*, 1984, v. 25, p. 155-188.

[2] Multiple crack problems for two bonded half planes in plane and antiplane elasticity // *Engineering Fracture Mechanics*, 1986, v. 25, p. 1-9.

[3] A survey of new integral equations in plane elasticity // *Engineering Fracture Mechanics*, 1995, v. 51, p. 97-134.

Chernykh K. F.

[1] *An Introduction to Modern Anisotropic Elasticity*. Begellhouse Inc. Publishers: New York, 1999.

Churchill R. V.

[1] *Introduction to Complex Variables and Applications*. London: McGrow-Hill, 1966.

Clements D. L.

[1] *Boundary Value Problems Governed by Second Order Elliptic Systems*. London: Pitman, 1981.

Cotterell B., Rice J. R.

[1] Slightly curved or kinked cracks // *Int. J. Fracture*, 1980, v. 16, p. 155-168.

Crouch S. L.

[1] Solution of plane elasticity problem by the displacement discontinuity method // *Int. J. Numerical Methods Engng.*, 1976, v. 10, p. 301-343.

Crouch S. L., Starfield A. M.

[1] *Boundary Element Methods in Solid Mechanics*. London: George Allen & Unwin, 1983.

Cruse T. A., Rizzo F. J.

[1] A direct fromulation and numerical solution of the general transient elasto-plastic problem // *J. Math. Anal. Appl.*, 1968, v. 22, p. 244-259.

Datsyshin A. P., Savruk M. P.

[1] Integral equations of the plane problem of crack theory // *Journal of Applied Mathematics and Mechanics (PMM)*, 1974, v. 38, p. 677-686.

[2] A periodic problem in the theory of cracks // *Mechanics of Solids*, 1974, v. 9, p. 123-129.

Denda M., Kosaka I.
[1] Dislocation and point-force-based approach to the special Green's function BEM for elliptic hole and crack problems in two dimensions // *Int. J. Numerical Methods in Engineering*, 1997, v. 40, p. 2857-2889.
Dobroskok A. A.
[1] Modeling of constitutive equations for a medium with joints and growing cracks // *Proc. XVIII Int. Conference "Mathematical Modeling in Solid Mechanics by BEM and FEM"* / A. N. Snitko (ed). St-Petersburg: St-Petersburg State University, 2000, v. 3, p. 225-300.
[2] On a new method for iterative calculation of crack trajectory // *Int. J. Fracture*, 2001, v. 11, p. L41-L46.
Dobroskok A. A., Linkov A. M.
[1] Complex symmetric boundary element method // *Proc. 13th ASCE Engineering Mechanics Division Specialty Conference* / N. P. Jones (ed). Baltimore: John Hopkins University, 1999 (CD-ROM).
Dobroskok A. A., Linkov A. M., Myer L., Roegiers J.-C.
[1] On a new numerical approach in micromechanics of soils and rocks // *Proc. 38th U. S. Rock Mechanics Symposium DC Rocks: "Rock Mechanics in the National Interest"* / J. Tinucci, Heasley (eds). Washington DC: Swets & Zeitlinger Lisse, 2001, p. 1185-1190.
England A. H..
[1] *Complex Variable Methods in Elasticity*. London: Wiley-Interscience, 1971.
Erdogan F.
[1] Complex function technique // *Continuum Physics, Continuum Mechanics of Single-Substance Materials* / A. C. Eringen (ed.), v. 2, part III, Chap. 3. Academic Press, New York, 1975, p. 523-603.
Erdogan F., Gupta G. D., Cook T. S.
[1] Numerical solution of singular integral equations // *Methods of Analysis and Solution of Crack Problems. Mechanics of Fracture* / G. C. Sih (ed.), Leyden: Noordhoff, v.1, p. 368-425.
Gahov F. D.
[1] *Boundary Value Problems*. Oxford: Pergamon Press, 1966.
Gauss K. F.
[1] *Briefwechsel Zwischen Gauss und Bessel*. Ed. G. F. Auwers. Leipzig, 1880. Reprint: New York: Hildsheim, 1975.
Gol'dshteyn R.V., Salganik R.L.
[1] Planar problem of curvilinear cracks in an elastic solid // *Mechanics of Solids*, 1970, v. 5, p. 54-64.
Gorgidze A. Ya., Rukhadze A. K.
[1] On the numerical solution of the integral equations of the plane theory of elasticity // *Soobshcheniya Gruz. Filiala A. N. S.S.S.R.*, 1940, v. 1, p. 255-258. (In Russian).
Green A. E., Zerna W.
[1] *Theoretical Elasticity*. Oxford: Clarendon Press, 1996.
Grekov M. A.
[1] On a method to solve plane elasticity problems for a half-plane with a cut // *Memorial Volume Devoted to V. V. Novoghilov* / N. S. Solomenko (ed). St-Petersburg: Sudostroenie, 1992, p. 171-179. (In Russian).
[2] *Plane Problems of the Crack Theory. Textbook*. St-Petersburg: St-Petersburg State University, 1997. (In Russian).
[3] Green's functions for periodic problems of elastic half-plane // *Mechanics of Solids*, 1998, v. 33, p. 142-146.
Grigolyuk E. I., Fil'shtinskii L. A.
[1] *Perforated Planes and Shells*. Moscow: Nauka, 1970. (In Russian).
[2] *Periodic Piece-wise Elastic Structures*. Moscow: Nauka, 1992. (In Russian).
Grigolyuk E. I., Kats V. E., Fil'shtinskii L. A.
[1] The doubly periodic problem of the theory of elasticity for a plane anisotropic medium // *Mechanics of Solids*, 1971, v. 6, p. 36-42.
Fairhurst C., Cook N. G. W.
[1] The phenomenon of rock splitting parallel to the direction of maximum compression in the neighbourhood of a surface // *Proc. 1st Congress Int. Soc. Rock Mechanics*, 1966, v. 1, p. 687-692.
Fil'shtinskii L. A.
[1] Interaction of doubly periodic system of rectilinear cracks in an isotropic medium // *Journal of Applied Mathematics and Mechanics (PMM)*, 1974, v. 38, p. 853-861.
[2] Elastic equilibrium of plane anisotropic medium weakened by arbitrary curvilinear cracks. Limit passage to the isotropic medium // *Mechanics of Solids*, 1976, v. 11, p. 78-84.

[3] Doubly periodic problem of elasticity theory for anisotropic medium with curvilinear cuts // *Mechanics of Solids*, 1977, v. 12, p. 98-105.

Germanovich L. N., Ring L. M, Carter B. J. et al.

[1] Simulation of crack growth and interaction in compression // *Proc. 8th Int. Congress on Rock Mechanics* / T. Fujii (ed). Rottedam: Balkema, 1995, v. 1, p. 219-226.

Hageman L. A., Young D. M.

[1] *Applied Iterative Methods*. New York: Academic Press, 1981.

Hartmann F.

[1] Elastostatics // *Progress in Boundary Element Methods* / C. A. Brebbia (ed). London: Pentech Press, 1981, v. 1, p. 84-167.

Hong H.-K., Chen J.-T.

[1] Derivation of integral equations of elasticity // *Journal of Engineering Mechanics*, 1988, v. 114, p. 1028-1044.

Hromadka II T. V., Lai C.

[1] *The Complex Variable Boundary Element Method in Engineering Analysis*. New York-Berlin: Springer Verlag, 1987.

Hui Ch.-Y., Mukherjee S.

[1] Evaluation of hypersingulr integrals in the boundary element method by complex variable techniques // *Int. J. Solids Structures*, 1997, v. 34, p. 203-221.

Ioakimidis, N. I.

[1] *General Methods for the Solution of Crack Problems in the Theory of Plane Elasticity* // Doctoral thesis at the National Technical University of Athens. Athens, 1976.

[2] Application of finite-part integrals to the singular equations of crack problems in plane and three-dimensional elasticity // *Acta Mechanica*, 1982, v. 45, p. 31-47.

Ioakimidis N. I., Theocaris P. S.

[1] Array of periodic curvilinear cracks in an infinite isotropic medium // *Acta Mechanica*, 1977, v. 28, p. 239-254.

[2] Doubly-periodic array of cracks in an infinite isotropic medium // *J. Elasticity*, 1978, v. 8, p. 157-169.

[3] The numerical evaluation of a class of generalized Lobatto-Jacobi numerical rule // *Int. J. Fracture*, 1978, v. 14, p. 469-484.

[4] Numerical solution of Cauchy type singular integral equations by use of Lobatto-Jacoby numerical integration rule // *Appl. Math.*, 1978, v. 23, p. 349-452.

[5] A system of curvilinear cracks in an isotropic elastic half-plane // *Int. J. Fracture*, 1979, v. 15, p. 299-309.

Isakhanov R. V.

[1] Differential boundary value problem by Riemann and its application in the theory of integro-differential equations // *Soobshcheniya Gruz. Filiala A. N. S.S.S.R.*, 1958, v. 20, p. 659-666. (In Russian).

[2] On a class of singular integro-differential equations // *Soviet Math., Doklady*, 1960, v. 1, p. 529-532.

Ivanov V. V.

[1] *The Theory of Approximate Methods and Their Application to the Numerical Solution of Singular Integral Equations*. Leyden: Noordhoff, 1976.

Kachanov M.

[1] Elastic solids with many cracks and related problems // *Advances in Applied Mechanics*. / J. Hutchinson, T. Wu. (eds). Academic Press, 1993, v. 30, p. 259-445.

Kalandiya A. I.

[1] On approximate solution of a class of singular equations // *Doklady Akademii Nauk SSSR*, 1959, v. 125, 713 –718. (In Russian, English translation available by the British Library-Lending Division, No RTS 8730, 1974).

Kane J. H.

[1] *Boundary Element Analysis in Engineering Continuum Mechanics*. New Jersey: Prentice Hall, 1994.

Kemeny J. M., Cook N. G. W.

[1] Micromechanics of deformation in rocks // *Toughness Mechanics in Quasi-Brittle Materials* / S. P. Shah (ed). Kluwer Acad. Publ.: Netherlands, 1991, p. 155-188.

Koiter W. T.

[1] An infinite row of collinear cracks in an infinite elastic sheet // *Ing. Arch.*, 1959, v. 28, p. 168-172.

[2] Stress distribution in an elastic sheet with a doubly-periodic set of equal holes // *Boundary Problems of Differential Equations*. Medison: Univ. Wisconsin Press, 1960, p. 191-213.

Kolosov G.V.

[1] *On an Application of Complex Function Theory to a Plane Problem of the Mathematical Theory of Elasticity.* Yuriev, 1909. (In Russian).

Korneychuk A. A

[1] Quadrature rules for singular integrals // *Numerical Methods of Solving Differential and Integral Equations and Quadrature Rules* / Moscow: Nauka, 1964, p. 64-74. (In Russian).

Kovneristov G. B.

[1] *Development of the Numerical Method of Potentials on the Basis of Interpolating Representations in Two-Dimensional Problems of Construction Mechanics* // Doctoral thesis at the Building University of Kiev, Kiev, 1991. (In Russian).

Kovneristov G. B., Rusinov I. A. et al.

[1] *Strength and contact deformability of reinforced concrete constructions.* Kiev: Budivelnik, 1991. (In Russian).

Krishnasamy, G., Schmerr, L. W. et al.

[1] Hypersingular boundary integral equations: some applications in acoustic and elastic wave scattering // *Trans. ASME*, 1990, v. 57, p. 404-414.

Kupradze V. D.

[1] *Potential Methods in the Theory of Elasticity.* Translated from Russian by Israel program for Scientific Translation, Jerusalem, 1965.

Kupradze V. D., Gegelia T. G., Basheleishvili M. O., Burchuladze T. V.

[1] *Three-Dimensional Problems of the Mathematical Theory of Elasticity.* Amsterdam- New York: North-Holland Publ. Co, 1979.

Levina Ts. O., Mikhlin S. G.

[1] To the evaluation of stress calculation in panel pillars // *Proc. Seismological Institute, Academy of Sciences USSR*, 1940, No 94, p. 3-35. (In Russian).

Ladopulos E. G.

[1] New aspects for the generalization of the Sokhotski-Plemelj formulae for the solution of finite-part singular integrals used in fracture mechanics // *Int. J. Fracture*, 1992, v. 54, p. 317-328.

Lee K. J.

[1] A boundary element technique for plane isotropic media using complex variable technique, // *Comput. Mech.*, 1993, v. 11, p. 83-91.

Lekhnitskii S. G.

[1] *Theory of Elasticity of an Anisotropic Elastic Body.* Holden-Day: San Francisco, 1963.

Linkov A. M.

[1] Integral equations in the theory of elasticity for a plane with cuts, loaded by a balanced system of forces // *Soviet Physics. Doklady*, 1975, v. 19, p. 718-720.

[2] On one integral relationship in the plane problem of elasticity theory // *Mechanics of Solids*, 1975, v. 10, p. 70-73.

[3] Problems of the elasticity theory for a plane with a finite number of curvilinear cuts // *Researches in Elasticity and Plasticity* / L. M. Kachanov (ed). Leningrad: Leningrad State University, 1976, v. 11, 3-11. (In Russian).

[4] Problems of the elasticity theory for a plane with periodic systems of cuts // *Researches in Elasticity and Plasticity* / L. M. Kachanov (ed). Leningrad: Leningrad State University, 1976, v. 11, p. 11-18. (In Russian).

[5] Integral equation for the plane problem of elasticity theory of a doubly periodic system of slots acted on by self-balancing loads // *Mechanics of Solids*, 1976, v. 11, p. 60-63.

[6] Plane problems of the static loading of piecewise homogeneous linearly elastic medium // *Applied Mathematics and Mechanics*, 1983, v. 47, p. 527-532.

[7] Boundary integral equation with a finite-part integral for the plane elasticity theory // *Problems of Mechanics and Control Processes: Academician V. V. Novojhilov, Scientist, Scholar, Citizen* / V. I. Zoubov (ed). Leningrad: Leningrad State University, 1990, v. 13, p. 103-107. (In Russian).

[8] Real and complex hypersingular integrals and integral equations in computational mechanics // *Demonstratio Mathematica*, 1995, v. 28, p. 759-769.

[9] Elasticity problems involving coupled half-planes // *J. Appl. Math. Mech.*, 1999, v. 63, p. 927-938.

Linkov A. M., Koshelev V. F.

[1] Complex BIE and BEM in periodic and double-periodic problems // *Proc. 13th ASCE Engineering Mechanics Conference* / N. P. Jones (ed). Baltimore: John Hopkins University, 1999 (CD-ROM).

[2] Complex variables BIE and BEM for a plane doubly periodic system of flaws // *J. Chinese Inst. Engineers*, 1999, v. 22, p. 709-720.

[3] Linkov A. M., Koshelev V. F. On numerical modeling of effective properties of blocky systems with cracks, inclusions and voids // *Proc. Int. Workshop on Fracture Mechanics and Advanced Engineering Materials* / Lin Ye, Yiu-Wing Mai (eds). Sydney: University of Sydney, 1999, p. 265-271.

Linkov A. M., Mogilevskaya S. G.

[1] Hypersingular integrals in plane problems of the theory of elasticity // *Applied Mathematics and Mechanics*, 1990, v. 54, p. 93-99.

[2] Complex variables hypersingular integrals and integral equations in plane elasticity problems // *Researches in Mechanics of Building Constructions and Materials* / V. P. Il'in (ed). St-Petersburg: St-Petersburg State Architecture and Building University, 1991, p. 17-34. (In Russian).

[3] Complex hypersingular integrals and integral equations in plane elasticity // *Acta Mechanica*, 1994, v. 105, p. 189-205.

[4] On the theory of complex hypersingular equations // *Computational Mechanics'95* / S. Atluri, G. Yagawa, G. T. Cruse (eds). Berlin - Heidelberg - New-York: Springer-Verlag, 1995, v. 2, p. 2836-2840.

[5] Complex hypersingular BEM in plane elasticity problems // *Singular Integrals in Boundary Element Methods* / V. Sladek, J. Sladek (eds). Southempton: Computational Mechanics Publications, 1999, p. 299-364.

Linkov A. M., Mogilevskaya S. G., Napier J. A. L.

[1] Multiple interacting curvilinear crack problems: a method of solution and numerical results // *Proc. 36th US Rock Mechanics Symposium (NYRocks '97)* / K. Kim (ed). Oxford: Elsevier, 1997, (CD-ROM).

Linkov A. M., Zoubkov V. V.

[1] Boundary integral equations in problems for jointed rocks // *Proc. 7-th Int. Conference on Computer Methods and Advances in Geomechanics* / G. Beer, J. R. Bookers, J. P. Carter (eds). Rotterdam: Balkema, 1992, p. 1747-1750.

Linkov A. M., Zoubkov V. V., Mogilevskaya S. G.

[1] *Complex Variables Integral Equations*. St-Petersburg: Institute for Problems of Mechanical Engineering (Russian Academy of Sciences), 1994, No 118. (In Russian).

Liu Z., Cook N. G. W., Myer L. R.

[1] Numerical studies relating to micro fracture in granular materials // *Proc. 1st North American Rock Mechanics Symposium* / P. P. Nilson, S. E. Laubach (eds). Rotterdam: Balkema, 1995, p. 63-74.

Maier G., Polizzotto C.

[1] A Galerkin approach to boundary element elastoplastic analysis // *Computational Methods in Applied Mechanics Engineering*, 1987, v. 60, p. 175-194.

Manjavidze G. F.

[1] On a class of singular integral equations with discontinuous coefficients // *Soobshcheniya Gruz. Filiala A. N. S.S.S.R.*, 1950, v. 11, p. 269-274. (In Russian).

[2] On a singular integral equation with discontinuous coefficients and its application in the theory of elasticity // *Prikladnaya Matematika i Mekhanika.*, 1951, v. 15, p. 279-296. (In Russian).

Maz'ya V., Nazarov S., Plamenevskij V.

[1] *Asymptotic Theory of Elliptic Boundary Value Problems in Singularly Perturbated Domains*. Birkhauser Verlag: Basel-Boston-Berlin, 2000, v. 1, 2.

Mikhlin S. G.

[1] Reduction of the fundamental problems of the plane theory of elasticity to Fredholm integral equations // *Doklady, A. N. SSSR (Comptes rendus de l'académie des sciences de l'U.R.S.S.)*, 1934, new ser., v. 1, p. 295-298.

[2] *Multidimensional Singular Integrals and Integral Equations*. Oxford, New York: Pergamon Press, 1965.

Mikhlin S. G., Morozov N. F., Paukshto M. V.

[1] *The Integral Equations of the Theory of Elasticity*. Stuttgart: Teubner, 1995.

Mogilevskaya S. G.

[1] The universal algorithm based on complex hypersingular integral equation to solve plane elasticity problems // *Comput. Mech.*,1996, v. 18, p. 127-138.

[2] The complex one-sided integrals of Cauchy and Hadamard and application to boundary element method // *Int. J Numerical and Analytical Methods in Geomechanics*, 1998, v. 22, p. 947-968.

[3] Complex singular and hypersingular integral equations for the system of openings, cracks and inclusions in a half-plane // *Proc. Internat. Symposium on Boundary Element Methods* / M. Bonnet (ed). Paris: Ecole Polytechnique, 1998, p. 153-154.

[4] Complex hypersingular integral equation for piece-wise homogeneous half-plane with cracks // *Int. J. Fracture*, 2000, v. 102, p. 177-204.

Mogilevskaya S. G., Linkov A. M

[1] Complex fundamental solutions and complex variables boundary element method // *Proc. IABEM-97 Workshop "Fundamental Solutions in Boundary Elements. Formulation and Integration"* / F. G. Benitez (ed). Seville: Univ. of Seville, 1997, p. 71-81.

[2] Complex fundamental solutions and complex variables boundary element method // *Computational Mechanics*, 1998, v. 22, p. 88-92.

Mogilevskaya S. G., Rothenburg L., Dusseault M. B.

[1] Interaction between a circular opening and fractures originating from its boundary in a piecewise homogeneous plane // *Int. J. Numerical and Analytical Methods in Geomechanics* // 2000, v. 24, p. 247-970.

Murakami Y.

[1] *Stress Intensity Factors Handbook.* Oxford-New York: Pergamon Press, 1990, v. 1, 2.

Muskhelishvili N. I.

[1] A new general method of solution of the fundamental boundary problems of the plane theory of elasticity // *Doklady, A. N. SSSR (Comptes rendus de l'académie des sciences de l'U.R.S.S.)*, 1934, v. 3, p. 7-11.

[2] A study of the new integral equations of the plane theory of elasticity // *Doklady, A. N. SSSR (Comptes rendus de l'académie des sciences de l'U.R.S.S.)*, 1934, v. 3, p. 73-77.

[3] On the numerical solution of the plane problems of the theory of elasticity // *Trudy Tbilissk. Mat. Inst.*, 1937, v. 1, 83-87. (In Georgian with short Russian summary).

[4] *Singular Integral Equations.* Groningen: Noordhoff, 1953.

[5] *Some Basic Problems of the Mathematical Theory of Elasticity.* Groningen: Noordhoff, 1975.

Napier J. A. L.

[1] Modelling of fracturing near deep level gold mine excavations using a displacement discontinuity approach // *Proc. 2nd Conf. "Mechanics of Jointed and Faulted Rock"* / H.-P. Rossmanith (ed). Rotterdam: Balkema, 1990.

Napier J. A. L, Pierce A. P.

[1] Simulation of extensive fracture formation and interruption in brittle materials // *Proc. 2nd Conf. "Mechanics of Jointed and Faulted Rock"* / H.-P. Rossmanith (ed). Rotterdam: Balkema, 1995, p. 63-74.

Natanzon V. Ya.

[1] On stresses in a tensioned plate with holes located in the check order // *Matematicheskii sbornik*, 1935, v. 42, p. 617-636. (In Russian).

Panasiuk V. V., Savruk M. P., Datsishin A. P.

[1] *Stress Distribution near Cracks in Plates and Shells.* Kiev: Naukova dumka, 1976. (In Russian)

Parton V. Z., Perlin P. I.

[1] *Integral Equations in Elasticity.* Moscow: Mir, 1982.

Pierce A. P., Spottiswoode S., Napier J. A. L.

[1] The spectral boundary element method: a new window on boundary elements in rock mechanics // *Int. J. Rock Mechanics and Mining Sciences & Geomechanics Abstracts*, 1992, v. 29, p. 379-400.

Potter D.

[1] *Computational Physics.* New York: Wiley-Interscience, 1973.

Rabotnov Y. N.

[1] *Solid Mechanics.* Moscow: Nauka, 1988. (In Russian).

Rice J.

[1] Mathematical methods in the mechanics of fracture // *Fracture* / H. Liebowitz (ed). New York-London: Academic Press, 1968, v. 2, p. 191-311.

Savruk M. P.

[1] Constructing integral equations for two-dimensional elasticity theory problems of a body with curvilinear cracks // *Soviet Materials Science*, 1977, v. 12, p. 682-683.

[2] System of curved cracks in an elastic body under different boundary conditions on their lips // *Soviet Materials Science*, 1979, v. 14, p. 641-649.

[3] Plane problems of the theory of elasticity for a multiply connected area with holes and cracks // *Soviet Materials Science*, 1980, v. 16, p. 432-437.

[4] *Two-dimensional Elasticity Problems for Bodies with Cracks.* Kiev: Naukova dumka, 1981. (In Russian).

[5] *Stress Intensity Factors in Bodies with Cracks.* Kiev: Naukova dumka, 1988. (In Russian).

Savruk M. P., Osiv P. N., Prokopchuk I. V.

[1] *Numerical Analysis in Plane Problems of the Crack Theory.* Kiev: Naukova dumka, 1989. (In Russian).

Savruk M. P., Timoshuk N. V.

[1] Singular integral equations of the plane problem of the theory of elasticity for an infinite piecewise-uniform body with cracks // *Soviet Materials Science,* 1984, v. 20, p. 573-579.

[2] The plane problem of the theory of elasticity for a piecewise-uniform semiinfinite plate with elastic inclusions and cracks // *Soviet Materials Science,* 1987, v. 23, p. 160-165.

Sherman D. I.

[1] On the new method of N. I. Muskhelishvili for the plane problem of the theory of elasticity // *Doklady, A. N. SSSR (Comptes rendus de l'académie des sciences de l'U.R.S.S.),* 1936, v. 1 (X), p. 201-206.

[2] On the solution of the plane static problem of the theory of elasticity for displacements, given on the boundary // *Doklady, A. N. SSSR (Comptes rendus de l'académie des sciences de l'U.R.S.S.),* 1940, v. 27, p. 911-913.

[3] On the solution of the plane static problem of the theory of elasticity for given external forces // *Doklady, A. N. SSSR. (Comptes rendus de l'académie des sciences de l'U.R.S.S.),* 1940., v. 28, p. 25-28.

[4] The mixed problem of the static theory of elasticity for plane multiply connected regions // *Doklady, A. N. SSSR (Comptes rendus de l'académie des sciences de l'U.R.S.S.),* 1940, v. 28, p. 309-310.

[5] On the problem of plain strain in non-homogeneous media // *Non-Homogeneity in Elasticity and Plasticity* / Oxford: Pergamon Press, 1959, p. 3-20.

Sokolnikoff I. S.

[1] *Mathematical Theory of Elasticity,* 2nd edn. New-York: McGrow-Hill, 1956.

Stevenson A. C.

[1] Some boundary problems of two-dimensional elasticity // *Philos. Mag.,* 1943, v. 347, p. 766-793.

[2] Complex potentials in two-dimensional elasticity // *Proc. Roy. Soc. Ser. A,* 1945, v. 184 (997), p. 129-179, 218-229.

Tang S., Kitching R

[1] Elastoplastic analysis of two-dimensional problems with hole by boundary element method using complex variables // *Int. J. Mech. Science,* 1993, v. 35, p. 577-586.

Tanaka M., Sladek V., Sladek J.

[1] Regularization techniques applied to boundary element methods // *Appl. Mech. Reviews,* 1994, v. 47, p. 457-499.

Tashkinov a. A., Wildemann V. E., Zaitsev A. V.

[1] Micro-mechanical models of stable damage evolution on post-critical stage and failure of composite materials // *Proc. 12th Europ. Conf. "Fracture from Defects* / M. W. Brown, K. J. Miller, E. R. de los Rios (eds). London: EMAS Publ., 1998, v. III, p. 1483-1488.

Theocaris P. S., Ioakimidis N. I.

[1] The inclusion problem in plane elasticity // *Quarterly J. Applied Me*chanics, 1977 v. 30, p. 437-448.

[2] A star-shaped array of curvilinear cracks in an infinite isotropic elastic medium // *J. of Applied Mechanics,* 1977, v. 44, p. 619-624.

[3] Numerical integration methods for the solution of singular integral equations // *Quarterly of Applied Mathematics,* 1977, v. 35, p. 173-183.

[4] On the Gauss-Jacobi numerical integration method applied to the solution of singular integral equations // *Bulletin of Calcutta Math. Soc.,* 1978, v. 71, p. 29-43.

[5] A remark on the numerical solution of singular integral equations and determination of stress intensity factors // *J. Engineering Math.,* 1978, v. 13, p. 213-222.

[6] A numerical method for the solution of plane crack problems in finite media // *Int. J. Math. and Math Sci.,* 1980, v. 3, p. 739-760.

[7] A generalized crack problem in plane elasticity // *Int. J. Engineering Sci.,* 1980, v. 18, p. 491-499.

Timoshenko S. P., Goodier J. N.

[1] *Theory of Elasticity,* 3rd edn. New York: McGrow-Hill, 1970.

Tsamasphyros G., Dimou G.

[1] Gauss quadrature rules for finite part integrals // *Int. J. for Numerical Methods in Engineering,* 1990, v. 30, p. 13-26.

Ugodchikov A. G., Khutorianski N. M.

[1] *Boundary Element Method in Solid Mechanics.* Kazan: Kazan State university, 1986. (In Russian).

Vekua N. P.

[1] *Generalized Analytic Functions.* London: Pergamon Press, 1962.

[2] *Systems of Singular Integral Equations*. Groningen: Noordhoff, 1967.

Wang J., Mogilevskaya S. G., Crouch S. L.

[1] A Galerkin boundary integral method for nonhomogeneous materials with cracks // *Proc. 38th U. S. Rock Mechanics symposium DC Rocks: "Rock Mechanics in the National Interest"* / J. Tinucci, Heasley (eds). Washington DC: Swets & Zeitlinger Lisse, 2001, p. 1453-1460.

Westergaard H. M.

[1] Bearing pressures and cracks // *J. Appl. Mech.*, 1939, v. 6, p. A49-A53.

Zoubkov V. V.

[1] On stressed state and stability of elastic blocks interacting on their interfaces // *Problems of the Crack Theory and Mechanics of Fracture. Researches on Elasticity and Plasticity* / N. F. Morozov (ed), 1986, p. 39-46. (In Russian).

Mechanics

SOLID MECHANICS AND ITS APPLICATIONS

Series Editor: G.M.L. Gladwell

69. P. Pedersen and M.P. Bendsøe (eds.): *IUTAM Symposium on Synthesis in Bio Solid Mechanics.* Proceedings of the IUTAM Symposium held in Copenhagen, Denmark. 1999 ISBN 0-7923-5615-2
70. S.K. Agrawal and B.C. Fabien: *Optimization of Dynamic Systems.* 1999 ISBN 0-7923-5681-0
71. A. Carpinteri: *Nonlinear Crack Models for Nonmetallic Materials.* 1999 ISBN 0-7923-5750-7
72. F. Pfeifer (ed.): *IUTAM Symposium on Unilateral Multibody Contacts.* Proceedings of the IUTAM Symposium held in Munich, Germany. 1999 ISBN 0-7923-6030-3
73. E. Lavendelis and M. Zakrzhevsky (eds.): *IUTAM/IFToMM Symposium on Synthesis of Nonlinear Dynamical Systems.* Proceedings of the IUTAM/IFToMM Symposium held in Riga, Latvia. 2000 ISBN 0-7923-6106-7
74. J.-P. Merlet: *Parallel Robots.* 2000 ISBN 0-7923-6308-6
75. J.T. Pindera: *Techniques of Tomographic Isodyne Stress Analysis.* 2000 ISBN 0-7923-6388-4
76. G.A. Maugin, R. Drouot and F. Sidoroff (eds.): *Continuum Thermomechanics.* The Art and Science of Modelling Material Behaviour. 2000 ISBN 0-7923-6407-4
77. N. Van Dao and E.J. Kreuzer (eds.): *IUTAM Symposium on Recent Developments in Non-linear Oscillations of Mechanical Systems.* 2000 ISBN 0-7923-6470-8
78. S.D. Akbarov and A.N. Guz: *Mechanics of Curved Composites.* 2000 ISBN 0-7923-6477-5
79. M.B. Rubin: *Cosserat Theories: Shells, Rods and Points.* 2000 ISBN 0-7923-6489-9
80. S. Pellegrino and S.D. Guest (eds.): *IUTAM-IASS Symposium on Deployable Structures: Theory and Applications.* Proceedings of the IUTAM-IASS Symposium held in Cambridge, U.K., 6–9 September 1998. 2000 ISBN 0-7923-6516-X
81. A.D. Rosato and D.L. Blackmore (eds.): *IUTAM Symposium on Segregation in Granular Flows.* Proceedings of the IUTAM Symposium held in Cape May, NJ, U.S.A., June 5–10, 1999. 2000 ISBN 0-7923-6547-X
82. A. Lagarde (ed.): *IUTAM Symposium on Advanced Optical Methods and Applications in Solid Mechanics.* Proceedings of the IUTAM Symposium held in Futuroscope, Poitiers, France, August 31–September 4, 1998. 2000 ISBN 0-7923-6604-2
83. D. Weichert and G. Maier (eds.): *Inelastic Analysis of Structures under Variable Loads.* Theory and Engineering Applications. 2000 ISBN 0-7923-6645-X
84. T.-J. Chuang and J.W. Rudnicki (eds.): *Multiscale Deformation and Fracture in Materials and Structures.* The James R. Rice 60th Anniversary Volume. 2001 ISBN 0-7923-6718-9
85. S. Narayanan and R.N. Iyengar (eds.): *IUTAM Symposium on Nonlinearity and Stochastic Structural Dynamics.* Proceedings of the IUTAM Symposium held in Madras, Chennai, India, 4–8 January 1999 ISBN 0-7923-6733-2
86. S. Murakami and N. Ohno (eds.): *IUTAM Symposium on Creep in Structures.* Proceedings of the IUTAM Symposium held in Nagoya, Japan, 3-7 April 2000. 2001 ISBN 0-7923-6737-5
87. W. Ehlers (ed.): *IUTAM Symposium on Theoretical and Numerical Methods in Continuum Mechanics of Porous Materials.* Proceedings of the IUTAM Symposium held at the University of Stuttgart, Germany, September 5-10, 1999. 2001 ISBN 0-7923-6766-9
88. D. Durban, D. Givoli and J.G. Simmonds (eds.): *Advances in the Mechanis of Plates and Shells The Avinoam Libai Anniversary Volume.* 2001 ISBN 0-7923-6785-5
89. U. Gabbert and H.-S. Tzou (eds.): *IUTAM Symposium on Smart Structures and Structonic Systems.* Proceedings of the IUTAM Symposium held in Magdeburg, Germany, 26–29 September 2000. 2001 ISBN 0-7923-6968-8

Mechanics

SOLID MECHANICS AND ITS APPLICATIONS

Series Editor: G.M.L. Gladwell

90. Y. Ivanov, V. Cheshkov and M. Natova: *Polymer Composite Materials – Interface Phenomena & Processes.* 2001 ISBN 0-7923-7008-2
91. R.C. McPhedran, L.C. Botten and N.A. Nicorovici (eds.): *IUTAM Symposium on Mechanical and Electromagnetic Waves in Structured Media.* Proceedings of the IUTAM Symposium held in Sydney, NSW, Australia, 18-22 Januari 1999. 2001 ISBN 0-7923-7038-4
92. D.A. Sotiropoulos (ed.): *IUTAM Symposium on Mechanical Waves for Composite Structures Characterization.* Proceedings of the IUTAM Symposium held in Chania, Crete, Greece, June 14-17, 2000. 2001 ISBN 0-7923-7164-X
93. V.M. Alexandrov and D.A. Pozharskii: *Three-Dimensional Contact Problems.* 2001 ISBN 0-7923-7165-8
94. J.P. Dempsey and H.H. Shen (eds.): *IUTAM Symposium on Scaling Laws in Ice Mechanics and Ice Dynamics.* Proceedings of the IUTAM Symposium held in Fairbanks, Alaska, U.S.A., 13-16 June 2000. 2001 ISBN 1-4020-0171-1
95. U. Kirsch: *Design-Oriented Analysis of Structures.* A Unified Approach. 2002 ISBN 1-4020-0443-5
96. A. Preumont: *Vibration Control of Active Structures.* An Introduction (2nd Edition). 2002 ISBN 1-4020-0496-6
97. B.L. Karihaloo (ed.): *IUTAM Symposium on Analytical and Computational Fracture Mechanics of Non-Homogeneous Materials.* Proceedings of the IUTAM Symposium held in Cardiff, U.K., 18-22 June 2001. 2002 ISBN 1-4020-0510-5
98. S.M. Han and H. Benaroya: *Nonlinear and Stochastic Dynamics of Compliant Offshore Structures.* 2002 ISBN 1-4020-0573-3
99. A.M. Linkov: *Boundary Integral Equations in Elasticity Theory.* 2002 ISBN 1-4020-0574-1

Kluwer Academic Publishers – Dordrecht / Boston / London

ICASE/LaRC Interdisciplinary Series in Science and Engineering

1. J. Buckmaster, T.L. Jackson and A. Kumar (eds.): *Combustion in High-Speed Flows.* 1994 ISBN 0-7923-2086-X
2. M.Y. Hussaini, T.B. Gatski and T.L. Jackson (eds.): *Transition, Turbulence and Combustion.* Volume I: Transition. 1994 ISBN 0-7923-3084-6; set 0-7923-3086-2
3. M.Y. Hussaini, T.B. Gatski and T.L. Jackson (eds.): *Transition, Turbulence and Combustion.* Volume II: Turbulence and Combustion. 1994 ISBN 0-7923-3085-4; set 0-7923-3086-2
4. D.E. Keyes, A. Sameh and V. Venkatakrishnan (eds): *Parallel Numerical Algorithms.* 1997 ISBN 0-7923-4282-8
5. T.G. Campbell, R.A. Nicolaides and M.D. Salas (eds.): *Computational Electromagnetics and Its Applications.* 1997 ISBN 0-7923-4733-1
6. V. Venkatakrishnan, M.D. Salas and S.R. Chakravarthy (eds.): *Barriers and Challenges in Computational Fluid Dynamics.* 1998 ISBN 0-7923-4855-9
7. M.D. Salas, J.N. Hefner and L. Sakell (eds.): *Modeling Complex Turbulent Flows.* 1999 ISBN 0-7923-5590-3

KLUWER ACADEMIC PUBLISHERS – DORDRECHT / BOSTON / LONDON

ERCOFTAC SERIES

1. A. Gyr and F.-S. Rys (eds.): *Diffusion and Transport of Pollutants in Atmospheric Mesoscale Flow Fields.* 1995 ISBN 0-7923-3260-1
2. M. Hallbäck, D.S. Henningson, A.V. Johansson and P.H. Alfredsson (eds.): *Turbulence and Transition Modelling.* Lecture Notes from the ERCOFTAC/IUTAM Summerschool held in Stockholm. 1996 ISBN 0-7923-4060-4
3. P. Wesseling (ed.): *High Performance Computing in Fluid Dynamics.* Proceedings of the Summerschool held in Delft, The Netherlands. 1996 ISBN 0-7923-4063-9
4. Th. Dracos (ed.): *Three-Dimensional Velocity and Vorticity Measuring and Image Analysis Techniques.* Lecture Notes from the Short Course held in Zürich, Switzerland. 1996 ISBN 0-7923-4256-9
5. J.-P. Chollet, P.R. Voke and L. Kleiser (eds.): *Direct and Large-Eddy Simulation II.* Proceedings of the ERCOFTAC Workshop held in Grenoble, France. 1997 ISBN 0-7923-4687-4
6. A. Hanifi, P.H. Alfredson, A.V. Johansson and D.S. Henningson (eds.) : *Transition, Turbulence and Combustion Modelling.* 1999 ISBN 0-7923-5989-5
7. P.R. Voke, N.D. Sandham and L. Kleiser (eds.) : *Direct and Large-Eddy Simulation III.* 1999 ISBN 0-7923-5990-9
8. B.J. Geurts, R. Friedrich and O. Métais (eds.) : *Direct and Large-Eddy Simulation IV.* 2001 ISBN 1-4020-0177-0

KLUWER ACADEMIC PUBLISHERS – DORDRECHT / BOSTON / LONDON

Mechanics

FLUID MECHANICS AND ITS APPLICATIONS

Series Editor: R. Moreau

24. R. Benzi (ed.): *Advances in Turbulence V.* 1995 ISBN 0-7923-3032-3
25. B.I. Rabinovich, V.G. Lebedev and A.I. Mytarev: *Vortex Processes and Solid Body Dynamics.* The Dynamic Problems of Spacecrafts and Magnetic Levitation Systems. 1994 ISBN 0-7923-3092-7
26. P.R. Voke, L. Kleiser and J.-P. Chollet (eds.): *Direct and Large-Eddy Simulation I.* Selected papers from the First ERCOFTAC Workshop on Direct and Large-Eddy Simulation. 1994 ISBN 0-7923-3106-0
27. J.A. Sparenberg: *Hydrodynamic Propulsion and its Optimization.* Analytic Theory. 1995 ISBN 0-7923-3201-6
28. J.F. Dijksman and G.D.C. Kuiken (eds.): *IUTAM Symposium on Numerical Simulation of Non-Isothermal Flow of Viscoelastic Liquids.* Proceedings of an IUTAM Symposium held in Kerkrade, The Netherlands. 1995 ISBN 0-7923-3262-8
29. B.M. Boubnov and G.S. Golitsyn: *Convection in Rotating Fluids.* 1995 ISBN 0-7923-3371-3
30. S.I. Green (ed.): *Fluid Vortices.* 1995 ISBN 0-7923-3376-4
31. S. Morioka and L. van Wijngaarden (eds.): *IUTAM Symposium on Waves in Liquid/Gas and Liquid/Vapour Two-Phase Systems.* 1995 ISBN 0-7923-3424-8
32. A. Gyr and H.-W. Bewersdorff: *Drag Reduction of Turbulent Flows by Additives.* 1995 ISBN 0-7923-3485-X
33. Y.P. Golovachov: *Numerical Simulation of Viscous Shock Layer Flows.* 1995 ISBN 0-7923-3626-7
34. J. Grue, B. Gjevik and J.E. Weber (eds.): *Waves and Nonlinear Processes in Hydrodynamics.* 1996 ISBN 0-7923-4031-0
35. P.W. Duck and P. Hall (eds.): *IUTAM Symposium on Nonlinear Instability and Transition in Three-Dimensional Boundary Layers.* 1996 ISBN 0-7923-4079-5
36. S. Gavrilakis, L. Machiels and P.A. Monkewitz (eds.): *Advances in Turbulence VI.* Proceedings of the 6th European Turbulence Conference. 1996 ISBN 0-7923-4132-5
37. K. Gersten (ed.): *IUTAM Symposium on Asymptotic Methods for Turbulent Shear Flows at High Reynolds Numbers.* Proceedings of the IUTAM Symposium held in Bochum, Germany. 1996 ISBN 0-7923-4138-4
38. J. Verhás: *Thermodynamics and Rheology.* 1997 ISBN 0-7923-4251-8
39. M. Champion and B. Deshaies (eds.): *IUTAM Symposium on Combustion in Supersonic Flows.* Proceedings of the IUTAM Symposium held in Poitiers, France. 1997 ISBN 0-7923-4313-1
40. M. Lesieur: *Turbulence in Fluids.* Third Revised and Enlarged Edition. 1997 ISBN 0-7923-4415-4; Pb: 0-7923-4416-2
41. L. Fulachier, J.L. Lumley and F. Anselmet (eds.): *IUTAM Symposium on Variable Density Low-Speed Turbulent Flows.* Proceedings of the IUTAM Symposium held in Marseille, France. 1997 ISBN 0-7923-4602-5
42. B.K. Shivamoggi: *Nonlinear Dynamics and Chaotic Phenomena.* An Introduction. 1997 ISBN 0-7923-4772-2
43. H. Ramkissoon, *IUTAM Symposium on Lubricated Transport of Viscous Materials.* Proceedings of the IUTAM Symposium held in Tobago, West Indies. 1998 ISBN 0-7923-4897-4
44. E. Krause and K. Gersten, *IUTAM Symposium on Dynamics of Slender Vortices.* Proceedings of the IUTAM Symposium held in Aachen, Germany. 1998 ISBN 0-7923-5041-3
45. A. Biesheuvel and G.J.F. van Heyst (eds.): *In Fascination of Fluid Dynamics.* A Symposium in honour of Leen van Wijngaarden. 1998 ISBN 0-7923-5078-2

Mechanics

FLUID MECHANICS AND ITS APPLICATIONS

Series Editor: R. Moreau

46. U. Frisch (ed.): *Advances in Turbulence VII.* Proceedings of the Seventh European Turbulence Conference, held in Saint-Jean Cap Ferrat, 30 June–3 July 1998. 1998 ISBN 0-7923-5115-0
47. E.F. Toro and J.F. Clarke: *Numerical Methods for Wave Propagation.* Selected Contributions from the Workshop held in Manchester, UK. 1998 ISBN 0-7923-5125-8
48. A. Yoshizawa: *Hydrodynamic and Magnetohydrodynamic Turbulent Flows.* Modelling and Statistical Theory. 1998 ISBN 0-7923-5225-4
49. T.L. Geers (ed.): *IUTAM Symposium on Computational Methods for Unbounded Domains.* 1998 ISBN 0-7923-5266-1
50. Z. Zapryanov and S. Tabakova: *Dynamics of Bubbles, Drops and Rigid Particles.* 1999 ISBN 0-7923-5347-1
51. A. Alemany, Ph. Marty and J.P. Thibault (eds.): *Transfer Phenomena in Magnetohydrodynamic and Electroconducting Flows.* 1999 ISBN 0-7923-5532-6
52. J.N. Sørensen, E.J. Hopfinger and N. Aubry (eds.): *IUTAM Symposium on Simulation and Identification of Organized Structures in Flows.* 1999 ISBN 0-7923-5603-9
53. G.E.A. Meier and P.R. Viswanath (eds.): *IUTAM Symposium on Mechanics of Passive and Active Flow Control.* 1999 ISBN 0-7923-5928-3
54. D. Knight and L. Sakell (eds.): *Recent Advances in DNS and LES.* 1999 ISBN 0-7923-6004-4
55. P. Orlandi: *Fluid Flow Phenomena.* A Numerical Toolkit. 2000 ISBN 0-7923-6095-8
56. M. Stanislas, J. Kompenhans and J. Westerveel (eds.): *Particle Image Velocimetry.* Progress towards Industrial Application. 2000 ISBN 0-7923-6160-1
57. H.-C. Chang (ed.): *IUTAM Symposium on Nonlinear Waves in Multi-Phase Flow.* 2000 ISBN 0-7923-6454-6
58. R.M. Kerr and Y. Kimura (eds.): *IUTAM Symposium on Developments in Geophysical Turbulence* held at the National Center for Atmospheric Research, (Boulder, CO, June 16–19, 1998) 2000 ISBN 0-7923-6673-5
59. T. Kambe, T. Nakano and T. Miyauchi (eds.): *IUTAM Symposium on Geometry and Statistics of Turbulence.* Proceedings of the IUTAM Symposium held at the Shonan International Village Center, Hayama (Kanagawa-ken, Japan November 2–5, 1999). 2001 ISBN 0-7923-6711-1
60. V.V. Aristov: *Direct Methods for Solving the Boltzmann Equation and Study of Nonequilibrium Flows.* 2001 ISBN 0-7923-6831-2
61. P.F. Hodnett (ed.): *IUTAM Symposium on Advances in Mathematical Modelling of Atmosphere and Ocean Dynamics.* Proceedings of the IUTAM Symposium held in Limerick, Ireland, 2–7 July 2000. 2001 ISBN 0-7923-7075-9
62. A.C. King and Y.D. Shikhmurzaev (eds.): *IUTAM Symposium on Free Surface Flows.* Proceedings of the IUTAM Symposium held in Birmingham, United Kingdom, 10–14 July 2000. 2001 ISBN 0-7923-7085-6
63. A. Tsinober: *An Informal Introduction to Turbulence.* 2001 ISBN 1-4020-0110-X; Pb: 1-4020-0166-5
64. R.Kh. Zeytounian: *Asymptotic Modelling of Fluid Flow Phenomena.* 2002 ISBN 1-4020-0432-X
65. R. Friedrich and W. Rodi (eds.): *Advances in LES of Complex Flows.* Prodeedings of the EUROMECH Colloquium 412, held in Munich, Germany, 4-6 October 2000. 2002 ISBN 1-4020-0486-9
66. D. Drikakis and B.J. Geurts (eds.) *Turbulent Flow Computation.* 2002 ISBN 1-4020-0523-7

Kluwer Academic Publishers – Dordrecht / Boston / London

GPSR Compliance

The European Union's (EU) General Product Safety Regulation (GPSR) is a set of rules that requires consumer products to be safe and our obligations to ensure this.

If you have any concerns about our products, you can contact us on ProductSafety@springernature.com

In case Publisher is established outside the EU, the EU authorized representative is:

Springer Nature Customer Service Center GmbH
Europaplatz 3
69115 Heidelberg, Germany

Batch number: 10615058

Printed by Printforce, the Netherlands